...NATIONAL TECHNO... ...UNIVERSITY
This Book is Donated...
PROF. WAI-KAI CHEN

Date:

Proceedings of the IX Winter Meeting on Low Temperature Physics

HIGH TEMPERATURE SUPERCONDUCTORS

SERIES ON PROGRESS IN HIGH TEMPERATURE SUPERCONDUCTIVITY

Editors-in-charge

Paul C. W. Chu (*Houston*)
D. C. Mattis (*Utah*)
K. K. Phua (*Nat'l Univ. of S'pore*)
C. N. R. (*Indian Inst. of Sc*)

S. Tanaka (*Tokyo*)
Yu Lu (*Beijing/ICTP*)
Zhao Zhongxian (*Acad Sinica, Beijing*)

Published

Vol. 1 — Proceedings of the Adriatico Research Conference on High Temperature Superconductors
(*eds. S. Lundqvist, E. Tosatti, M. Tosi and Yu Lu*)

Vol. 2 — Proceedings of the Beijing International Workshop on High Temperature Superconductivity
(*eds. Z. Z. Gan, G. J. Cui, G. Z. Yang and Q. S. Yang*)

Vol. 3 — Proceedings of the Drexel International Conference on High Temperature Superconductivity (*eds. S. M. Bose and S. D. Tyagi*)

Vol. 4 — Proceedings of the 2nd Soviet-Italian Symposium on Weak Superconductivity (*eds. A. Barone and A. Larkin*)

Forthcoming

Vol. 6 — The Physical Properties of High Temperature Superconductors — review volume (*D M Ginsberg*)

Vol. 7 — Structural and Chemical Aspects of High Temperature Superconductors (*ed. C. N. R. Rao*)

Vol. 8 — World Congress on Superconductivity (*eds. C. G. Burnham and R. Kane*)

Vol. 9 — Latin American Conference on High Temperature Superconductivity (*R Nicholsky*)

Vol. 10 — Macroscopic Theories of Superfluidity & Superconductivity — review volume (*A A Sobyanin*)

Vol. 11 — Volume on High Temperature from Russia (*A Larkin*)

Vol. 12 — High Temperature Superconductivity and other related topics — 1st Asia Pacific Conference on Condensed Matter Physics (*K K Phua et. al.*)

Vol. 13 — Applications of High Temperature Superconductors — review volume (*C Y Huang*)

Vol. 14 — Combined Proceedings on "Mini Workshop on Mechanism of High Temperature Superconductivity" and "Towards the Theoretical Understanding of High Temperature Superconductors" (*S Lundqvist et. al.*)

Progress in High Temperature Superconductivity — Vol. 5

Proceedings of the IX Winter Meeting on Low Temperature Physics

HIGH TEMPERATURE SUPERCONDUCTORS

Vista Hermosa, Mor., México 10-15 January 1988

Editors:
J. Heiras
R. A. Barrio
T. Akachi
J. Tagüeña

Instituto de Investigaciones en Materiales
Universidad Nacional Autónoma de México, México.

World Scientific
Singapore • New Jersey • Hong Kong

Published by

World Scientific Publishing Co. Pte. Ltd.
P.O. Box 128, Farrer Road, Singapore 9128

U. S. A. office: World Scientific Publishing Co., Inc.
687 Hartwell Street, Teaneck NJ 07666, USA

Proceedings of the IX Winter Meeting on Low Temperature Physics
HIGH TEMPERATURE SUPERCONDUCTORS

Copyright © 1988 by World Scientific Publishing Co Pte Ltd.

All rights reserved. This book, or parts thereof, may not be reproduced in any form or by any means, electronic or mechanical, including photocopying, recording or any information storage and retrieval system now known or to be invented, without written permission from the Publisher.

ISBN 9971-50-583-5
    9971-50-584-3 (pbk)

Printed in Singapore by JBW Printers & Binders Pte. Ltd.

IX WINTER MEETING ON LOW TEMPERATURE PHYSICS

"HIGH TEMPERATURE SUPERCONDUCTORS"

Hotel Hacienda Vista Hermosa

Morelos, México

January 10 - 15, 1988

ORGANIZING COMMITTEE:

- Guillermo Aguilar Sahagún
- Tatsuo Akachi Miyazaki
- Rafael Barrio Paredes
- Roberto Escudero Derat
- Jesús Heiras Aguirre
- Julia Tagüeña Parga

S P O N S O R S :

- UNIVERSIDAD NACIONAL AUTONOMA DE MEXICO:
  - Programa Universitario de Investigación Sobre Superconductores de Alta Temperatura de Transición.
  - Centro de Instrumentos.
  - Dirección General de Asuntos del Personal Académico.
  - Dirección General de Intercambio Académico.
  - Dirección General de Servicios Auxiliares.
- CONSEJO NACIONAL DE CIENCIA Y TECNOLOGIA
- GRUPO CONDUMEX.
- INSTITUTO DE INVESTIGACIONES ELECTRICAS.
- REUNIONES DE INVIERNO, A.C.
- SUBSECRETARIA DE EDUCACION SUPERIOR E INVESTIGACION CIENTIFICA-SEP.
- THE THIRD WORLD ACADEMY OF SCIENCES.

ACKNOWLEDGEMENTS

The organizing committee wishes to express its debt to the central authorities of the Universidad Nacional Autónoma de México whose support made, this meeting to a great extent possible.

Special thanks are due to the Programa Universitario de Investigación de Superconductores de Alta Temperatura Crítica de Transición. The technical assistante of Leticia Baños, José de J. Camacho, Angel Diez, Candelaria Flores, Amelia Hernández, Violeta Leyton, Patricia Murillo and Carmen Vázquez is fully acknowledged.

FOREWORD

This meeting is the ninth of an uninterrupted series of annual international meetings on Low Temperature Physics organized by our Institute. The main purpose of this series has been to stimulate the direct exchange of ideas between the Mexican scientific community and the rest of the world on current topics of research in this field.

It was decided to devote the winter meeting this year to High-Tc Superconductivity, in view of the extremely important discovery of ceramic superconductors at the end of 1986, and of the research on this topic carried out in México since early 1987.

We are certain that the present volume reflects the high quality of the papers presented and that its content will retain its relevance for a long time, despite the rapid development of the field. Key results, some of them presented for the first time at this meeting are: the appearance of new superconductors which do not contain rare-earths, as the B_i compound; the impressive enhancement of critical currents in bulk materials; the important role of antiferromagnetism in these ceramics; the subtle differences between the magnetic behaviour of these materials and the spin-glass properties; the detailed study of the macroscopic magnetic properties of the La ceramics, and the remarks about the possible rôle of Andreiev reflexions in the transport properties which lead to the undestanting of tunneling results. We knew about extremely important experimental results on Hall Effect, upper critical field measurements and compositional phase diagrams of superconducting ceramics. There were also outstanding reviews of different aspects of the field, such as the rôle of defects, the enhancement of Tc, revealing historical points and the importance of determining the oxygen content.

This volume is conveniently organized in two sections, separating the invited papers from the contributed papers presented in a poster session. The invited papers are divided into four thematic sections, loosely covering the main topics discussed in the course of the meeting, not only during the formal presentations but also in workshops and panel sessions. Some of the topics touched on in these workshops, such as industry-research links and prospects for High-Tc in Latinamerican countries, are not recorded here. The cartoons that can be found at the beginning of each section represent a humoristic view of one of our students.

We hope that the present volume will meet our expectations of its relevance and usefulness.

The Editors

February 1988.

CONTENTS

FOREWORD vii

INVITED PAPERS

CHEMICAL ASPECTS

MIXED VALENCE COPPER OXIDES: RELATIONSHIPS BETWEEN
CRYSTAL CHEMISTRY AND SUPERCONDUCTIVITY 5
   B. Raveau, C. Michel, A. Maignan, M. Hervieu,
   and J. Provost

THE ROLE OF OXYGEN IN $YBa_2Cu_3O_{7-\delta}$ 18
   J.B. Goodenough and A. Manthiram

NEW MATERIALS EXHIBITING SUPERCONDUCTIVITY ABOVE 90 K 28
   C.Y. Huang

THE EFFECTS OF OXYGEN STOICHIOMETRY AND OXYGEN ORDERING
ON SUPERCONDUCTIVITY IN $Y_1Ba_2Cu_3O_{9-x}$ 38
   R. Beyers, E.M. Engler, P.M. Grant, S.S.P. Parkin, G. Lim,
   M.L. Ramirez, K.P. Roche, J.E. Vazquez, V.Y. Lee,
   R.D. Jacowitz, B.T. Ahn, T.M. Gür and R.A. Huggins

THE SEARCH FOR HIGH TEMPERATURE SUPERCONDUCTIVITY 44
   C. Politis

CERAMIC MATERIALS AND HIGH-T_c SUPERCONDUCTIVITY:
THE 1:2:3 COMPOUNDS 54
   R. Escudero

PHYSICAL PROPERTIES

PHYSICAL PROPERTIES OF HIGH T_c OXIDES 63
   S. Uchida

HOLE CONCENTRATION COMPENSATION EFFECT IN $Ln_{1+x}Ba_{2-x}Cu_3O_{7-\delta}$
AND ANISOTROPIC UPPER CRITICAL FIELD OF $LnBa_2Cu_3O_{7-\delta}$ 73
   K. Takita

TUNNELING SPECTROSCOPY OF HIGH TEMPERATURE SUPERCONDUCTORS 82
   A.L. de Lozanne

MICROSTRUCTURE AND PROPERTIES OF CERAMIC $Ba_2YCu_3O_7$ 88
   G.J. Fisanick, S. Nakahara, S. Jin, T.F. Tiefel,
   R.C. Sherwood, M. Yan, R. Moore and R.B. van Dover

MACROSCOPIC VIEW OF THE HIGH T_c SUPERCONDUCTIVITY 102
F. de la Cruz, L. Civale, E. Osquiguil, H. Safar,
R. Decca, D.A. Esparza and C. D'Ovidio

MÖSSBAUER SPECTROSCOPY IN HIGH T_c SUPERCONDUCTORS 113
R.W. Gómez

STRUCTURAL PROPERTIES AND DEFECTS

METASTABILITY PROPERTIES OF GAUGE GLASSES AND THEIR
RELEVANCE TO HIGH TEMPERATURE SUPERCONDUCTORS 123
J.V. José

STRUCTURAL DEFECTS IN THE 1:2:3 PHASES OF HIGH T_c
SUPERCONDUCTORS.. 135
D. Ríos-Jara

CRYSTALLINE STRUCTURE IN THE Cu BASAL PLANE AND
MICROSTRUCTURE IN SUPERCONDUCTING $YBa_2Cu_3O_{7-y}$ 145
C. Varea, and A. Robledo

ON THE STRUCTURAL CHARACTERISTICS OF THE HIGH-T_c
SUPERCONDUCTORS $YBa_2Cu_3O_{7-x}$ AND $HoBa_2Cu_3O_{7-x}$ 154
R. Pérez, J.G. Pérez-Ramírez, J. Reyes-Gasga,
U. Oseguera, J. Cruz and M. José-Yacamán

SUPERCONDUCTIVITY AND STABILITY IN Y-Ba-Sr-Cu-O COMPOUNDS.......... 161
E. Carrillo, A. Mendoza, G. Pacheco, J. L. Albarrán,
J. Fuentes, L. Martínez and E. Orozco

NORMAL MODES OF VIBRATION IN HIGH T_c SUPERCONDUCTORS
A PICTORIAL VIEW 167
A. Calles, E. Yépez, A. Salcido,
J.J. Castro and A. Cabrera

MICROSTRUCTURE CHARACTERIZATION OF SUPERCONDUCTIVE
MIXED PHASES 172
J.M. Domínguez, C. Falcony, O. Guzmán, P. del Angel,
J. Omaña and A. Montoya

MECHANISMS OF SUPERCONDUCTIVITY

THE ELECTRONIC STRUCTURE AND SUPERCONDUCTIVITY
OF THE HIGH T_c COPPER OXIDES 183
M. Schluter

PROPERTIES OF HIGH T_c PHONON-EXCITON SUPERCONDUCTORS 196
J.P. Carbotte

ANTIFERROMAGNETIC INTERACTIONS IN HIGH-T_c SUPERCONDUCTORS 206
R.A. Barrio

SUPERCONDUCTOR-NORMAL METAL INTERFACES IN HIGH T_c
SUPERCONDUCTORS 219
   R. Nicolsky

CONTRIBUTED PAPERS

THE ORTHORHOMBIC TO TETRAGONAL PHASE TRANSITION
IN THE $Er_{1-x}La_xBa_2Cu_3O_y$ SYSTEM 237
   L. Govea, R. Escudero, D. Ríos-Jara, C. Piña,
   C. Wang and R.A. Barrio

SOLID SOLUTION IN THE SYSTEM $(Gd_yYb_{1-y})Ba_2Cu_3O_{7-x}$ 243
   G. Pacheco, Ma. de L. Chávez, and C. Piña

SUPERCONDUCTIVITY OF Gd-X-Ba-Cu-Oxides (X=Yb,La,Y) 247
   M. de L. Chávez, D. López, G. Pacheco and C. Piña

EVIDENCE OF STRUCTURAL CHANGES NEAR T_c IN A
$YBa_2Cu_{3-x}Fe_xO_\delta$ SUPERCONDUCTOR... 251
   R. Gómez, S. Aburto, V. Marquina, M.L. Marquina,
   M. Jiménez, C. Quintanar, T. Akachi, R. Escudero
   R.A. Barrio and D. Ríos-Jara

THERMOPOWER OF $Y_{(1-x)}Pr_xBa_2Cu_3O_{7-\delta}$ ($0 \leq x \leq 1$) 256
   A.P. Goncalves, I.C. Santos, E.B. Lopes,
   R.T. Henriques, M. Almeida and L. Alcacer

CRYSTAL FIELD INTERACTION AND MAGNETIC ORDER IN $GdBa_2Cu_3O_7$ 260
   M.T. Causa, C. Fainstein, G. Nieva, R. Sánchez,
   L.B. Steren, M. Tovar, R. Zysler, D.C. Vier,
   S. Schultz, S.B. Oseroff, Z. Fisk and J.L. Smith

SOME REMARKS ON EPR STUDIES OF Y-Ba-Cu-O COMPOUNDS 264
   G. Aguilar, H. Murrieta, J. Ramírez,
   T. Akachi, R.A. Barrio and R. Escudero

MECHANICAL CHARACTERIZATION OF $YBa_2Cu_3O_{7-x}$ 268
   L. Martínez, J.L. Albarrán, S. Valdés and J. Fuentes

STUDIES ON THE ELECTRONIC STRUCTURE OF $YBa_2Cu_3O_{7-x}$ 275
   C. de Teresa, P. de la Mora and J. Keller

ANALYSIS OF JAHN-TELLER DISTORTIONS IN $YBa_2Cu_3O_7$ COMPOUND 279
   A. Calles, E. Yépez, A. Salcido, J.J. Castro and A. Cabrera

STUDY OF THE SUPERCONDUCTORS CERAMICS OF THE TYPE
$Yb_2Ba_4Cu_6O_x$, $YbGdBa_4Cu_6O_x$ and $Gd_2Ba_4Cu_6O_x$ 284
   C. Piña, A. Montoya, P. Bosch and R. Escudero

VIBRONIC STATES AS THE POSSIBLE ORIGIN OF HIGH T_c
SUPERCONDUCTIVITITY 289
  J. Keller

AUTHOR INDEX..293

INVITED PAPERS

CHEMICAL ASPECTS

SAMPLE PREPARATION

SINGLE CRYSTAL TAKING OXYGEN

MIXED VALENCE COPPER OXIDES : RELATIONSHIPS BETWEEN CRYSTAL CHEMISTRY AND SUPERCONDUCTIVITY

B. Raveau, C. Michel, A. Maignan, M. Hervieu and J. Provost.

Laboratoire de Cristallographie et Sciences des Matériaux
I.S.M.Ra. Bd du Maréchal Juin. 14032 CAEN Cedex. FRANCE.

ABSTRACT : The influence of different factors - mixed valence of copper, coordination of this element, low dimensionality of the structure, phase transition and non-stoichiometry- upon superconductivity is analyzed for the two families of higher Tc superconductors La_2CuO_4 and $YBa_2Cu_3O_7$-type.

The mechanisms which govern superconductivity in copper oxides are far to be understood in spite of the numerous investigations carried out over the world since the discovery of this property (1) in $La_{2-x}Ba_xCuO_{4-y}$ (2). The key-words which are involved in different questions can be summarized as follows : mixed valence of copper, coordination resulting from the Jahn-Teller effect of this element, dimensionality of the structure, phase transition, oxygen stoichiometry deviation, effect of substitutions and especially of magnetic impurities. The recent studies concerning the two families of superconductors, La_2CuO_4 -type and $YBa_2Cu_3O_7$ -type, bring information about the relationships between superconductivity and those factors.

La_2CuO_4 - type structures

The examination of the electron transport properties of the oxides $La_{2-x}Sr_xCuO_{4-x/2+\delta}$ (3) shows that the mixed valence of copper $Cu(II)$-$Cu(III)$ is necessary for the existence of a semi-metallic or metallic behaviour of these materials : $La_{1.85}Sr_{0.15}CuO_4$ is metallic whereas $LaSrCu^{II}O_{3.5}$, which exhibits only $Cu(II)$, is a poor semiconductor. The importance of the mixed valence of copper which allows a delocalisation of the holes over the "copper oxygen framework" to be achieved is demonstrated for several other oxides such as $La_4BaCu_5O_{12+\delta}$ (4) or $La_{8-x}Sr_xCu_8O_{20-\epsilon}$ (5) which are also metallic. All those oxides have their structure closely related to the perovskite but, among them, only one is a superconductor - $La_{2-x}Sr_xCuO_{4-x/2+\delta}$. The comparison of the structure (Fig. 1) of this latter oxide (6) to those of the other copper oxides shows its particular character : it is indeed built up from single oxygen deficient single perovskite layers separated by insulating SrO-type layers. This low dimensionality of the copper-oxygen framework has to be opposed to the three dimensional character of the structures of $La_4BaCu_5O_{12+\delta}$ (7) or $La_{8-x}Sr_xCu_8O_{20-\varepsilon}$ (8) which are not superconductors in spite of their metallic properties. These observations suggest that two factors -mixed valence of copper and low-dimensionality of the structure - are absolutely necessary for the existence of superconductivity. Thus, these experimentals features are in agreement with the theory developped by Labbé and Bok (9) who explain the superconductivity in terms of a degenerate logarithmic singularity in the electronics states.

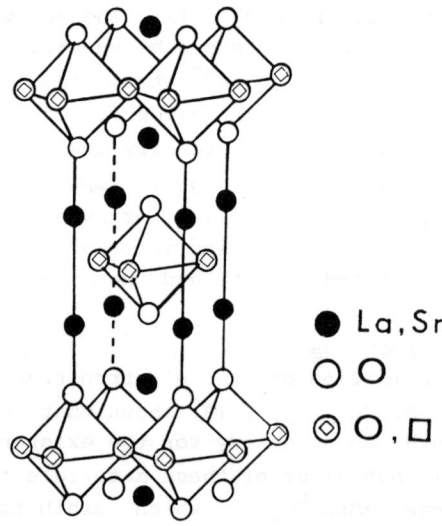

Fig.1. Structure of K_2NiF_4-type superconductor $La_{2-x}Sr_xCuO_{4-x/2+\delta}$

A remarkable feature concerning those oxides deals with the existence of a phase transition versus x or versus temperature. For instance La_2CuO_4 (x = 0) is orthorhombic at room temperature, and becomes only tetragonal beyond 233 °C. This orthorhombic symmetry, which corresponds in fact to a monoclinic distortion of the tetragonal cell, tends to disappear as x increases. Kang et al (10) have shown that the superconductivity appears near the boundary between the two domains, orthorhombic and tetragonal, and that Tc is maximum at the boundary. Thus it seems that the phase transition influences the superconducting properties of these oxides, since Tc decreases in the tetragonal domain. This phenomenon is interpreted by Friedel (4) in terms of a lattice or spin modulation.

Nevertheless, the recent observation for La_2CuO_4 of superconducting properties even in the bulk (12-13) shows that the relationships between symmetry and superconductivity are not really understood for this structural type. On the opposite the fact that "La_2CuO_4" can be prepared either as a semiconductor at low temperature, or as superconductor with a critical temperature of 37 K is easily explained in terms of oxygen non-stoichiometry. Starting from the notion of Schottky defects it is easy to understand that for a molar ratio La : Cu = 2 one can obtain either a Cu(II) oxide $La_{2-2\varepsilon}Cu^{II}_{1-\varepsilon}O_{4-4\varepsilon}$ or a mixed valence oxide $La_{2-2\varepsilon}Cu^{II,III}_{1-\varepsilon}O_{4-\varepsilon'}$ ($\varepsilon' < 4\varepsilon$). The first formulation, which can be obtained by synthesis of the oxide in air and quenching, will correspond to a semiconducting behaviour, whereas annealing in air during a long time, or under an oxygen flow, or under a high oxygen pressure leads to the second formulation which presents superconductivity. However in this latter case one must distinguish the bulk superconductor obtained at high pressure for which ε' tends towards zero, from the surface superconductor obtained at normal oxygen pressure for which ε' is not negligeable compared to 4ε. Changing the ratio La: Cu can favour the appearance of superconductivity. This has been shown by using an excess of copper oxide with respect to the nominal composition La_2CuO_4: the curve $\rho(T)$ exhibits indeed in this case a transition at 37K and a much lower resistivity for the composition "$La_2O_3:1.04CuO$". Such a phenomenon is interpreted by a simultaneous deficiency of lanthanum and oxygen in the structure leading to the "semiconductor" $La_{1.92}Cu^{II}O_{3.88}$, whose oxidation allows the limit superconductor $La_{1.92}Cu^{II,III}O_4$ to be synthesized. In the case of an excess of lanthanum with respect to the nominal composition La_2CuO_4, two antagonist effects will be involved which can be understood by considering the formula $La_2Cu_{1-\delta}Cu^{II}O_{4-\delta}$. The anionic deficiency favours the oxidation of Cu(II) into Cu(III), leading to the limit superconductor $La_{2-\delta}Cu^{II}_{1-\epsilon}{:}Cu^{III}_\epsilon O_4$; but on the opposite the copper deficiency ($\delta \gg \epsilon$) tends to break the Cu-O chains, leading to a decrease of Tc as shown by the experimental results.

$YBa_2Cu_3O_7$ type oxides.

The 92K-superconductor $YBa_2Cu_3O_7$ (14-18) confirms the role of the mixed valence of copper in the superconducting properties of those oxides. It is remarkable that the molar ratio CuIII:CuII equal to 0.5 in this compound is about three times greater than the maximum value obtained for $La_{1.85}Sr_{0.15}CuO_4$ ($\simeq 0.18$) which exhibits a much lower critical temperature. The second factor -low dimensionality of the structure- which seems to play a role in superconductivity is also observed for $YBa_2Cu_3O_7$. The structure of this oxide (Fig. 2) is indeed built up of $[Cu_3O_7]_\infty$ triple layers of corner-sharing CuO_5 pyramids and CuO_4 square planar groups, whose cohesion is ensured by yttrium planes. The low-dimensionality of the structure is in fact more complex than for La_2CuO_4 -type oxides. One can indeed distinguish $[CuO_{2.5}]_\infty$ pyramidal layers which are connected through $[CuO_2]_\infty$ files of CuO_4 groups parallel to $\vec{b}$. These infinite files of square planar groups, which are isolated one from each other and connected to the pyramidal layers by a much larger Cu-O distances (2.35Å against 1.95Å), give to this structure an unidimensional character. The presence of such $[CuO_2]_\infty$ chains implies for this oxygen deficient perovskite an orthorhombic symmetry.

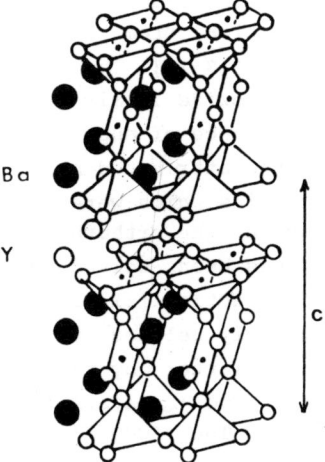

Fig.2. Structure of the orthorhombic superconductor $YBa_2Cu_3O_{7-\delta}$

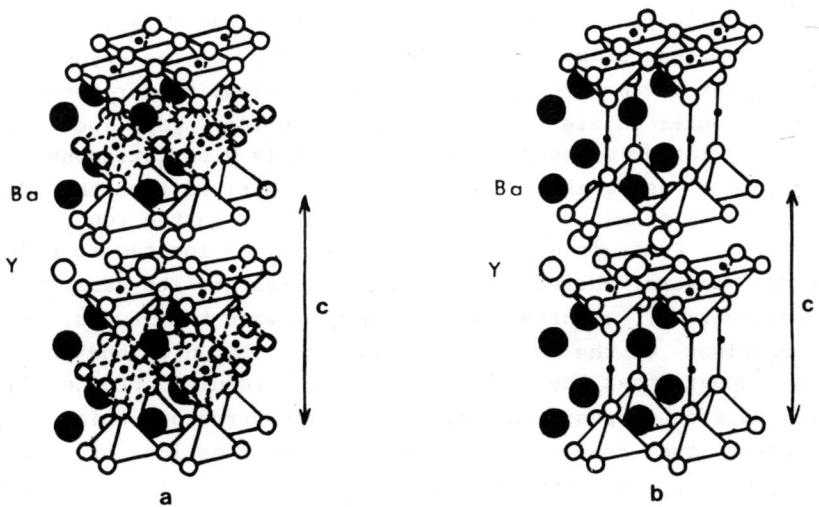

Fig.3. a) Structure of the tetragonal phase $YBa_2Cu_3O_{6.25}$
b) Structure of the tetragonal oxide $YBa_2Cu_3O_6$

Like for the first series of superconductors, $YBa_2Cu_3O_7$ exhibits a phase transition, from orthorhombic to tetragonal, by increasing temperature [19-20]. The tetragonal phase can be isolated by quenching the oxide in air from 950°C down to room temperature. The disappearance of superconductivity for this symmetry was as a first step attributed by several authors to a breaking of the $[CuO_2]_\infty$ chains which were thought to be at the origin of superconductivity in the orthorhombic form. The structure of this phase (Fig.3a) has been solved by several authors by X-ray and neutron powder diffraction [21-23]. This oxygen deficient perovskite is very closely related to the orthorhombic form : it only differs from the latter by the distribution of the oxygen atoms belonging to the $[CuO_2]_\infty$ layers, located between the pyramidal $[CuO_{2.5}]_\infty$ layers. The $[CuO_2]_\infty$ chains of square planar groups, running along $\vec{b}$ in the orthorhombic form, are replaced by planes of strongly oxygen deficient CuO_6 octahedra in the tetragonal form. Moreover the oxygen vacancies are statistically distributed in the basal plane of the octahedra. However this "transition" is in fact rather complex, since the change of symmetry is accompanied by an oxygen loss, so that the Cu(III): Cu(II) ratio may decrease dramatically. This decrease of the oxygen content sets out the problem of the copper coordination for the compositions intermediate between $YBa_2Cu_3O_7$ and $YBa_2Cu_3O_6$. The structure of this latter compound (Fig.3b), which is also tetragonal can be described as formed of $[Cu_3O_6]_\infty$ layers derived from the $[Cu_3O_7]_\infty$ layers by elimination of the oxygen atoms in the $[CuO_2]_\infty$ layers (24-26). Thus, in such a structure Cu(II) is located in the pyramidal layers, whereas Cu(I), which presents its usual two-fold coordination, ensures the cohesion between the layers. From the comparison of the two limit structures, $YBa_2Cu_3O_7$ and $YBa_2Cu_3O_6$, it appears that the neutron diffraction results obtained for intermediate compositions lead to a mean structure which does not reflect the reality. The recent investigation of the oxide $YBa_2Cu_3O_{6.50}$ (27) confirms this point of view. This composition, which corresponds to the mean oxidation state two of copper, was obtained by quenching in air the superconductor $YBa_2Cu_3O_{6.9}$ from 950°C down to room temperature. The refinement from neutron data leads to the structure of Fig.3a with a high oxygen deficiency corresponding to the formulation $YBa_2Cu_3O_{6.25}$ in agreement with previous results. The important features concerning the neutron data deals with the high background which suggests the presence of an amorphous phase, and the high anisotropic thermal factor of some oxygen atoms which indicates a rather high degree of disorder. The electron microscopy study of this composition shows that the crystal chemistry is more complex than deduced from powder data. The electron diffraction patterns show indeed that the crystallization varies from one grain to the next, ranging from a very poor to a quite good quality. A large majority of crystals exhibit different streaks along the $\vec{c}$ axis (Fig. 4a) with h0l or hhl spots arranged in continuous arcs (Fig. 4b). Correlatively the low resolution images of these crystals show the existence of numerous defects such as variation of the layer spacing, misorientation between different areas and presence of amorphous layers (Fig. 4c).

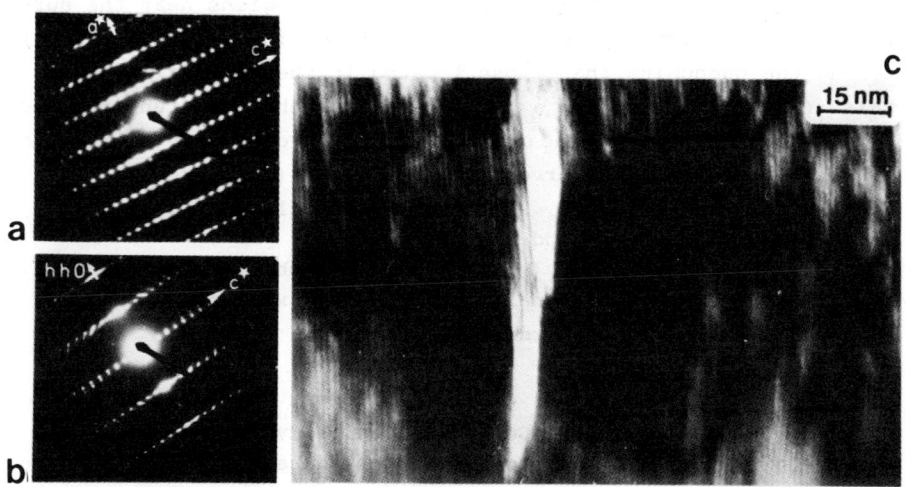

Fig.4. a) h0l electron diffraction pattern with diffuse streaks parallel to $\vec{c}*$, b) hhl electron diffraction pattern showing arcs, c) corresponding low resolution image.

The high resolution electron microscopy investigation confirms that crystals can be strongly disturbed and that most of them are coated with an amorphous layer of some ten angström thickness. One important point concerns the variations of image contrast from one crystal to the other or from one part of the crystal to the other, (Fig. 5) which are in favour of a significant variation of the oxygen content, as confirmed by the

systematic simulation of the through focus series (see some examples Fig. 6), taking into account the fact that a change of oxygen distribution implies a displacement of the cations. These results clearly demonstrate that the intermediate compositions between "O_7" and "O_6" cannot be characterized by only two oxidation states of copper Cu(II)-Cu(III) or Cu(II) - Cu(I) for $0 < \delta < 0.5$ and $0.5 < \delta < 1$ respectively. For instance in the present case the formulation $YBa_2 Cu^{II}_{2.5} Cu^{I}_{0.5} O_{6.25}$ cannot be retained since it would imply a coordination smaller than four for a part of divalent copper, which is not likely. On the opposite the inhomogeneity of the structure observed by HREM, and especially diffusion streaks observed along $\vec{c}$ are in favour of the coexistence in the same particle of Cu(III)-Cu(II) and Cu(II)-Cu(I) regions, leading to the following formulation : $YBa_2 [Cu^{II}_2 Cu^{III} O_7]_{0.25} [Cu^{II}_2 Cu^{I} O_6]_{0.75}$. This apparent . "disproportionation" of Cu(II) into Cu(III) and Cu(I) can be explained by the method of preparation, the regions near the surface of the particles being more easily oxidized (Cu(III)-Cu(II)) during the quenching than those located in the core of the crystals (Cu(II)-Cu(I)). Such a distribution of the different regions can also be applied to the orthorhombic superconducting form, and especially to that observed by Cava et al (28) for oxygen contents lower than "$O_{6.5}$" so that the formulation $YBa_2 Cu_3 O_{7-\delta}$, should be better understood in the form
$YBa_2 [Cu^{II}_2 Cu^{III} O_7]_{1-\delta} [Cu^{II}_2 Cu^{I} O_6]_{\delta}$. The orthorhombic form should only differ from the tetragonal one by the existence of infinite chains $[CuO_2]_\infty$ perfectly ordered.

The existence of extended defects in the orthorhombic superconducting phase $YBa_2 Cu_3 O_{7-\delta}$ has already been described elsewhere (29-35) and will not be discussed here. It must just be recalled that all the crystals exhibit microtwinning, due to the transition orthorhombic to tetragonal and that ceramics do not exhibit only perfect crystals but that numerous defects are observed, showing regions characterized either by an excess of oxygen or by an oxygen substoichiometry. All these extended defects could have an influence on the superconducting properties of this material.

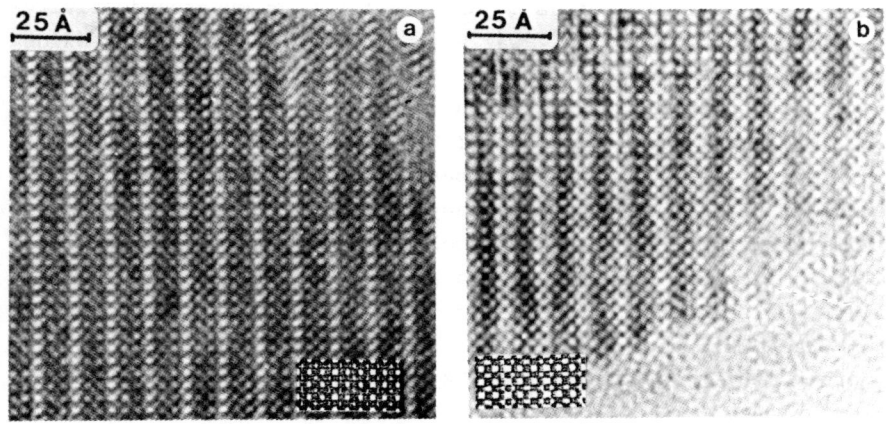

Fig.5. [100] high resolution images showing different oxygen-content areas : a) poor b) rich ones. Insert images calculated with Df= -150Å, t = 30Å, corresponding to the formulation :
a) $YBa_2Cu_3O_{6.2}$ and b) $YBa_2Cu_3O_7$.

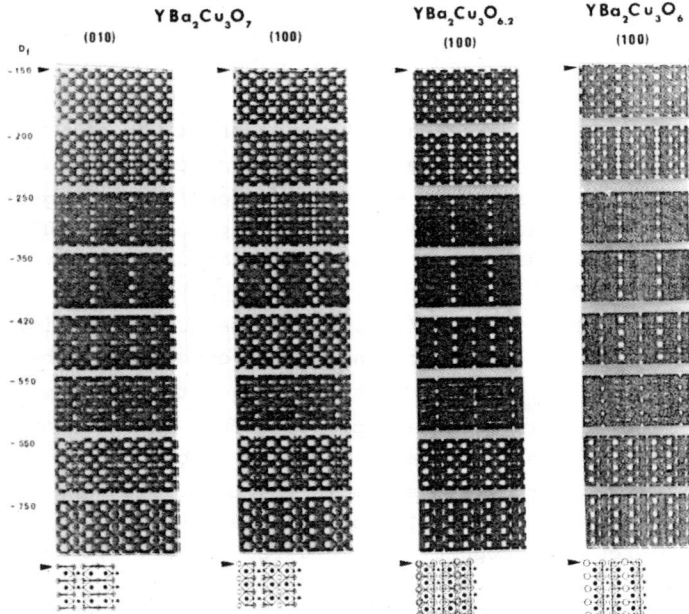

Fig.6. Through focus simulated images : voltage v = 200kV, spherical aberration constant Cs = 0.8mm, semi-angle of beam convergence α/2 = 0.8 mrad, focus spread Df = 150Å, 111 beams, thickness ≃ 30Å.

The possible influence of symmetry upon superconductivity in the oxide $LaBa_2Cu_3O_{7-\delta}$ was also recently studied. This oxide was found to be a superconductor at temperatures ranging from 60K to 75K in spite of its tetragonal symmetry (36-37). However a recent microscopy investigation (38) shows that this superconducting phase exhibits only an apparent tetragonality. Besides rigorously tetragonal crystals, many crystals are characterized by a distortion of the tetragonal cell. The deviation from tetragonality which can be rather weak, is shown by the presence of twins. Moreover two types of domains are observed which result from the distortion of the cell and form the setting up of the superstructures ($\vec{c}$ axis) along the three equivalent perpendicular orientations.

Substitution reactions in $YBaCu_3O_{7-\delta}$ bring interesting information about mechanisms of superconductivity in this structure. The substitution of yttrium ions by magnetic rare earth ions without changing Tc (39) has shown that these cations were very weakly involved in superconductivity contrary to the $[Cu_3O_7]_\infty$ layers. On the opposite substitution of copper for different cations leads to a significant decrease of superconductivity (40-41) in agreement with the breaking of Cu-O chains. However some of those effects are unexpected such as the fact that Tc is less affected by introduction of the magnetic ions Fe^{3+} than by Zn^{2+}. An explanation to this phenomenon could be that Fe^{3+} ions can be distributed between pyramidal sites and $[CuO_2]_\infty$ files, whereas Zn^{2+} more rarely takes the pyramidal coordination. Consequently the $[CuO_2]_\infty$ files would be much more disturbed by Zn^{2+} than by Fe^{3+} (displacement of the oxygen atoms between the pyramidal layers). Such a hypothesis suggests the prominent role of the $[CuO_2]_\infty$ rows of square planar groups in superconductivity with respect to the pyramidal layers. The recent study of the replacement of copper by palladium in $YBa_2Cu_3O_{7-\delta}$ is in agreement with this point of view (42). In the oxide $YBa_2Cu_{2.6}Pd_{0.4}O_{6.5}$ Pd(II) partially replaces copper in an ordered manner, i.e. in the CuO_4 groups of the $[CuO_2]_\infty$ rows. Contrary to $YBa_2Cu_3O_{7-\delta}$ this oxide is no more a superconductor, although being orthorhombic. Thus it appears that Pd (II) kills the superconductivity in spite of its

$4d^8$ configuration similar to that of Cu(Ⅲ). This suggests that Cu(Ⅲ) has a particular configuration which would be $3d^9 L$ (with holes in the oxygen band) rather than $3d^8$ in agreement with Bianconi et al (43). Moreover, it seems from this study that Cu(Ⅲ) would be mainly located in the $[CuO_2]_\infty$ rows and that those latter rows of square planar groups would be mainly responsible for superconductivity.

REFERENCES

1. J.G. BEDNORZ and K.A. MULLER Z. Phys. **B64**, 189 (1986).
2. C. MICHEL and B. RAVEAU, **21**, 407 (1984).
3. N. NGUYEN, F. STUDER and B. RAVEAU, J. Phys. Chem. Solids, **44**, 389 (1983).
4. C. MICHEL, L. ER-RAKHO and B. RAVEAU, Mat. Res. BULL. **20**, 667 (1985).
5. C. MICHEL L. ER-RAKHO and B. RAVEAU, J. Phys. Chem. Solids, in press.
6. N. NGUYEN, J. CHOISNET, M. HERVIEU and B. RAVEAU, J. Solid State Chem. **39**, 120, (1981).
7. C. MICHEL, L. ER-RAKHO, M. HERVIEU, J. PANNETIER and B. RAVEAU, J. Solid State Chem, **68**, 143 (1987).
8. L. ER-RAKHO, C. MICHEL and B. RAVEAU, J. Solid State Chem. in press.
9. J. LABBE and J. BOK, Europhysics Letters **3**, 1225 (1987).
10. W. KANG, G. COLLIN, M. RIBAULT, J. FRIEDEL, D. JEROME, J.M. BASSAT, J.P. COUTURES and Ph. ODIER, J. Physique.
11. J. FRIEDEL J. Physique **48**, 1787 (1987).
12. J. BEILLE, R. CABANEL, C. CHAILLOUT, B. CHEVALIER, G. DEMAZEAU, F. DESLANDES, J. ETOURNEAU, P. LEJAY, C. MICHEL, J. PROVOST, B. RAVEAU, A. SULPICE, J.L. THOLENCE and R. TOURNIER, C.R. Acad.Sc. **304** Ⅱ, 1097 (1987).
13. J. BEILLE, B. CHEVALIER, G. DEMAZEAU, F. DESLANDES, J. ETOURNEAU, O. LABORDE, C. MICHEL, P. LEJAY, J. PROVOST, B. RAVEAU, A. SULPICE, J.L. THOLENCE, Physica B under press.
14. M.K. WU, J.R. ASHBURN, C.D. TORNG, P.H. HOR, R.L. MENG, L. GAO, Z.J.HUANG, Y.Q. WANG and C.W. CHU, Phys. Rev. Lett., **58**, 908, (1987).
15. C. MICHEL, F. DESLANDES, J. PROVOST, P. LEJAY, R. TOURNIER, M. HERVIEU and B. RAVEAU, C.R. Acad. Sc. **17** Ⅱ, 1050 (1987).

16. R.J. CAVA, B. BATTLOG, R.B. VANDOVER, D.W. MURPHY, S. SUNSHINE, T. SIEGRIST, J.P. REMEIKA, E.A. RIETMAN, S. ZAHURAK and G.P. ESPINOSA Phys. Rev. Lett. , **58** 1676 (1987).
17. Y. LEPAGE, W.R. MC KINNON, J.M. TARASCON, L.H. GREENE, G.W. HULL and D.M. HWANG, Phys.Rev., **B35**, (1987), 7245.
18. J.J.CAPPONI, C. CHAILLOUT, A.W. HEWAT, P. LEJAY, M. MAREZIO, N. NGUYEN, B. RAVEAU, J.L. SOUBEYROUX, J.L. THOLENCE and R. TOURNIER, Europhysics Lett.,12, (1987),1301.
19. M.A. BENO, L. SODERHOLM, D.W. CAPONE, D.G. HINKS, J.D. JORGENSEN, J.K. SCHULLER, C.V. SEGRE, K. ZHANG, J.D. GRACE, Appl. Phys. Lett. to be published.
20. G. ROTH, D. EWERT, G. HEGER, C. MICHEL, M. HERVIEU, B. RAVEAU, F. D'YVOIRE and A. REVCOLEVSHI, Zeitschrift für Physik, to be published.
21. F. IZUMI, H. ASANO, T. ISHIGAKI, E. TAKAYAMA, Y. UCHIDA, N. WATANABE and T. NISHIKAWA, Japn. J. Appl. Phys., **26**, 649 (1987).
22. F. IZUMI, H. ASANO, T. ISHIGAKI, E. TAKAYAMA, Y. UCHIDA and N. WATANABE, Japn. J. Appl. Phys. **26**, 1214 (1987).
23. J.D. JORGENSEN, M.A. BENO, D.G. HINKS, L. SODERHOLM, K.J. VOLIN, R.L. HITTERMAN, J.D. GRACE, I.K. SCHULLER, C.V. SEGRE, K. ZHANG and M.S. KLEEFISCH, Phys. Rev., **B36**, 3608 (1987).
24. A. SANTORO, S. MIRAGLIA, F. BEECH, D.W. MURPHY, L.F. SCHNEEMEUER and J.V. WASZCAK, Mat. Res. Bull. under press.
25. G. ROTH, B. RENKER, G. HEGER, M. HERVIEU, B. DOMENGES and B. RAVEAU, Z. für Phys. under press.
26. J.J. CAPPONI, C. CHAILLOUT, M. MAREZZIO, High Tc superconductor RCP, Meeting. Bordeaux. April 1987.
27. B. DOMENGES, M. HERVIEU, V. CAIGNAERT, B. RAVEAU, JL. THOLENCE a,d R. TOURNIER, J. Micr. et Spectr. Electroniques, under press.
28. B. BATLOGG, R.J. CAVA, C.H. CHEN, G. KOUROUKLIS, W. WEBER, A. JAYARAMAN, A.E. WHITE, K.T. SHORT, E.A. RIETMAN, L.W. RUPP, D. WERDER and S.M. ZAHURAK, International Workshop on Novel Mechanisms of Superconductivity. Berkeley. June 1987. Ed. S.A. Wolf. and V.Z. Kresin p. 652.
29. M. HERVIEU, B. DOMENGES, C. MICHEL, G. HEGER, J. PROVOST and B. RAVEAU, Phys. Rev. **B36**, 3920 (1987).

30. M. HERVIEU, B. DOMENGES, C. MICHEL and B. RAVEAU, Europhysics Letters **4(2)**, 205 (1987).
31. B. DOMENGES, M. HERVIEU, C. MICHEL and B. RAVEAU, Europhysics Letters, **4(2)**, 211 (1987).
32. M. HERVIEU, B. DOMENGES, C. MICHEL, J. PROVOST and B. RAVEAU, J. Solid State Chem., in press.
33. G. VAN TENDELOO, H.W. ZANDBERGEN and S. AMELINCKX, Solid State Commun, **63**, 603 (1987).
34. G. VAN TENDELOO, H.W. ZANDVERGEN and S. AMELYNCKX, Solid State Comm., **63**, 603 (1987).
35. H.W. ZANDBERGEN, G. VAN TENDELOO, T. OKABE and S. AMELINCKX, Submitted to Phys. Statu Solidi (A).
36. C. MICHEL, F. DESLANDES, J. PROVOST, P. LEJAY, R. TOURNIER, M. HERVIEU and B. RAVEAU, C.R. Acad. Sc. **304** II, 1169 (1987).
37. R. YOSHIZAKI, H. SAWADA, T. IWAZUMI, Y. SAITO, Y. ABE, H. IKEDA, I. NAKAI, International Workshop on Novel Superconductivity, Berkeley. June 1987. Edit. S.A. Wolf AND V.Z. Kresin p. 1089.
38. M. HERVIEU, B. DOMENGES, J. PROVOST, F. DESLANDES and B. RAVEAU., Angewandte Chem., under press.
39. J.M. TARASCON, L.H. GREENE, B.G. BAGLEY, W.R. MC KINNON, P. BARBOUX and G.W. HULL, International Workshop on Novel Superconductivity. Berkeley. June 1987. Edit. S.A. Wolf AND V.Z. Kresin, p. 705.
40. Y. MAENO and T. FUJITA, International Workshop on Novel Superconductivity. Berkeley. June 1987. Edit. S.A. Wolf and V.Z. Kresin, p. 1073.
41. S.B. OSEROF, D.C. VIER, J.F. SMYTH, C.T. SALLING, S. SCHULTZ, Y. DALICHAOUCH, B.N. LEE, M.B. MAPLE, Z. FISK, J.D. THOMPSON, J.L. SMITH and E. ZIRNGIEBL, International Workshop on Novel Superconductivity. Berkeley. June 1987. Edit. S.A. Wolf and V.Z. Kresin, p. 679.
42. Y. LALIGAN, A. LEBAIL, G. FEREY, M. HERVIEU, B. RAVEAU, A. SULPICE, R. TOURNIER, J. de Physique, submitted.
43. A. BIANCONI, J. BUDNICK, A.M. FLANK, A. FONTAINE, P. LAGARDE, A. MARCELLI, H. TALENTINO, B. CHAMBERLAND, G. DEMAZEAU, C. MICHEL and B. RAVEAU, Phys. Letters A, in press.

THE ROLE OF OXYGEN IN $YBa_2Cu_3O_{7-\delta}$

John B. Goodenough and A. Manthiram
Center for Materials Science & Engineering, ETC 5.160
The University of Texas at Austin, Austin, Texas 78712

ABSTRACT

Five observations concerning the role of oxygen in the high-T_c superconductors $YBa_2Cu_3O_{7-\delta}$ are emphasized: (1) An internal electric field parallel to the c-axis constrains the redox reactions associated with the intercalation/disintercalation of oxygen primarily to the $[CuO_{3-\delta}]^{3-}$ layers located between Ba^{2+}-ion layers. (2) Intercalation of O^{2-} ions into the tetragonal $YBa_2Cu_3O_6$ phase results in an ordering onto one of the basal-plane axes of the Cu(1) layer, the b-axis of the orthorhombic phase. (3) Ordering between and within b-axis chains gives rise to discrete phases predicted to occur at $O_{6.875}$, $O_{6.75}$, $O_{6.5}$, $O_{6.25}$, and $O_{6.125}$. (4) An equilibrium oxidation state for the intercalation layer depends not only on the oxygen partial pressure and temperature, but also on the cations of the structure. (5) At higher oxidation states of the intercalation layer, any oxygen atom on the a-axis interacts with a near-neighbor oxygen on the b-axis to trap out holes in a peroxide ion $(O_2)^{2-}$.

INTRODUCTION

The system $YBa_2Cu_3O_{7-\delta}$, $0.04 \leq \delta \leq 1$, has the tetragonal structure of Fig. 1(a) in the limit O_6 (i.e. $\delta = 1$) [1]; it has the orthorhombic structure of Fig. 1(b) in the ideal limit O_7 (i.e. $\delta = 0$) [2]. By varying the temperature in the interval 350 - 950 °C and the partial pressure of oxygen over the range $0 \leq p_{O_2} \leq 1$ atm, it is possible to adjust the equilibrium oxygen concentration over the entire range $O_{6.96}$ to O_6 ($0.04 \leq \delta \leq 1$) [3]. The mobile oxygen is rapidly inserted into/extracted from the Cu(1) planes at temperatures above 350 °C; at room temperature the oxygen is essentially immobile, so non-equilibrium oxygen concentrations can be quenched in. The full range of oxygen compositions $0.04 \leq \delta \leq 1$ has been obtained at room temperature; the structure was found [3] to change smoothly with oxygen concentration through an orthorhombic-tetragonal transition at $O_{6.27}$, see Fig. 2. At lower oxygen concentrations (O_6 to $O_{6.27}$), the inserted oxygen atoms occupy the Cu(1)-bridging positions on the two equivalent axes of the tetragonal structure; at higher oxygen concentrations ($O_{6.27}$ to $O_{6.96}$) the inserted oxygen are ordered onto one of these axes, the b-axis of the orthorhombic structure.

On raising the temperature in air or O_2, the $O_{6.96}$ composition loses oxygen, but the O_6 composition initially picks up oxygen. Fig. 3 illustrates the weight change due to oxygen insertion on heating an $O_{6.18}$ sample in 1 atm O_2 at 1 °C/min; the equilibrium oxygen concentration is approached below 400 °C, and a continued increase in temperature results in a loss of oxygen as the equilibrium oxygen concentration decreases. On cooling from 920 °C at 1 °C/min, the sample regains oxygen, but the time at temperature is never sufficient for the full equilibrium concentration to be reached [3]. High-temperature x-ray data of an oxidized sample shows an orthorhombic-tetragonal transition occurs smoothly, but rather abruptly, near 600 °C in air [4]; in 1 atm O_2, it occurs near 680 °C [5]. The c-axis oxygens--O(1) of Fig. 1--neighboring the Cu(1) atoms are not disordered by either temperature or changing oxygen concentration [6].

On the other hand, the superconducting transition temperature T_c does not vary smoothly with oxygen concentration. Cava et al. [7] have shown that T_c varies stepwise from near 90 K at $O_{6.96}$ to about 60 K in the vicinity of $O_{6.67}$ and to a lower temperature ($T_c \approx 20K$) below $O_{6.5}$. Moreover, the transitions are sharp and the room-temperature resistivity is lower in the center of a plateau region; the transitions are broad at intermediate compositions. These data clearly indicate the presence of some additional ordering not readily detectable with neutron or x-ray diffraction.

This chemical flexibility raises several important questions that can be addressed experimentally.

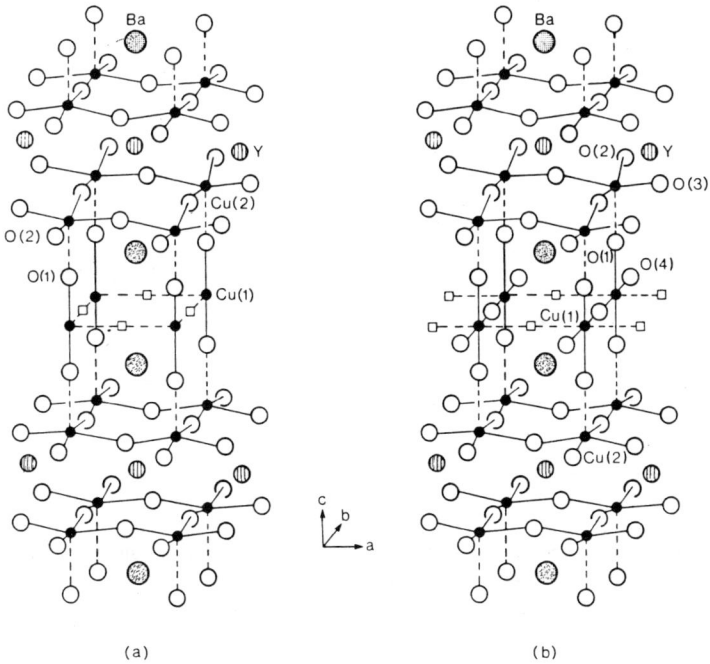

FIG. 1. Structure of (a) tetragonal $YBa_2Cu_3O_6$ and (b) orthorhombic, ideal $YBa_2Cu_3O_7$.

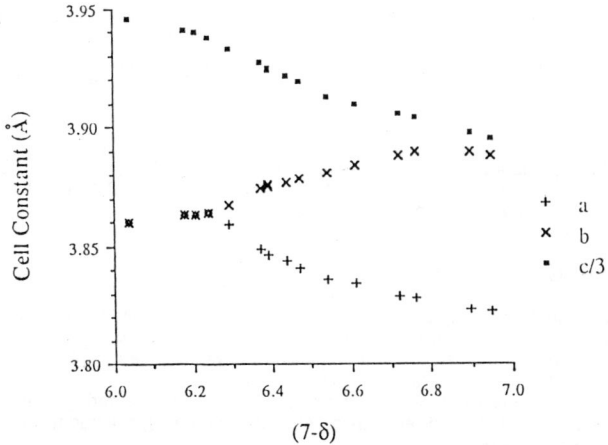

FIG. 2. Variation of room-temperature lattice parameters with oxygen concentration in the system $YBa_2Cu_3O_{7-\delta}$, after Swinnea (3).

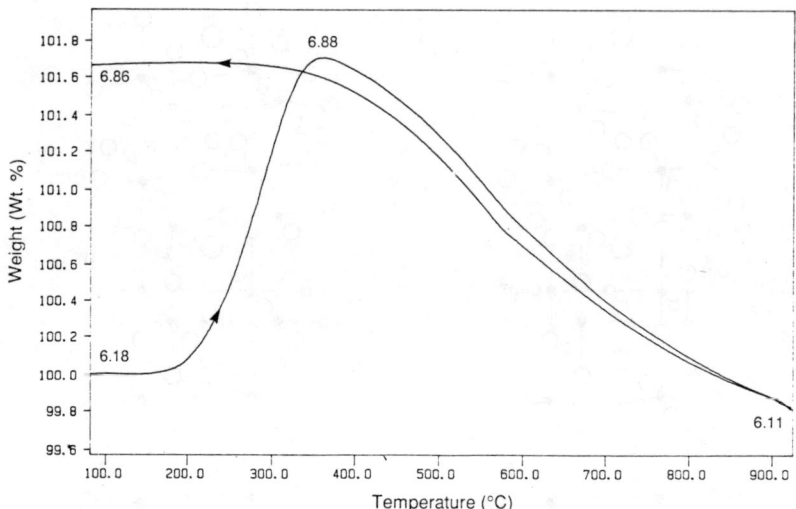

FIG. 3. TGA curve for a metastable, room-temperature $YBa_2Cu_3O_{6.18}$ sample heated in 1 atm O_2 at 1 °C/min. Numbers refer to oxygen content.

THE REDOX REACTION

The structure of Fig. 1 represents an intergrowth of layers of different charge. As a result, an internal electric field is established parallel to the c-axis that determines not only the location of the oxygen vacancies in the structure, but also the region of the crystal where redox reactions occur on insertion/extraction of oxygen.

The crystallographic feature that builds in an internal electric field is the stabilization of eightfold coordination at the Y^{3+} ion instead of the twelvefold coordination found in the perovskite structure. In the tetragonal $YBa_2Cu_3O_6$ structure of Fig. 1(a), the Cu(1) atoms have a linear O-Cu-O coordination characteristic of Cu(I) species; fivefold-coordinated Cu(2) atoms have the formal valence Cu(II). The Cu(2) atoms order antiferromagnetically, the Cu(1) atoms carry no magnetic moment [8].

Introduction of oxygen into the Cu(1) plane must oxidize primarily the Cu(1) atoms, not the Cu(2), if the internal electric field is to be minimized. Therefore insertion/extraction of oxygen into/from the Cu(1) plane introduces a redox reaction between Cu(1) and inserted oxygen that is essentially constrained to the Cu(1) plane and its near-neighbor oxygen O(1), which means that this layer retains its formal charge $[Cu(1)O_{3-\delta}]^{3-}$ for all values of δ.

OXYGEN ORDERING

A remarkable feature of Fig. 2 is the appearance of ordering onto the b-axis of the inserted oxygen with less than 15 percent occupancy of the Cu(1)-bridging oxygen sites. Two factors contribute to the ordering, the electrostatic repulsion between oxide ions and the preference of formal-valence Cu(II) for square-coplanar anion coordination.

Two types of additional oxygen ordering are possible, intrachain ordering of oxygen vacancies and interchain ordering of fully oxidized chains. At a composition $O_{6.5}$, intrachain ordering within a b-axis chain may be represented as

$$- \square \text{-Cu-O-Cu-} \square \text{-Cu-O-Cu-} \square -$$

This ordering would double the periodicity along the b-axis and probably along the a-axis as well; the Cu(1) atoms would be uniformly oxidized to Cu(II) and have only three coplanar oxygen each. Interchain ordering, on the other hand, is illustrated in Fig. 4 for $O_{6.5}$; it would give a doubling of the periodicity only along the a-axis. It involves a disproportionation of the Cu(1) atoms into "Cu(III)" within a complete chain, each having square-coplanar coordination, and Cu(I) with linear, twofold oxygen coordination. Electron-diffraction data [9] show a doubling only along the a-axis, which indicates that interchain ordering is stabilized.

Electron diffraction data [9] also indicate an ordering of oxygen vacancies in $O_{6.875}$, and of inserted oxygen in $O_{6.125}$, that produce periodicities in the a-b plane corresponding to $2\sqrt{2}a_c \times 2\sqrt{2}a_c$, where a_c is a Cu-Cu distance. These basal-plane periodicities are uncorrelated along the c-axis.

These data can be harmonized with the Gibbs phase rule applied to the observation [7] of steps in T_c vs composition by the preliminary phase diagram of fig. 5.

The solid curve in Fig. 5 is an experimental thermogravimetric oxygen analysis (TGA) like that of Fig. 3 taken in an O_2 atm at 1 °C/min. The dashed orthorhombic-tetragonal transition temperature T_t is estimated from the room-temperature transition at $O_{6.27}$ and the equilibrium transition in 1 atm air near 680 °C. The shaded areas represent, schematically, immiscibility domes between ordered phases.

In the oxidized phases, the oxygen vacancies are attracted to twin-plane boundaries, as is illustrated in Fig. 6, so as to inhibit formation of close oxide-ion pairs across such a boundary. The orthorhombic phase contains a high density of twin planes of the type illustrated, but the density is not great enough to attract all the oxygen vacancies at the equilibrium concentration $O_{6.96}$. Therefore ordering into b-axis chains prevents near-neighbor contact between inserted oxygen over the entire compositional range $O_{6.96}$ to O_6.

FORMAL VALENCES

The formal valences Y^{3+} and Ba^{2+} are not at issue. However, the assignment of formal valences to the other atomic species does require definition.

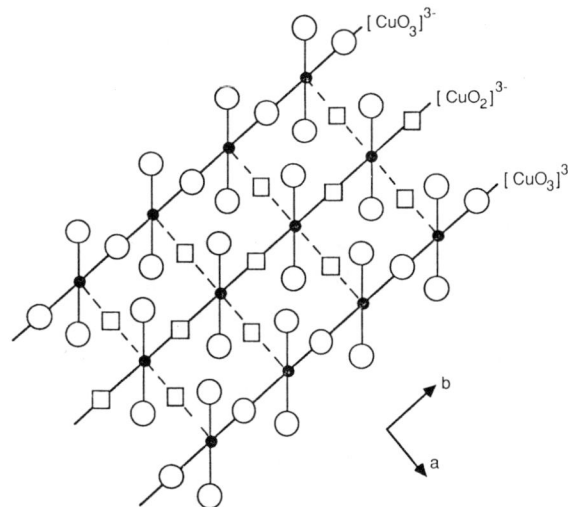

FIG. 4. Interchain ordering in a-b plane for $O_{6.5}$.

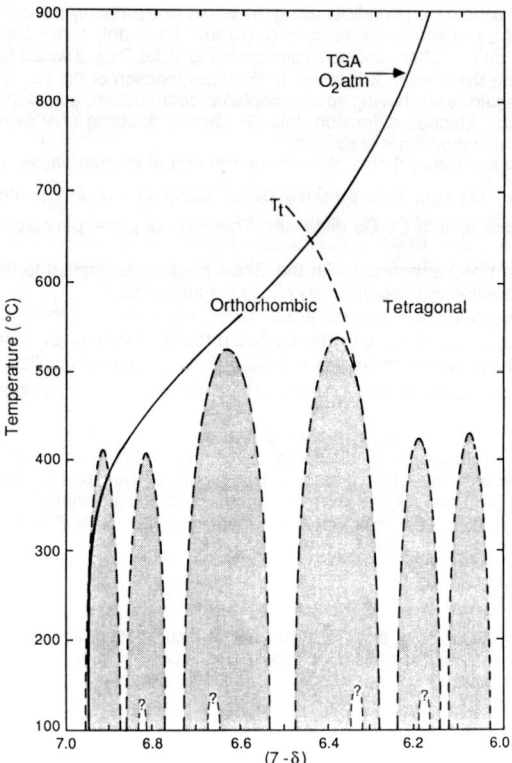

FIG. 5. Schematic phase diagram for the system $YBa_2Cu_3O_{7-\delta}$ and the TGA curve, solid line, for O_2 atm at 1 °C/min.

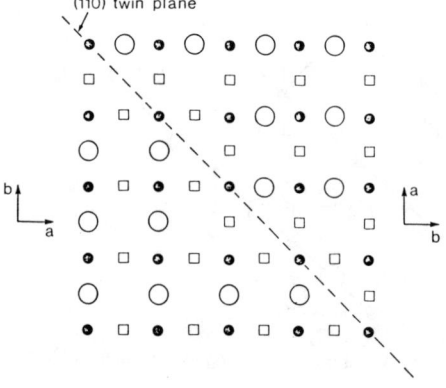

FIG. 6. Orthorhombic $YBa_2Cu_3O_{7-\delta}$ fully oxidized except at twin planes.

The covalent hybridization between Cu-3d and O-2p orbitals is strong. The copper d_{z^2} and $d_{x^2-y^2}$ orbitals (x,y,z axes are the crystallographic axes) σ-bond with the nearest-neighbor oxygen atoms; they are orthogonal to the O-p_π orbitals (we neglect the displacement of the Cu(2) atoms from the oxygen plane). Conversely, the Cu-3d orbitals d_{xy}, d_{yz}, d_{zx} π-bond with the nearest-neighbor oxygen atoms; they are orthogonal to the O-p_σ and O-s orbitals. So long as the Madelung energy stabilizes the O-2p orbitals relative to the cation 3d orbitals, the antibonding states retain the symmetry of the cation 3d orbitals after covalent mixing, and a formal valence at the cation designates the number of electrons or holes in these antibonding states whether they remain "localized" or are made itinerant by interatomic interactions.

What makes the copper oxides a special case is the positioning of the relative energies of the Cu-3d and O-$2p_\sigma$ orbitals. The $Cu^{3+/2+}$ redox couple lies above the O-$2p_\sigma$ energy, so covalent mixing raises the energy of the σ-bonding, copper crystal-field orbitals and lowers the energy of the corresponding O-$2p_\sigma$ orbitals. The antibonding σ* band is therefore composed of orbitals of primarily Cu-3d parentage; the square-pyramidal and square-coplanar symmetries at Cu(II) and Cu(III) atoms show this to be the case. On the other hand, the O-$2p_\pi$ energies may lie above the π-bonding Cu-3d energies; in this event, covalent mixing would lower the energy of the π-bonding copper crystal-field orbitals and raise the energy of the corresponding O-$2p_\pi$ orbitals. The antibonding π* band would thus be composed of orbitals of primarily O-$2p_\pi$ parentage. Nevertheless, this situation need not create a potential ambiguity in the definition of formal valence so long as the Fermi energy lies above the top of the π* band. This situation appears to hold so long as the oxidation involves only a $Cu^{2+/+}$ redox couple and not the $Cu^{3+/2+}$ redox couple. Therefore we refer to the Cu(I) and Cu(II) valence states in association with oxide ions O^{2-} wherever the operative redox couple is $Cu^{2+/+}$.

On the other hand, any correlation splitting between the $Cu^{2+/+}$ and $Cu^{3+/2+}$ redox couples would drop the energy of the $Cu^{3+/2+}$ couple below the top of the π* band [10]; and holes in the π* bands may be associated with orbitals of primarily O-$2p_\pi$ parentages. In this situation, assignment of formal valences to the Cu(1) atoms and their near-neighbor oxygen atoms becomes ambiguous, which is why the notation "Cu(III)" was used above for the Cu(1) atoms in completed chains. It would be more meaningful to use the notation $(CuO_{3-\delta})^{3-}$ to designate the formal charge on the chain without specifying the distribution of holes in σ* and π* bands.

An experimental test of the formal-valence ambiguity would be the trapping out of holes at oxygen-atom clusters. A peroxide ion $(O_2)^{2-}$, for example, traps two holes in its antibonding orbitals. Trapping out of holes on oxygen-atom clusters would cause a transfer of electrons to the σ* bands. With sufficient electron transfer, a formal valence on the copper may be meaningfully restored.

The apparent absence of peroxide-ion formation in $YBa_2Cu_3O_{6.96}$--as normally prepared from tetragonal $YBa_2Cu_3O_{6+\varepsilon}$, $\varepsilon < 0.25$, by oxygen insertion--demonstrates that there is no peroxide formation so long as the copper retain a coplanar coordination. We may conclude from this observation that the terminal O(1) oxygen atoms of a b-axis chain do not participate in O-O bonding; presumably they retain a formal valence O^{2-}. On the other hand, the Cu(1) bridging oxygen are prevented from making direct contact with one another by the ordering that occurs. The inserted, bridging oxygen are disordered only at low oxidation levels, $(7-\delta) < 6.5$, where formal valence states are meaningful.

If π*-band holes are associated with orbitals of primarily $2p_\pi$ character on Cu(1)-bridging oxygen, this situation leads to the following testable prediction:

Where the oxygen concentration in a $YBa_2Cu_3O_{7-\delta}$ phase--or in a cation-substituted isostructural phase--is high enough to create π*-band holes, then these holes will be trapped out as peroxide ions $(O_2)^{2-}$ wherever two Cu(1)-bridging oxygen atoms make direct contact with one another.

Two types of experiments have been performed that support this prediction: low-temperature preparations of tetragonal phases containing higher oxygen concentrations, $(7-\delta) > 6.5$, and cation substitutions that induce oxygen concentrations in excess of O_7.

LOW-TEMPERATURE PREPARATIONS

In order to prepare $YBa_2Cu_3O_{7-\delta}$ phases in particle sizes < 0.5 μm, it is necessary to decompose precursors mixed in stoichiometric proportions at temperatures too low for sintering. This procedure has led to tetragonal phases having an oxygen content $(7-\delta) > 6.5$ [11,12]. In view of the rapid diffusion of oxide ions O^{2-} inserted into $YBa_2Cu_3O_{7-\delta}$, the synthesis of a tetragonal phase having $(7-\delta) = 6.7$ at 780 °C [12] clearly indicates that the Cu(1) bridging oxygen may have a quite different character if present in high concentration and also disordered on the a-axis and b-axis bridging sites.

As normally prepared, the system $YBa_2Cu_3O_{7-\delta}$ undergoes an orthorhombic-tetragonal transition at a $(7-\delta) < 6.5$, an oxidation state where formal valence states are still meaningful. At higher oxygen concentrations, ordering of the bridging oxygen prevents their contact even at twin planes.

In a low-temperature preparation, on the other hand, the bridging oxygen are initially present in a disordered manner, which allows direct contact between them. If the initial concentration is high, $(7-\delta) > 6.5$, then O-2p holes may be trapped at oxygen clusters such as a peroxide $(O_2)^{2-}$ ion. A molecular species would be much less mobile than a monomeric O^{2-} ion, so the oxygen is retained to higher temperatures. Transfer of electrons back to the σ^* bands reduces the copper to a Cu(II) valence state, and the system is semiconducting rather than superconducting.

Transformation of a tetragonal, semiconducting phase prepared at temperatures $T \leq 780$ °C to the orthorhombic, superconducting phase can be achieved by first removing the "paired" oxygen--at temperatures T > 810 °C in air or, to prevent sintering, at 750 °C in N_2--and then, by reannealing in O_2 at 400 °C, reinserting monomeric oxygen atoms. The monomeric oxygen order--as in a conventional preparation--before their concentration is high enough for $O-2p_\pi$ holes to be introduced [12].

CATION SUBSTITUTIONS

Three types of cation substitutions have been investigated: (1) isovalent substitutions for Y or Ba, (2) aliovalent substitutions for Y or Ba, and (3) substitutions for copper.

Isovalent substitutions for Y or Ba.

Substitution of the smaller Ln^{3+} lanthanides for Y^{3+} in $YBa_2Cu_3O_{7-\delta}$ has a negligible effect on the superconducting transition temperature $T_c \approx 90$ K [13,14]. In view of the sensitivity of T_c to the oxygen concentration, this observation leads to two conclusions: (1) the equilibrium oxygen concentration is not appreciably changed by these substitutions and (2) the rare-earth ions are isolated from the Cooper pairs responsible for superconductivity, which is consistent with confinement of the redox reactions primarily to the Cu(1) planes.

On the other hand, substitution of the non-magnetic Sr^{2+} ion for Ba^{2+} causes T_c to decrease gradually with increasing Sr^{2+}-ion concentration [15] even though the oxygen concentration remained independent of x for a given thermal treatment [16].

Aliovalent substitutions for Y or Ba.

Investigations of M^{4+}-ion substitutions for Y^{3+} need to be repeated and refined.

An investigation [17] of the system $Y_{1-x}Ca_xBa_2Cu_3O_{7-\delta}$ ($0 \leq x < 0.3$) showed that, after annealing in oxygen in the range $350 < T < 450$ °C, the oxygen concentration of the Cu(1) layer varies as $(CuO_{3-\delta_0-0.5x})^{(3-x)-}$ for all x, where δ_0 is the equilibrium value for x = 0 at a particular temperature and partial pressure of oxygen. The "formal average copper oxidation state" is thus seen to be independent of x; only the oxygen-vacancy concentration $[V_O]$ within the $(CuO_{3-\delta})^{3-}$ chains is changing. Measurement of T_c gave a decrease with x that varied as

$$(\partial T_c/\partial(0.5x))_{[h]=1.9} = (\partial T_c/\partial[V_O])_{[h]=1.9} \approx -180 \text{ K} \qquad (1)$$

for a fixed hole concentration $[h] \approx 1.9$, i.e. $\delta_0 \approx 0.05$.

An investigation [18] of La substitution for Ba in the system $YBa_{2-x}La_xCu_3O_{7\pm\delta}$ ($0 \leq x < 0.5$) showed a similar tendency to retain compositional domains over which the "formal average copper oxidation state" remains constant; the oxygen concentration of the Cu(1) layers varies as $(CuO_{3-\delta_0-0.5x})^{(3+x)-}$ for a given temperature and partial pressure of oxygen. Three domains could be distinguished: Domain I ($0 \leq x < 0.05$) is narrow; both T_c and the orthorhombic b/a ratio increase with increasing x, and a $\delta_0 \approx 0.05$ is similar to that found in the $Y_{2-x}Ca_xBa_2Cu_3O_{7-\delta}$ system. In this domain the oxygen atoms remain monomeric. Domain III ($0.10 \leq x < 0.5$) is extensive; both T_c and the orthorhombic b/a ratio decrease with increasing x, see Fig. 7, and a $\delta_0 \approx 0.15$ indicates that the equilibrium oxygen concentration is changed. Domain III can therefore represent a second crystallographic phase with Domain II ($0.5 \leq x < 0.1$) corresponding to a two-phase region separating the phases in Domain I and Domain III. A straightforward interpretation of the properties of the phase of Domain III can be obtained by introducing the possibility of peroxide formation between a-axis and b-axis oxygen. With this assumption, the Cu(1) layer is represented formally as

$$[Cu^{n+}O^{2-}_{3\pm\delta-y}(O_2)^{2-}_{0.5y}]^{(3+x)-}$$

$$n = 3-y-x\pm2\delta \qquad (2)$$

where a $|\delta_0| = 0.15$ makes $n = 2.7-y$. If x_0 is the critical concentration for the generation of peroxide ions, then $y = (x-x_0)$ and

$$n = (2.7+x_0) - x \qquad (3)$$

In the high-T_c superconducting copper oxides, the appearance of superconductivity correlates well with a formal oxidation state on the copper that is greater than Cu^{2+}. Extrapolation of T_c vs x, Fig. 8, gives $T_c = 0$ at $x_c = 0.77$. Substitution of $x_c = 0.77$ for x at $n = 2$ in the above expression for n gives $x_0 = 0.07$, which is in the middle of the two-phase region, Domain II. Were the samples perfectly homogeneous, the upper bound of the $y = 0$ phase would, according to the calculation, be $O_{6.98}$, which is consistent with what is permissible according to the sparse data for the twin-plane density.

Substitutions for copper

Numerous investigations of cation substitutions for copper have been reported. All such substitutions decrease T_c whether or not the substituent carries an atomic magnetic moment.
An investigation [19] of the compounds representing the limiting substitution of Co or Fe for Cu in $YBa_2Cu_3O_{7-\delta}$ gives additional evidence of peroxide (or oxygen clustering) formation where a-axis and b-axis oxygen make contact.

COOPER PAIRING

The identification of oxygen-atom clustering, or peroxide-ion formation, between inserted oxygen atoms at higher oxidation states signals the presence of holes in orbitals of primarily O-2p character. As pointed out above, such a situation should be anticipated first for the π-bands. An important correlation splitting between the $Cu^{2+/+}$ and $Cu^{3+/2+}$ levels would cause the Fermi energy to move into the π^* band of the $[Cu(1)O_{3-\delta}]^{3-}$ layer [10]. This situation leads to the hole configuration

$$\pi^* p_\sigma^* 2(1-\delta')-p$$

for the π^* and σ^* bands of fully oxidized ($CuO_{3-\delta'}$) chains; where p is the number of holes per formula unit in the π^* bands. Note that an increase in the hole concentration p in the π^* bands of primarily O-2p parentage results in a corresponding increase in the electron concentration in the σ^* bands of primarily Cu(1)-3d parentage. This reciprocal relationship is capable of introducing polarization fluctuations that suppress mechanisms competitive with superconductivity--eg. charge-density waves or magnetic order--and enhance the Bardeen-Cooper-Schrieffer (BCS) pairing potential

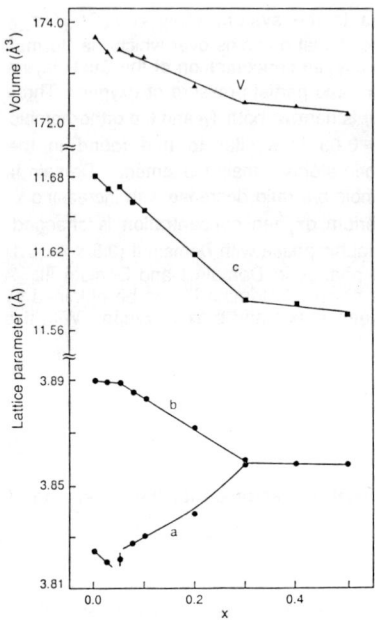

FIG. 7.
Variation of lattice parameters and unit-cell volume with composition x of oxygen-annealed $YBa_{2-x}La_xCu_3O_{7\pm\delta}$.

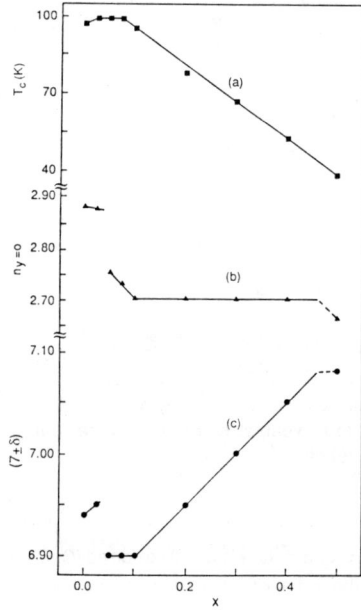

FIG. 8.
Variation with composition of (a) T_C, (b) formal oxidation state of Cu(1) were there no oxygen clustering (y = 0), and (c) the total oxygen content $O_{7\pm\delta}$ for oxygen-annealed $YBa_{2-x}La_xCu_3O_{7\pm\delta}$.

$$V_{BCS} = V_C - U \tag{4}$$

where V_C is the Cooper-pair attractive potential and U is the electrostatic repulsive energy between paired electrons.

The enhancement arises from the fact that the relative strengths of the π and σ components of a Cu-O bond vary with the bond distance R, which makes $p = p(R)$. Therefore, coupling of a pair of σ^* electrons to a phonon may be enhanced by an associated increase in p, thus increasing V_C. Moreover, the interrelationship between σ^* electrons and π^* holes would serve to reduce the electrostatic repulsion between the paired electrons, thus decreasing U. Coupling between b-axis chains may occur via the $Cu(2)O_2$ layers.

The band calculation of de Groot et al. [20] indicates the presence of "valence" fluctuations that can enhance T_C, and their view of the O-2p character of the states at the top of the π^* band appears to be similar to the one presented here. Yu et al. [10] have also emphasized the role of charge fluctuations between σ^* and π^* bands, but without introducing an O-2p parentage for the states at the top of the π^* band.

We thank the Robert A. Welch Foundation, Houston, Texas, for partial support of this research.

REFERENCES

1. J.S. Swinnea and H. Steinfink, J. Mat. Res. 2, 424 (1987).
2. W.I.F. David, W.T.A. Harrison, J.M.F. Gunn, O.Moze, A.K. Soper, P.Day, J.D. Jorgensen, D.G. Hinks, M.A. Beno, L. Soderholm, D.W. Capone II, I.K. Schuller, C.U. Segre, K. Zhang, J.D. Grace, Nature 327, 310 (1987).
3. A. Manthiram, J.S. Swinnea, Z.T. Sui, H. Steinfink, and J.B. Goodenough, J. Am. Chem. Soc. 109, 6667 (1987).
4. M.O. Eatough, D.S. Grimley, B. Morosin, E.L. Venturimi, Appl. Phys. Lett. 51, 367 (1987).
5. P.P. Fereitas and T.S. Plaskett, Phys. Rev. B36, 5723 (1987).
6. J.D. Jorgensen, M.A. Beno, D.G. Hinks, L. Solderholm, K.J. Volin, R.L. Hitterman, J.D. Grace, I.K. Schuller, C.U. Segre, K. Zhang, M.S. Kleefisch, Phys. Rev. B36, 3608 (1987).
7. R.J. Cava, B. Batlogg, C.H. Chen, E.A. Rietman, S.M. Zahurak, D. Werder, Nature 329, 423 (1987); Phys. Rev. B36, 5719 (1987).
8. J.M. Tranquada, D.E. Cox, W. Kunnmann, H. Moudden, G. Shirane, M. Suenaga, P. Zolliker, D. Vaknin, S.K. Sinha, M.S. Alvarez, A.J. Jacobson, and D.C. Johnston, Phys. Rev. B (in press).
9. M.A. Alario-Franco, J.J. Capponi, C. Chaillout, J. Chenavas, M. Marezio, Proc. Mat. Res. Soc. 1987 Fall Meeting, Boston, MA, Nov. 30-Dec. 5, 1987.
10. J. Yu, S. Massidda, and A.J. Freeman, Phys. Lett. A 122, 203 (1987).
11. C.C. Torardi, E.M. McCarron, M.A. Subramanian, M.S. Horowitz, J.B. Michel, A.W. Sleight, and D.E. Cox, in Chemistry of High-Temperature Superconductors, ACS Symposium Series 351, edited by D.L. Nelson, M.S. Whittingham, and T.F. George, (Am. Chem. Soc., Washington, D.C., 1987), p. 152.
12. A. Manthiram and J. B. Goodenough, Nature 329, 701 (1987).
13. P.H. Hor, R.L.Meng, Y.Q. Wang, L. Gao, Z.J. Huang, J. Bechtold, K. Forster, and C.W. Chu, Phys.Rev. Lett. 58, 1891 (1987).
14. J.M. Tarascon, L. H. Greene, B.G. Bagley, W.R. McKinnon, P. Barboux, and G.W. Hull, in Novel Superconductivity, edited by S.A. Wolf and V. Kreshnin, (Plenum Press, N.Y. London, 1987), p. 705.
15. B.W. Veal, W.K. Kwok, A. Umezawa, G.W. Crabtree, J.D. Jorgensen, J.W. Downey, L.J. Nowicki, A.W. Mitchell, A.P. Panlikas, C.H. Showers, Appl. Phys. Lett. 51, 279 (1987).
16. A. Manthiram, unpublished
17. A. Manthiram, S.J. Lee, and J.B. Goodenough, J. Solid State Chem. (in press).
18. A. Manthiram, X.X. Tang, and J.B. Goodenough, Phys. Rev. B (in press).
19. Y.K. Tao, J.S. Swinnea, A. Manthiram, J.S. Kim, J.B. Goodenough and H. Steinfink, J. Mat. Res. (in press).
20. R.A. de Groot, H. Gutfreund, and M. Weger, Solid State Commun. 63, 451 (1987).

NEW MATERIALS EXHIBITING SUPERCONDUCTIVITY ABOVE 90 K

C. Y. HUANG
Lockheed Missiles & Space Company, Inc.
Research & Development Division
Palo Alto, CA 94304-1191

ABSTRACT

A brief history of the discovery of the 90 K superconductors will be presented. Some salient properties (magnetism, Mössbauer, and electron spin resonance) are reviewed. We also discuss some recent results of super-high-temperature superconductors.

INTRODUCTION

Aware of the metallic character of ceramic samples, $BaLa_4Cu_5O_{13.4}$, synthesized by Michel and Raveau [1], and knowing that Cu^{2+} is a Jahn-Teller ion in many cases, Bednorz and Muller [2] in early 1986 measured the electrical resistance, R, of $Ba_{0.75}La_{4.25}Cu_5O_4(3-y)$ down to liquid-helium temperature. They found that electrical resistance dropped sharply around 30 K and became 0 below ~12 K. Based on the onset, they also suggested that the material could be superconducting as high as 30 K. The high-temperature superconductivity at ~30 K was quickly confirmed independently by the Tokyo [3] and the Houston [4] groups in late 1986. The Tokyo group [5] also determined that $(La_{0.85}Ba_{0.15})_2 CuO_{4-\delta}$ has a single-layer K_2NiF_4 perovskite structure, which they suggested is responsible for superconductivity at ~30 K. By using a SQUID magnetometer, Zirngiebl et al. [6] found that $(La_{0.9}Ba_{0.1})_2 CuO_{4-\delta}$ is a type-II superconductor with a lower critical field $H_{c1} = 3.7 \times 10^{-2}$ T, an upper critical field $H_{c2} = 7.3$ T, a coherence length $\xi = 6.7$ nm, and a Ginzburg-Landau parameter $\kappa \sim 11$ at 2 K. In addition, the Houston group investigated the high-pressure effect of this ceramic superconductor and found that the superconducting transition temperature, T_c, increased at 1 K/kbar, which is close to that for $EuMo_6S_8$ [7]. A superconducting onset temperature, T_{co}, as high as 57 K at 12 kbar was achieved. This exciting high-pressure result prompted the Houston-Alabama group to replace Ba with smaller Sr to simulate the pressure effect. These authors [8] determined that $(La_{0.9}Sr_{0.1})_2 CuO_{4-\delta}$ had an onset temperature $T_{co} \simeq 42.5$ K at ambient pressure. Several other groups [9-13] also observed similar results. The dc magnetization of $(La_{0.9}Sr_{0.1})_2CuO_{4-\delta}$ was first observed by Zirngiebl, et al. [14]. They found that the sample is also a type-II superconductor with H_{c1} ~ 4×10^{-2} T, H_{c2} ~ 20 T, ξ = 4.1 nm, and κ ~ 15 at 2 K.

In the course of investigating the La-Ba-Cu-O system, the Houston group also discovered that some mixed phase samples exhibited an onset of a resistance drop in the resistance-temperature curve at ~70 K, indicating a possible superconducting transition temperature around 70 K. The Beijing group [15] subsequently reported a similar observation.

From observation of the atomic radii and the inter-atomic distances forming the perovskite structure, the Alabama group [16] determined that $Y_{1.2}Ba_{0.8}CuO_{4-\delta}$ is the right composition for high T_c, because its inter-atomic distance is close to that of $(La_{0.9}Sr_{0.1})_2CuO_{4-\delta}$, based on the estimation from ionic radii. They discovered that the electrical resistance of the sample became zero even above 77 K. In order to confirm this finding, they prepared another sample, which took only 30 min to react in their oven; they again found that the T_c was indeed above 80 K [17]. In fact, the second sample showed sharper transition than the first one even though the second one had been reacted in the oven for only 30 min. Subsequently, the Houston group and Wu reproduced this result jointly in Houston. In order to confirm this result, they also measured the ac magnetic susceptibility and showed that the sample had a large (24%) Meissner effect [17]. However, they found the effect of high pressure on T_c to be small [18]. The discovery of the high T_c superconductor definitely set off a new era of intense superconductivity research worldwide.

The Houston-Alabama group was aware that $Y_{1.2}Ba_{0.8}CuO_{4-\delta}$ was a mixed-phase sample. It was later found that orthorhombic $YBa_2Cu_3O_{6+\delta}$ is the phase responsible for high-temperature superconductivity [19-22]. It was quickly recognized that Y can be replaced by any rare earth, and several groups [23-28] independently found that all $LBa_2Cu_3O_{6+\delta}$ (L = Y, all rare earths) with the orthorhombic structure have a T_c above 90 K.

In this paper we review some salient properties of these new high-temperature superconductors. In addition, we report some results of recent research in super-high-temperature superconductors.

90 K SUPERCONDUCTORS
(a) Mixed-phase $Y_{1.2}Ba_{0.8}CuO_{4-\delta}$

Even though zero resistance was important for identifying superconductivity, it was also pertinent to measure dc magnetic susceptibility, χ. Figure 1 shows the data for one of the earliest $Y_{1.2}Ba_{0.8}CnO_{4-\delta}$ samples [29]. DC susceptibility, χ, starts to become negative at ~90 K, demonstrating that $Y_{1.2}Ba_{0.8}CuO_{4-\delta}$ was indeed a superconductor with T_c ~ 90 K. While investigating the Meissner effect, we also observed some unexpected switching phenomena [30], which we attributed to the granular nature of the samples. In addition, when a sample was cooled below T_c in zero field, the magnetization responded to a magnetic field according to the fractional exponential time dependence as in the case of a typical spin glass [31]. We thus attributed the switching and nonexponential phenomena to the frustration effect in the granular samples as is the case in a spin glass [31,32]. We also measured the magnetic field dependence of the electrical resistance up to 20 T using the high-field magnets available at the Francis Bitter National Magnet Laboratory at the Massachusetts Institute of Technology. We estimated H_{c2} to be ~160 T, which yields $\xi(0)$ ~ 14 Å [33].

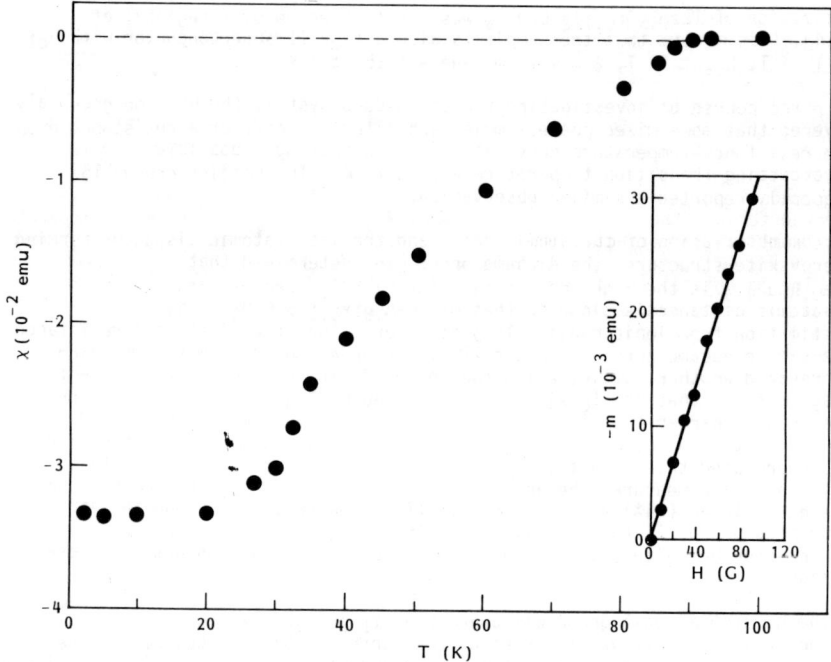

Figure 1. Temperature dependence of the dc magnetic susceptibility for $Y_{1.2}Ba_{0.8}CuO_{4-\delta}$. The inset shows the linear relation between the magnetic moment and the field, H [29].

(b) "123" Single-phase superconductors

Magnetic properties. As described in the Introduction, the 90 K superconductivity of $Y_{1.2}Ba_{0.8}CuO_{4-\delta}$ arose from the orthorhombic $YBa_2Cu_3O_{6+\delta}$ "123" phase. The fact that Y can be replaced by all rare earths with T_c all above 90 K, puzzled researchers in the field. In particular, one rare earth, Gd, completely destroys superconductivity, even when only 1% is introduced into an ordinary superconductor, but $GdBa_2Cu_3O_{6+\delta}$ also has a T_c of ~90 K. The independence of T_c on the rare earth suggested that rare earth ions interact only very weakly with superconducting electrons. This conclusion led us to measure the magnetic susceptibility of some of these superconductors ($ABa_2Cu_3O_{6+\delta}$, A = Y, Eu, Gd, Sm) above T_c. They are all paramagnetic. In particular, for $GdBa_2Cu_3O_{6+\delta}$ the magnetic susceptibility above T_c [34] can be fitted to the Curie-Weiss law: $\chi = C/(T + \theta)$ where C is the Curie constant, which can be accounted for by the moments of Gd^{3+} ions, and $\theta (= 6$ K) is the Curie-Weiss temperature. The sign of θ suggests that the Gd^{3+} spins are coupled antiferromagnetically. Indeed, the antiferromagnetic order in $GdBa_2Cu_3O_{6+\delta}$ sets in at the Néel temperature $T_N = 2.22$ K, as verified by the heat capacity measurements [35,24] and the neutron scattering study [36]. In fact, for insulating GdBa$_2$Cu$_3$O$_{6.5}$, T_N is still 2.2 K, suggesting the participation of the electrons in the deeper levels in the antiferromagnettic ordering of Gd spins [37].

Figure 2 depicts the field dependencies of the magnetization, M, for a GdBa$_2$Cu$_3$O$_{6+\delta}$ sample measured at various temperatures [34]. All show peaks at ~120 G. Above ~200 G, the slope, dm/dH, is much reduced. This reduction suggests that parts of the sample (presumably the grain boundaries) lose the superconducting shielding effect at ~120 G. Similar peaking has also been observed for all these ceramic superconductors. Practically, at low fields, all these samples are perfect diamagnets.

The magnetization at very high magnetic fields [38] also has been measured with a vibration sample magnetometer using a Bitter magnet at the Francis Bitter National Magnet Laboratory. Figure 3 exhibits the field dependence of the magnetic moment, m, at 1.48 K. As clearly demonstrated, the magnetic moment saturates at high fields (~10 T). The presence of the hysteresis indicates the presence of superconductivity at very high magnetic fields. (Zero resistance has been observed for fields up to 23 T.) Therefore, superconductivity and very nearly parallel Gd^{3+} spins can coexist at low temperatures. Figure 4 shows the derivative, dm/dH, of the up-sweep curve of Fig. 3. The anomaly around 2 T indicates that there is a phase transition from the "canted phase" to the paramagnetic state [38].

ESR Measurements [39]. Electron spin resonance (ESR) at 9 GHz was employed to study YBa$_2$Cu$_3$O$_{6+\delta}$. The large signal might have originated from Cu^{2+} of the Y$_2$BaCuO$_5$ impurity phase. For good samples, no Cu^{2+} ESR signal was

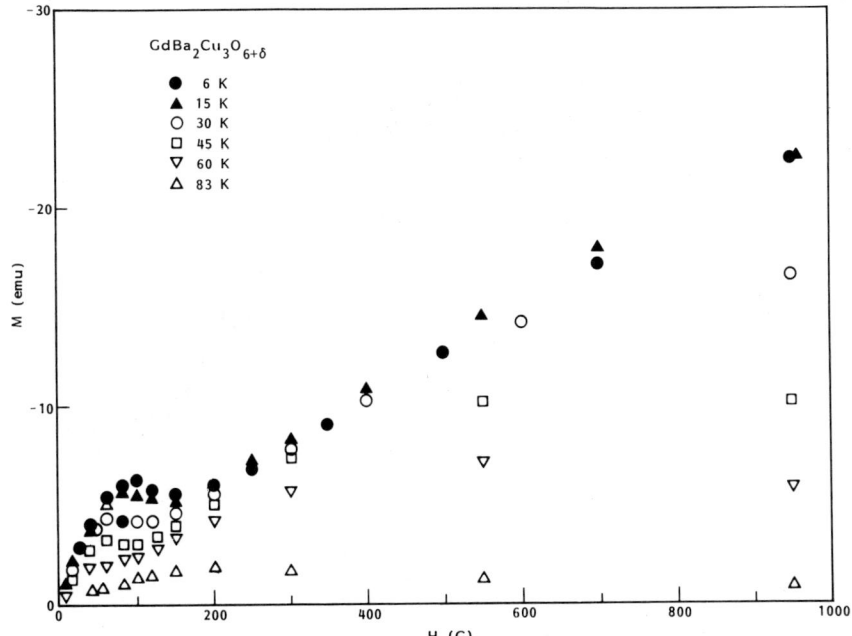

Figure 2. Field dependence of the magnetization for GdBa$_2$Cu$_3$O$_{6+\delta}$ at various temperatures [34].

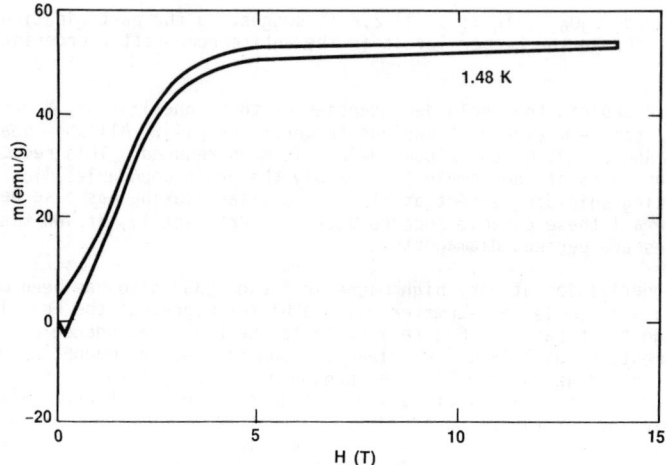

Figure 3. The magnetic field dependence of the magnetic moment, M, for $GdBa_2Cu_3O_{6+\delta}$ at 1.48 K. The lower (upper) curve is the up- (down-) sweep [38].

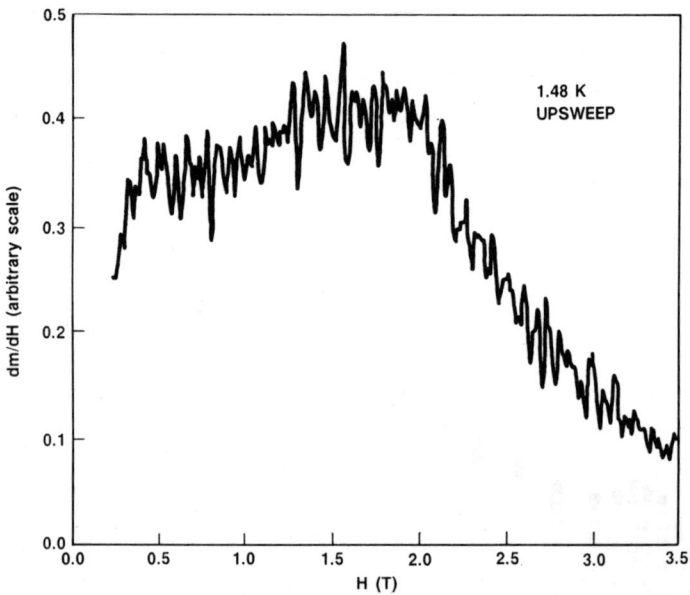

Figure 4. The derivative of the magnetic-moment up-sweep curve for $GdBa_2Cu_3O_{6+\delta}$ at 1.48 K [38].

observed. However, below T_c, an unexpected peak appeared at low fields (~50 G). This absorption signal was presumably related to the losses owing to the flux penetration at fields greater than H_{c1} (~50 G.)

The Gd^{3+} ESR in $(Y_{0.99}Gd_{0.01})Ba_2Cu_3O_{6+\delta}$ also has been made at 9 GHz. The Gd^{3+} ions substitute the positions of Y^{3+}. The g-value of the Gd^{3+} ESR is shifted from that of a free Gd^{3+} ion by as much as ~0.1, indicating that a Gd^{3+} ion interacts with neighboring electrons contributed by Cu. This result contradicts the assumption that rare earth ions do not interact with superconducting electrons, which originate from Cu. For further studies along this line, single crystals are desirable.

Mössbauer Spectroscopy. ^{151}Eu Mössbauer spectroscopy is easy to perform. For this reason, $EuBa_2Cu_3O_{6+\delta}$ has been investigated, employing $^{151}SmF_3$ as a source of the 21.6-keV γ-radiation [40]. No evidence of Eu^{2+} could be detected from 4.2 K to 300 K. Within experimental accuracy, Eu are all in the trivalent (Eu^{3+}) state. The temperature dependence of the integrated intensity of the spectrum varies normally according to a Debye density of vibrational states with a Debye temperature $\theta_D = 280 \pm 5$ K.

To observe the vibrations at the Cu sites, ^{119}Sn was doped into the sample, substituting for Cu. The spectra show that the temperature dependence of $EuBa_2Cu_{2.98}Sn_{0.02}O_{6+\delta}$ cannot be fitted to a Debye-Waller temperature dependence, but it displays a deviation below ~110 K, indicating a precursor effect that vibrational modes of the Cu-O chains soften before the onset of superconductivity. This result, along with the observation of the isotope effect [42], seems to support the idea that electron pairing in these novel high-temperature superconductors involves optical phonons of the Cu-O chains.

SUPER-HIGH-TEMPERATURE SUPERCONDUCTORS

Immediately after the discovery of the 90 K superconductor, researchers [43] determined that a change of the oxygen contents could yield a higher T_c (~98 K). In February 1987, Chu [44] announced the observation of up to a ~50% drop in the electrical resistance, R, of several Y-, Sc-, and La-based multi-phase Ba-Cu oxide samples at ~ 240 K. The resistance drop reported did not survive the thermal cycles for magnetic diagnoses. Later, the Berkeley group [45] reported similar anomalies in some Y-based oxides. In the absence of both "zero" resistivity and evidence of the Meissner effect, the two criteria for superconductivity, the resistance drops have been taken only as an indication of the possible existence of superconductivity at these unusual high temperatures. To provide strong evidence for superconductivity, Chen et al. [46] detected the reverse ac Josephson signal in mixed-phased Y-based Ba-Cu oxide samples below 240 K. To date, many groups have reported T_c above ~150 K [44-49]. In this paper, we report the observation of a sharp drop of resistance to "zero" around 230 K and a sudden appearance of the Meissner component at the same temperature in annealed $EuBa_2Cu_3O_{6+\delta}$. These two measurements suggest the occurrence of superconductivity at 230 K.

A conventional $EuBa_2Cu_3O_{6+\delta}$ sample was synthesized at the University of Houston. The sample was then tested resistively and determined to be a 90 K superconductor. The sample was then heat treated in vacuum and argon atmospheres at Lockheed Missiles & Space Company, Inc. (LMSC). A conventional four-probe technique was employed to measure the electrical resistance, R, of the sample. Figure 5 displays the results obtained at LMSC [49]. The onset of the sharp drop in resistance (from ~1 ohm) occurs at ~238 K, and the resistance reaches "zero" (<10^{-4} ohm, our experimental resolution) at ~228 K. The midpoint of

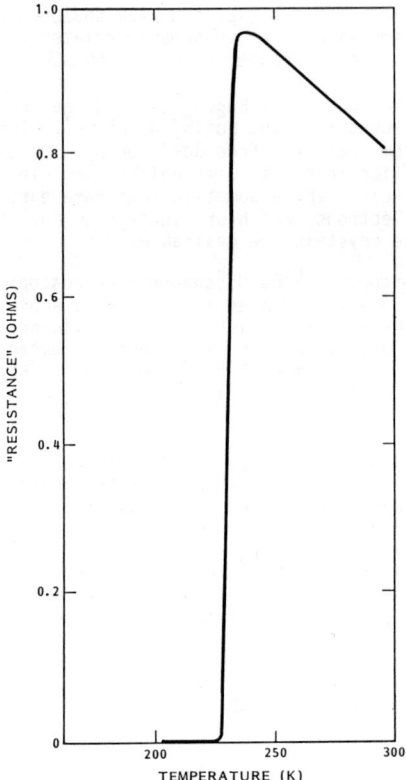

Figure 5. The electrical resistance, R, versus temperature [49].

the drop occurs at 230 K. A similar sharp drop was observed the following day. This sharp drop was interpreted in terms of a superconducting transition. However, the "zero-resistance" phenomenon at 230 K disappeared after several thermal cycles, even though a kink at ~230 K appeared.

To prove superconductivity, we had to measure the magnetization using a SQUID magnetometer. The magnetization is positive above ~ 90 K, but it is proportional to the applied field. On cooling, the magnetization increases with decreasing temperature as expected for a paramagnet. At 230 K, there is a discontinuity in dM/dT, and the magnetization becomes less than the value one would expect. This reduction in the magnetization suggests that the diamagnetic component due to the Meissner effect sets in. This onset temperature coincides with the occurrence of the resistance drop, as shown in Fig. 5. The appearance of both "zero" resistance and increased diamagnetism strongly suggests the onset of superconductivity at ~230 K. However, the change is small (a few parts in 10^4). The LMSC personnel also have shown that the diamagnetic component persists through many thermal cycles over more than 30 weeks even at 1 T. Hence, the diamagnetic component is believed to be long-lived, not short-lived as reported by others. Even though the transition temperature (~230 K) observed here is close to 240 K, the phase exhibiting this 230 K superconductivity may or may not be the same phase reported in references 46-48.

In the future, determining if a substance is a high-temperature superconductor should involve seven criteria:

(1) Phase coherence
(2) R = 0 (resistivity < 10^{-17} ohm-cm)
(3) At least a few percent of negative magnetization
(4) Identification of the phase
(5) Long lifetime of the phase
(6) The sample can be reproduced by other researchers
(7) Critical current exists

The proof of phase coherence as made in reference 46 is probably the most useful, because it does not require R = 0 and the sudden change of magnetization that might be interpreted in terms of a phase transition. However, in terms of practicality, the most important criterion is obviously the identification of the super-high-temperature phase.

CONCLUSIONS

In this paper we review some results of recent research in high-temperature superconductors. Most importantly, we lack an understanding of the origin of superconductivity, and we must also solve mechanical as well as electrical problems associated with these ceramic materials. More microscopic measurements are required. As for super-high-temperature superconductors, by means of the Chandrasekhar-Clogston limit, we speculate that a T_c as high as 1000 K could be obtained. The phase should be different from the perovskite structure and might consist only of copper, oxygen, and nitrogen. Therefore, to obtain super-high-temperature superconductors, methods different from those employed to prepare perovskite ceramics may be necessary.

ACKNOWLEDGMENTS

The author thanks C. W. Chu, P. H. Hor, R. B. Frankel, C. J. Lobb, R. L. Meng, M. K. Wu, C. S. Ting, L. Dries, P. Boolchand, and J. C. Ho for stimulating discussions. This work is supported by Lockheed Independent Research funds.

REFERENCES

1. C. Michel and B. Raveau. Rev. Chim. Miner., 21, 407 (1984); Mat. Res. Bull., 20, 667 (1985).
2. J. G. Bednorz and K. A. Muller. Z. Phys., B64, 189 (1986).
3. S. Uchida, H. Takagi, K. Kitagawa, and S. Tanaka. Jpn. J. Appl. Phys., 26, L1 (1987).
4. C. W. Chu, P. H. Hor, R. L. Meng, L. Gao, Z. J. Huang, and Y. Q. Wang. Phys. Rev. Lett., 58, 405 (1987).
5. H. Takagi, S. Uchida, K. Kitazawa, and S. Tanaka. Jpn. J. Appl. Phys., 26, L123 (1987). However, the relevance of the K_2NiF_4 structure to the occurrence of the 30 K superconductivity was first suggested by J. G. Bednorz, M. Takashige, and K. A. Muller. Europhys. Lett., 3, 379 (1987).
6. E. Zirngiebl, J. D. Thompson, C. Y. Huang, P. H. Hor, R. L. Meng, and C. W. Chu, unpublished.
7. C. Y. Huang, Critical Rev. in Sol. Stat. Phys. Mat. Sci., 12, 75 (1984).
8. M. K. Wu, P. H. Hor, R. L. Meng, and C. W. Chu, unpublished.
9. R. J. Cava, R. B. Van Dover, B. Batlogg, and E. A. Rietman. Phys. Rev. Lett., 58, 408 (1987).
10. Z. X. Zhau, L. Q. Chen, C. G. Cui, Y. Z. Huang, J. X. Liu, G. H. Chen, S. L. Li, S. Q. Guo, and Y. Y. He. Kexue Tongbau, 32, 522 (1987).
11. J. M. Tarascon, L. H. Greene, W. R. McKinnon, G. W. Hull, and T. H. Geballe. Science, 235, 1373 (1987).
12. D. W. Capone, D. G. Hinks, and J. D. Jorgensen. Appl. Phys. Lett., 50, 543 (1987).
13. D. Gubser, R. A. Hein, S. H. Lawrence, M. S. Osofsky, D. J. Schrodt, L. E. Toth, and S. A. Wolf. Phys. Rev., B35, 5350 (1987).
14. E. Zirngiebl, J. D. Thompson, C. Y. Huang, P. H. Hor, R. L. Meng, C. W. Chu, and M. K. Wu. Appl. Phys. Commun., 7, 1 (1987).
15. Z. X. Zhao, et al. People's Daily, January 16, 1987.
16. M. K. Wu, private communication.
17. M. K. Wu, J. R. Ashburn, C. J. Tong, P. H. Hor, R. L. Meng, Z. J. Huang, Y. Q. Wang, and C. W. Chu. Phys. Rev. Lett., 58, 908 (1987).
18. C. W. Chu, P. H. Hor, R. L. Meng, L. Gao, Z. J. Huang, Y. Q. Wang, M. K. Wu, J. R. Ashburn, and C. J. Tong. Phys. Rev. Lett., 58, 911 (1987).
19. R. M. Hazen, L. W. Finger, R. J. Angel, C. T. Prewitt, N. L. Ross, H. K. Mao, C. G. Hadichiacos, P. H. Hor, R. L. Meng, and C. W. Chu. Phys. Rev., B35, 7238 (1987).
20. T. Siegrist, S. Sunshine, D. W. Murphy, R. J. Cava, and S. M. Zahurak. Phys. Rev., B35, 7137 (1987).
21. P. M. Grant, R. Beyers, E. M. Engler, G. Lim, S.S.P. Parkin, M. L. Ramirez, V. Y. Lee, H. Nazal, J. E. Vasquez, and R. J. Savoy. Phys. Rev., B35, 7242 (1987).

22. M. A. Beno, L. Soderholm, D. W. Capone II, D. G. Hinks, J. D. Jorgensen, I. K. Schuller, C. U. Segre, K. Zhang, and J. D. Grace. Appl. Phys. Lett., $\underline{51}$, 57 (1987).
23. P. H. Hor, R. L. Meng, Y. Q. Wang, L. Gao, Z. J. Huang, J. Bechtold, K. Foster, and C. W. Chu. Phys. Rev. Lett., $\underline{58}$, 1891 (1987).
24. Z. Fisk, J. D. Thompson, E. Zingiebl, J. L. Smith, and S. W. Cheong. Sol. State Commun., $\underline{62}$, 743 (1987), preprint.
25. S. Hasoya, S. I. Shamoto, M. K. Onoda, and M. Sato. Jpn. J. Appl. Phys., $\underline{26}$, L325 (1987).
26. D. W. Murphy, S. Sunshine, R. B. Van Dover, R. J. Cava, B. Batlogg, S. M. Zahurak, and L. F. Schneemeyer. Phys. Rev. Lett., $\underline{58}$, 1888 (1987).
27. C. W. Chu has informed the author that the superconductors with L = Ce, Pr, Tb have been synthesized by a Russian group.
28. A Chinese group in Shen-Yang has made $CeBa_2Cu_3O_{6+\delta}$ superconducting at 90 K.
29. P. H. Hor, R. L. Meng, C. W. Chu, E. Zirngiebl, J. D. Thompson, and C. Y. Huang, unpublished.
30. P. H. Hor, R. L. Meng, C. W. Chu, M. K. Wu, E. Zirngiebl, J. D. Thompson, and C. Y. Huang, Nature, $\underline{326}$, 669 (1987).
31. C. Y. Huang. J. Magn. Magn. Mat., $\underline{51}$, 1 (1985).
32. C. Ebner and D. Stroud. Phys. Rev., $\underline{B31}$, 165 (1985).
33. P. H. Hor, R. L. Meng, J. Z. Huang, C. W. Chu, and C. Y. Huang. Appl. Phys. Commun., $\underline{7}$, 129 (1987).
34. P. H. Hor, R. L. Meng, C. W. Chu, C. Y. Huang, and R. B. Frankel, to be published.
35. J. C. Ho, P. H. Hor, R. L. Meng, C. W. Chu, and C. Y. Huang. Sol. Stat. Commun., $\underline{63}$, 711 (1987).
36. D. M. Paul, H. A. Mook, B. C. Sales, L. A. Boatner, J. R. Thompson, M. Mostoller, and S. W. Hewat, preprint.
37. J. C. Ho, C. Y. Huang, P. H. Hor, R. L. Meng, and C. W. Chu, to be published.
38. C. Y. Huang, Y. Shapira, P. H. Hor, R. L. Meng, and C. W. Chu. Modern Phys. Lett., November 1987.
39. C. Y. Huang, Z. J. Huang, P. H. Hor, R. L. Meng, and C. W. Chu, to be published.
40. P. Boolchand, R. N. Enzweiler, I. Zitkovsky, R. N. Meng, P. H. Hor, C. W. Chu, and C. Y. Huang. Sol. Stat. Commun., $\underline{63}$, 521, (1987).
41. P. Boolchand, R. N. Enzweiler, I. Zitkovsky, D. McDaniel, R. L. Meng, P. H. Hor, C. W. Chu, and C. Y. Huang, preprint.
42. K. J. Leary, H. C. Zur Loye, S. W. Keller, T. A. Faltens, W. K. Hain, J. N. Michaels, and A. M. Stacy. Phys. Rev. Lett., $\underline{59}$, 1236 (1987).
43. C. W. Chu, P. H. Hor, R. L. Meng, L. Gao, Z. J. Huang, Y. Q. Wang, J. Bechtold, D. Campbell, M. K. Wu, J. Ashburn, and C. Y. Huang, preprint.
44. C. W. Chu. National Science Foundation news release and University of Houston news release, February 16, 1987; New York Times, March 10, 1987.
45. L. C. Bourne, M. L. Cohen, W. N. Creager, M. F. Crommie, A. M. Stacy, and A. Zettl. Phys. Lett., $\underline{120A}$, 494 (1987).
46. J. T. Chen, L. E. Wenger, C. J. McEwan, and E. M. Logothetis. Phys. Rev. Lett., $\underline{58}$, 1972 (1987).
47. S. R. Ovshinsky, R. T. Young, D. D. Allred, G. DeMaggio, and G. A. Van den Leeden. Phys. Rev. Lett., $\underline{58}$, 2579 (1987).
48. H. Ihara et al. 18th Int. Low Temp. Phys. Conf., Kyoto, Japan, Aug 1987.
49. C. Y. Huang, L. Dries, P. H. Hor, R. L. Meng, and C. W. Chu. Nature, $\underline{328}$, 403 (1987).

THE EFFECTS OF OXYGEN STOICHIOMETRY AND OXYGEN ORDERING ON SUPERCONDUCTIVITY IN $Y_1Ba_2Cu_3O_{9-x}$

R. Beyers, E.M. Engler, P.M. Grant, S.S.P. Parkin, G. Lim, M.L. Ramirez, K.P. Roche, J.E. Vazquez, V.Y. Lee, and R.D. Jacowitz, IBM Almaden Research Center, 650 Harry Road, San Jose, CA 95120-6099

B.T. Ahn, T.M. Gür, and R.A. Huggins, Department of Materials Science and Engineering, Stanford University, Stanford, CA 94305

ABSTRACT

This paper reports on the structures and properties of $Y_1Ba_2Cu_3O_{9-x}$ samples prepared in precisely controlled oxygen environments using a solid-state ionic technique. By titrating out oxygen at low temperatures, orthorhombic $Y_1Ba_2Cu_3O_{9-x}$ samples were prepared with oxygen contents below 6.50. Resistivity and magnetometry studies indicated that these reduced, orthorhombic samples were marginally superconducting, with their superconductivity probably arising from local regions of higher oxygen content.

INTRODUCTION

Early on, it was recognized that the properties of oxide superconductors depend critically on processing conditions [1]. Using a variety of characterization techniques, numerous groups reached qualitatively similar conclusions regarding the importance of controlling the oxygen content and oxygen order for optimal superconducting properties. Quantitatively, however, the results of different groups vary considerably. For example, Johnston et al. [2] found samples prepared by interdiffusing mixtures of $Y_1Ba_2Cu_3O_6$ and $Y_1Ba_2Cu_3O_7$ in sealed silica tubes to be tetragonal and not superconducting below $Y_1Ba_2Cu_3O_{6.5}$, whereas Cava et al. [3] found samples prepared by a zirconium gettering technique to be orthorhombic with a large superconducting volume fraction at $Y_1Ba_2Cu_3O_{6.30}$. One cause for these discrepancies is that the techniques used to prepare oxide samples thus far do not allow for precise control of all of the important processing variables — namely, the annealing temperature, oxygen partial pressure, and quench rate [4]. Herein, we describe a solid-state ionic method for preparing oxide samples with well-defined processing histories. Using this method, we have prepared and examined $Y_1Ba_2Cu_3O_{9-x}$ samples with oxygen contents at and below 6.50.

EXPERIMENT

The solid-state ionic cell used consisted of a closed quartz jacket with two interconnected chambers, as shown schematically in Fig. 1. Each chamber had its own furnace for independent temperature control. One chamber contained the $Y_1Ba_2Cu_3O_{9-x}$ sample. A nearby thermocouple monitored the thermal history of the sample, including the quenching process. The other chamber housed a yttria-stabilized-zirconia (YSZ) solid electrolyte tube, closed at one end, with porous platinum electrodes deposited on the inner and outer walls at the closed end. The interior of the YSZ tube was exposed to ambient air, to act as an oxygen reference electrode.

The YSZ tube served two purposes. First, passing a known current through the electrolyte allowed quantitative titration of oxygen into and out of the chamber. Second, measuring the open circuit voltage across the electrolyte enabled the oxygen partial pressure inside the chamber to be accurately measured. The YSZ tube chamber was operated at a fixed temperature of 850 °C to obtain high ionic conductivity and thus rapid oxygen transport.

Samples with reduced oxygen content were prepared in the following manner. $Y_1Ba_2Cu_3O_{7.00}$ starting material (oxygen content derived from iodometric titration [5]) was placed in an alumina boat in the chamber and heated to 500°C. During the temperature ramp up, enough oxygen was pumped out of the chamber to ensure that the total pressure remained below 1 atm. At 500 °C, oxygen was titrated out of the chamber for 2 - 3 days. The sample remained

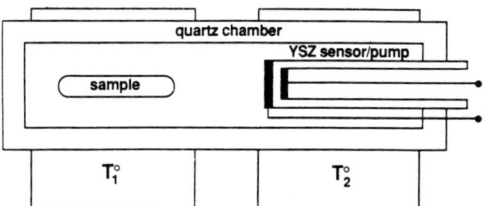

Figure 1. Schematic of the apparatus.

at 500 °C for an additional day after the desired oxygen content was attained. The sample was then either air quenched to 25 °C by removing the furnace around the sample compartment or slowly cooled by stepping down the sample furnace temperature to 25 °C over a 28 hour period.

The oxygen content in the reduced samples was derived from the known starting composition and the measured weight change in the sample (total sample weight ~10 grams). This value was checked by performing thermogravimetric analysis and iodometric titration on small portions of each sample. Based on the results of these three methods, we estimate that the quoted oxygen contents are accurate to within ±0.02 oxygens per formula unit.

X-ray diffraction and transmission electron microscopy (TEM) were used for structural characterization of the samples. Resistivity, thermopower, and susceptibility measurements were used to determine corresponding changes in superconducting properties.

RESULTS

The x-ray diffraction patterns, resistivity versus temperature plots, and diamagnetic shielding data for samples air quenched from 500°C with oxygen contents between 6.30 and 6.50 are shown in Figs. 2, 3, and 4, respectively.

Air quenched $Y_1Ba_2Cu_3O_{6.30}$

X-ray diffraction indicated that the $Y_1Ba_2Cu_3O_{6.30}$ sample was tetragonal [Fig. 2(a)]. Electron diffraction studies of individual crystals also found the material to be tetragonal. No twins were observed in the microstructure, as expected for tetragonal crystals. A few crystals, however, did exhibit a feint, tweed-like microstructure, as if on the verge of transforming to the orthorhombic structure. Resistively, the sample never went superconducting [Fig. 3(a)]. Magnetometry data, however, indicated approximately 0.2% of the sample was superconducting below ~30 K (Fig.4).

Air quenched $Y_1Ba_2Cu_3O_{6.41}$

Both x-ray and electron diffraction indicated that the $Y_1Ba_2Cu_3O_{6.41}$ sample was orthorhombic. Moreover, the electron diffraction patterns exhibited weak superlattice spots halfway between reflections along the a* axis, corresponding to a doubling of the a axis in the unit cell. Similar spots along the b* axis have been reported by Cava et al. in $Y_1Ba_2Cu_3O_{6.72}$ samples [3]. The twin density in this sample was similar to that in $Y_1Ba_2Cu_3O_{7.00}$ material. Dips occurred at 80, 50, and 30 K in the resistivity data [arrowed in Fig. 3(b), see below]. The sample did not reach zero resistance until 4 K. Diamagnetic shielding in the sample was only 2% of that expected for a perfect diamagnet, indicating that most of the sample was not superconducting (Fig. 4).

Air quenched $Y_1Ba_2Cu_3O_{6.50}$

The diffraction studies indicated that the orthorhombic distortion in this sample was greater than that in the $Y_1Ba_2Cu_3O_{6.41}$ sample, as expected [Fig. 3(c)]. Thermopower, resistivity, and susceptibility measurements all found a T_c of 37 K for this sample. The diamagnetic shielding in this sample was about 20% of that expected for a perfect diamanget (Fig. 4).

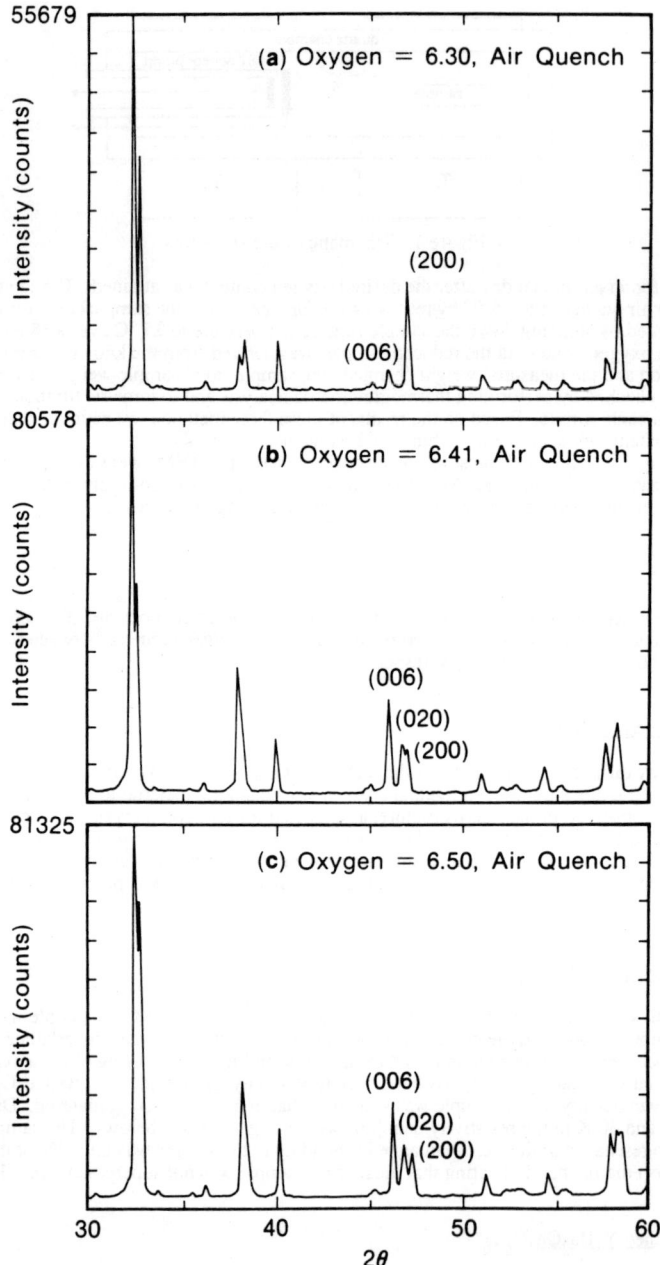

Figure 2. X-ray diffraction patterns for samples air quenched from 500°C with oxygen contents of (a) 6.30, (b) 6.41, and (c) 6.50.

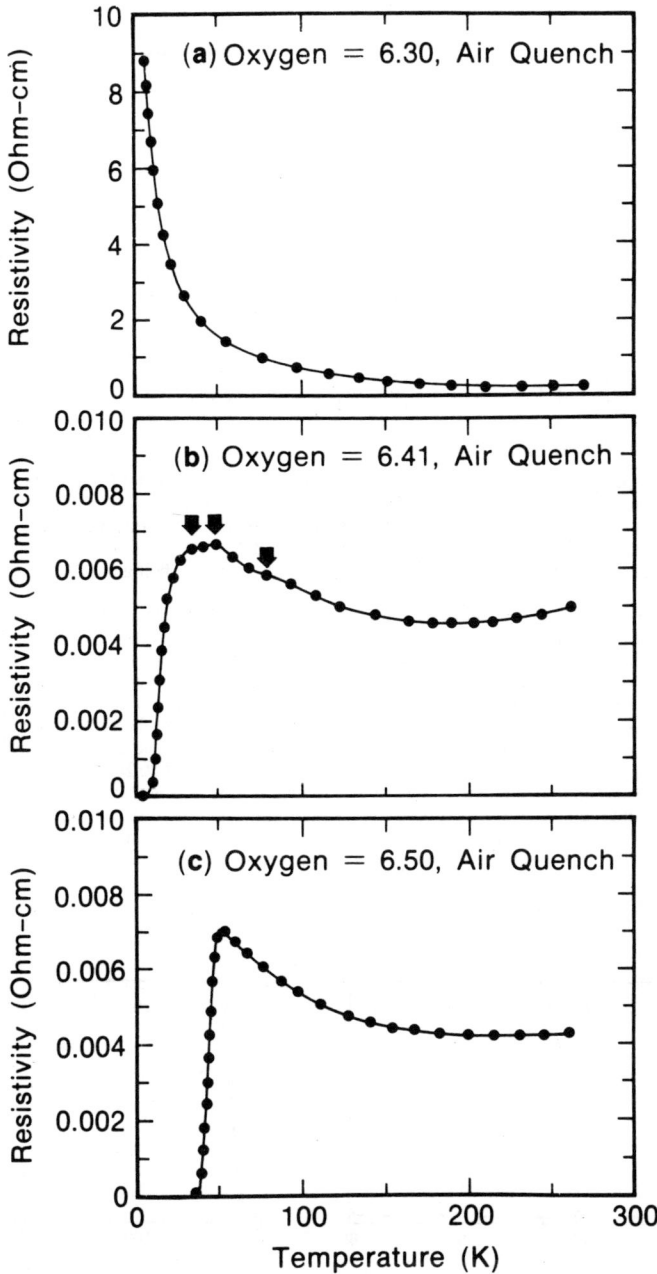

Figure 3. Resistivity versus temperature plots for samples air quenched from 500°C with oxygen contents of (a) 6.30, (b) 6.41, and (c) 6.50.

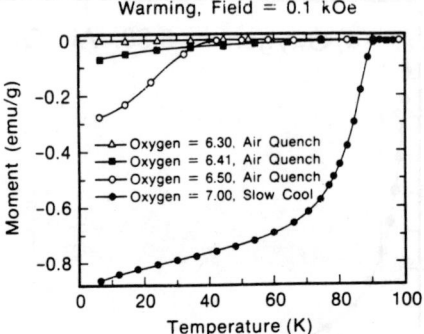

Figure 4. Diamagnetic shielding data for air quenched, reduced oxygen content $Y_1Ba_2Cu_3O_{9-x}$ samples and slowly cooled $Y_1Ba_2Cu_3O_{7.00}$ starting material.

Slow cooled $Y_1Ba_2Cu_3O_{6.41}$

The x-ray diffraction pattern and resistivity versus temperature plot for a $Y_1Ba_2Cu_3O_{6.41}$ sample that was slowly cooled from 500 °C to 25 °C over a 28 hour period are shown in Fig. 5. The (200) and (020) diffraction peaks are sharper and more clearly separated than those in the air quenched $Y_1Ba_2Cu_3O_{6.41}$ sample [compare Figs. 5(a) and 2(b)], indicating greater oxygen ordering in the slowly cooled sample. (This conclusion is currently being checked with neutron diffraction.)

Both resistivity and susceptibility measurements, however, indicated that the slowly cooled material had a lower transition temperature than the air quenched material. The slowly cooled sample had a higher resistivity over the entire temperature range studied and it did not reach zero resistance down to 4 K [Fig 5(b)]. Figure 6 compares the shielding data in the two samples. (Note the vertical scale is greatly expanded compared to Fig. 4.) The air quenched sample contained regions that were superconducting up to 80 K. We interpret the slope changes in both the shielding and resistivity data of the air quenched sample [Figs. 6 and 2(b)] as evidence for local regions in the material with critical temperatures of 30, 50, and 80 K. It is reasonable to suggest that local inhomogeneities in the oxygen content of this sample gave rise to these superconducting regions. We are currently investigating whether these inhomogeneities arose from insufficient equilibration time at 500 °C or were caused by stress induced diffusion during the quench. By comparison, the shielding data for the slowly cooled sample indicated only trace amounts of material with a critical temperature of 50 K (not observable in Fig. 6) and somewhat larger regions with a critical temperature of 10 K (arrowed in Fig. 6).

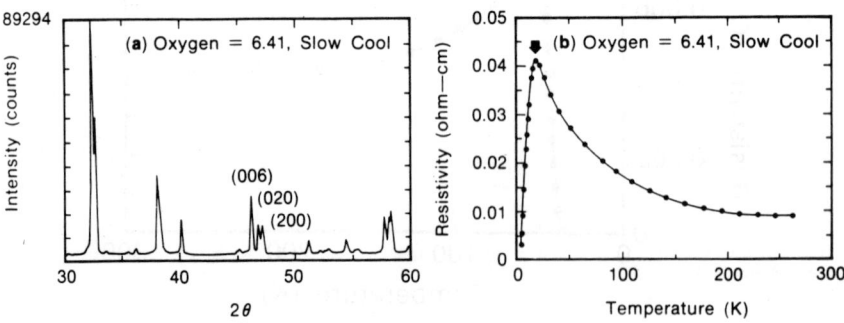

Figure 5. (a) X-ray diffraction pattern and (b) resistivity versus temperature plot for a slowly cooled $Y_1Ba_2Cu_3O_{6.41}$ sample.

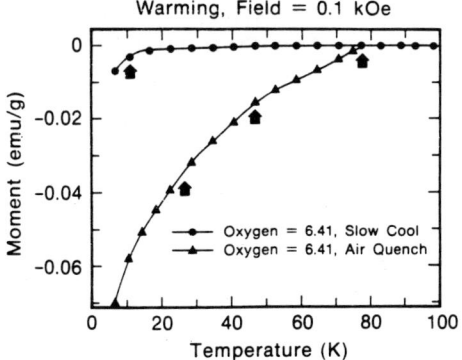

Figure 6. Enlarged plot of shielding data for slow cooled $Y_1Ba_2Cu_3O_{6.41}$ and air quenched $Y_1Ba_2Cu_3O_{6.41}$ samples. The arrows point out slope changes. Corresponding slope changes in the resistivity data are arrowed in Figs. 2(b) and 5(a).

CONCLUSIONS

(1) Like Cava et al. [3], we find that orthorhombic $Y_1Ba_2Cu_3O_{9-x}$ material with oxygen content below 6.5 can be prepared by low-temperature removal of oxygen from $Y_1Ba_2Cu_3O_{7.00}$. For oxygen removal at 500 °C, however, we find that the orthorhombic phase cannot be retained down to $Y_1Ba_2Cu_3O_{6.30}$. Lower titration temperatures may extend the oxygen stoichiometry range of the orthorhombic phase.

(2) Our results indicate that oxygen inhomogeneities are essential for observing small amounts of superconductivity in $Y_1Ba_2Cu_3O_{9-x}$ samples with oxygen contents below 6.5. This conclusion supports the view that a high oxidation state in the copper-oxygen network, whether on a local scale (inhomogeneities in $Y_1Ba_2Cu_3O_{<6.5}$ samples) or on a global scale ($Y_1Ba_2Cu_3O_{>6.5}$ samples), is essential for high-temperature superconductivity.

(3) The solid-state ionic method described here enables complete control of both the thermal history of the sample and its oxygen environment. It can be readily used for well-defined studies of other oxide samples of interest, including single crystals, thin films, and the many $(R.E._xY_{1-x})(A_yBa_{2-y})(B_zCu_{3-z})O_{9-\partial}$-type derivatives that have been made. Moreover, the same apparatus can be used for quantitative studies of phase equilibria in these oxide systems[6].

ACKNOWLEDGEMENTS

Financial support from the U.S. Department of Energy under subcontract LBL-4536310 is gratefully acknowledged.

REFERENCES

1. P.M. Grant, R.B. Beyers, E.M. Engler, G. Lim, S.S.P. Parkin, M.L. Ramirez, V.Y. Lee, A. Nazzal, J.E. Vazquez, and R.J. Savoy, Phys. Rev. B **35**, 7242 (1987).
2. D.C. Johnston, A.J. Jacobson, J.M. Newsam, J.T. Lewandowski, D.P. Goshorn, D. Xie, and W.B. Yelon, ACS Symposium Series **351**: Chemistry of High Temperature Superconductors, 136 (1987).
3. R.J. Cava, B. Batlogg, C.H. Chen, E.A. Rietman, S.M. Zahurak, and D. Werder, Phys. Rev. B **36**, 5719 (1987).
4. For example, see R. Beyers, G. Lim, E.M. Engler, V.Y. Lee, M.L. Ramirez, R.J. Savoy, R.D. Jacowitz, T.M. Shaw, S. La Placa, R. Boehme, C.C. Tsuei, Sung I. Park, M.W. Shafer, and W.J. Gallagher, Appl. Phys. Lett. **51**, 614 (1987).
5. D.C. Harris and T.A. Hewston, J. Solid St. Chem. **69**, 182 (1987).
6. B.T. Ahn, T.M. Gür, R.A. Huggins, R. Beyers, and E.M. Engler, these proceedings.

THE SEARCH FOR HIGH TEMPERATURE SUPERCONDUCTIVITY

C. POLITIS

Kernforschungszentrum Karlsruhe, Institut für Nukleare Festkörperphysik,
Postfach 36 40, 7500 Karlsruhe, FRG,
and
University of California, San Diego, Dept. of Electrical & Computer Engineering,
La Jolla, Mail-Code C-014, CA 92093, USA

Recently, following the discovery of the high-T_c superconductivity in mixed-valence copper oxides, we are living in a "gold rush" towards superconductors with surprising high critical temperatures. After the break-through concerning the transition temperature in the superconducting state a new break-through without rare earths will open soon the door for new superconductors. Here is a report about the search for high temperature superconductivity and some future aspects with respect to the constitution and substitution of the synthesized high-T_c superconductors.

Introduction

Although the phenomenon of superconductivity was discovered as early as more than 75 years ago, research on superconductivity has grown in importance for some decades. The reason for these ongoing activities lies in the relevance of superconductivity to application in the engineering field. Therefore, since superconductivity was discovered, materials have been searched which are characterized by the highest possible transition temperature. Very frequently this promissory goal which attracts each research worker had remained a dream.

The search for high-temperature superconductors stimulated contributions to fabricating by synthesis a great number of important and novel materials as well as to introduce them in the engineering field. This applies to several hundreds of new alloys and intermetallic phases and compounds and, likewise, to novel material classes such as amorphous alloys, layered structures and intercalation compounds. The research activities have shown that superconductivity is a rather common phenomenon among the metals and alloys so that the question "why do some metallic elements become superconducting?" could be reformulated "why do not all become superconducting[1]?" The transition temperatures of the elements are in the range of several mK to about 9.2 K. However, there is no correlation between the transition temperature and other characteristic features such as the melting temperature which would permit an ad hoc distinction to be made between superconductors and non-superconductors[2]. Elements which may occur with different crystal structures, by their transitions temperatures, which are dependent on the crystal structures, provide an indication to the effect that superconductivity is rather dependent on the arrangement of atoms. But the crystalline structure is not a prerequisite of superconductivity. It has even been demonstrated that the amorphous state which is attained, e. g., by cryo-condensation from the vapor phase[3,4] or by extreme quenching from the liquid state[5], might be superconducting with relatively high transition temperatures.

During a rather long period the search for materials with high transition temperatures had been concentrated on compounds of the transiton metals such as NbC_xN_y with T_c = 18 K[6], intermetallic phases such as A 15 compounds; V_3Si with T_c = 17 K[7], Nb_3Sn with T_c = 18 K[8], Nb_3 ($Al_{0.8}Ge_{0.2}$) with T_c = 20.5 K[9] and thin films of Nb_3Ge with T_c = 23.2 K[10]. There had been no lack of extrapolations, calculations and speculations concerning extremely metastable high temperature pressure phases such as cubic B 1-MoN and A 15-Nb_3Si. The "magic boundary" of 23.2 K (see Fig. 1) existing for more than 13 years as a possible asymptotic barrier had to be crossed; theoretical computations had predicted for stoichiometric δ-MoN a transition temperature of 29.3 K[11] and practical extrapolation a value of 25 K[12]. The best value which could currently be achieved in the experiment is 16 K for δ-MoN_{1-x}[13]. For A 15-Nb_3Si 38 K had been expected[14], whereas in the experiment only 18 to 19 K were achieved[15,16]. This is the reason why experimenters had been searching for a long time to detect superconductors with a new material as the basis which, actually, had not been usual among the superconductors, namely multi component oxides. For some oxides[17] of the composition $Me_{1-x}^{I,II}WO_3$ (with $Me^{I,II}$ = Li, Na, K, Rb, Cs and Ba, $0 \leq x \leq 1$) T_c's between 1.1 and 5.4 K were found. For the ternary oxide $Li_{1+x}Ti_{2-x}O_4$ ($0 < x < \frac{1}{3}$) critical temperatures T_c = 14 K were measured[18]. During the period in question (1973 until 1975) these values had been surprisingly high for an oxide material. Approximately at the same time (1975) another oxide material with the composition $BaPb_{1+x}Bi_xO_3$ ($0.05 \leq x \leq 0.35$), a perowskite like crystal structure, and a critical temperature of 13 K was discovered [19]. In the two oxides mentioned before the valencies are subjected to variations from Ti^{3+} to Ti^{4+} and Bi^{3+} to Bi^{5+}, respectively, which was considered to account for this relatively high T_c value. However, during the following decade all efforts failed to achieve still higher values of T_c by substitutions in $Li_{1+x}Ti_{2-x}O_3$ and in $BaPb_{1-x}Bi_xO_3$. On the other hand, a series of sensational news about materials with high T_c's such as CuCl and TiB_2 did not stand the test of critical examination [20] This may be the reason why the article by J. G. Bednorz and K. A. Müller, published in the September 1986 issue of Z. Physik B, Condensed Matter, entitled "Possible High T_c Superconductivity in the Ba-La-Cu-O System"[21], was not paid with commensurate attention[22].

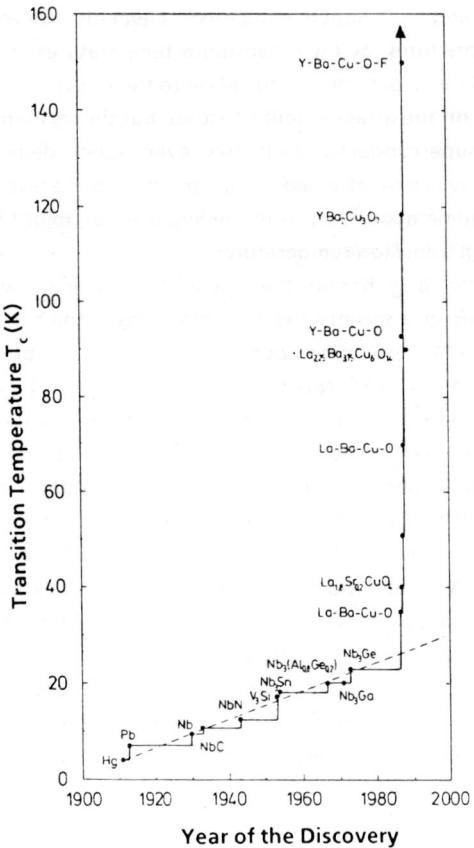

Fig. 1. The development of the transition temperatures since the discovery of superconductivity in 1911 to this date. Only the respective peak materials have been entered. Until 1986 the curve took an almost linear course with an increase of 0,3 K per year so that with T_C extrapolated the value of 30 K would have to be expected to be achieved not earlier than after the year 2000. The development in 1986/87 means for superconductivity a scientific and technical break-through the implications of which cannot yet be predicted.

The article reported for the first time about superconductivity above 30 K. The onset was measured by these authors at a temperature (T_c onset) of 35 K. The "zero" value of resistivity ($T_{c,o}$) was attained at 10 K. The composition of these new high-temperature oxide superconductors was indicated to be $Ba_xLa_{5-x}Cu_5O_{5(3-y)}$ ($x = 0.75$ and $1, y > 0$). On December 4, 1986, on the occasion of the Materials Research Society

(MRS) meeting in Boston, C. W. Chu, at the end of his lecture on $BaPb_{1-x}Bi_xO_3$, briefly mentioned the extraordinary results obtained by J. G. Bednorz and K. A. Müller. K.Kitazawa of Tokyo University immediately thereafter in a contribution to the discussion confirmed these results made on the basis of magnetic susceptibility and resistivity measurements[23,24]. One week later also other groups in China, the U.S.A., Japan and the Federal Republic of Germany started activities related to the La-Ba-Cu-O system with a view to confirm this results or to supplement them. On December 15, 1986 C. W. Chu of the University of Houston reported about superconductivity in the La-Ba-Cu-O system at 40 K [25].

In late December 1986 superconductivity was measured in Beijing on multiphase $La_{4.5}Ba_{0.5}Cu_5O_{5(3-y)}$ and on multiphase $La_{4.75}Sr_{0.25}Cu_5O_{5(3-y)}$ at 48.6 K[26]. On January 7, 1987 the Institute of Physics of the Chinese Academy of Sciences in Beijing reported about superconductivity measured at 70 K on a lanthanum mixed oxide[26] Independent of each other, groups of research workers of the University of Houston[25] and Tokyo[27], the Tohoku University[28], AT & T Bell Labs[29], the Argonne National Laboratory[30] and the Kernforschungszentrum Karlsruhe[31] within six weeks after the MRS meeting reported about the substitution of Sr for Ba in the oxide compound $La_{2-x}Ba(Sr)_xCuO_4$ ($0.1 \leq x \leq 0.3$) which yielded even higher transition temperatures of more than 40 K. For instance, at Karlsruhe a superconducting transition at 42 K and zero resistivity at 34 K were achieved as early as in January 1987 with the composition $La_{1.8}Sr_{0.2}Cu_4$[31]. On February 6, 1987 C. W. Chu reported about superconductivity with the transition occurring at 93 K without giving initially details of the composition of the material. This round the initiated clock activities geared worldwide to find the substance used by C. W. Chu and co-workers. The fact, however, that it was superconductive above the boiling temperature of liquid nitrogen (77 K) led to bustling activities displayed all over the world which were aimed to find out from the great number of potential oxide compounds the material in question and its composition. This was the starting shot for a unique scientific race in the history of physics and of modern engineering as a whole; it could be also termed a "scientific gold-fever" which has continued to persist until this day.

Several weeks immediately before and after publication (March 2, 1987) of the article by C. W. Chu and coworkers entitled "Superconductivity at 93 K in a New Mixed Phase Y-Ba-Cu-O Compound System at Ambient Pressure"[32] several groups, completely independent of each other, reported also about the application of superconductivity above 90 K and above 100 K, respectively, in the Y-Ba-Cu-O system[33-35].

Constitution and Superconducting properties

The quarternary compound $La_{2-x}Me_xCuO_{4-y}$ (Me: Mg, Ca, Sr or Ba) is an oxygen deficient structure of the K_2NiF_4 type. Substituting Mg, Ca, Sr or Ba for La brings about a wide range of stable homogeneity, depending on the temperature, oxygen pressure and conditions of cooling, e. g. $0 \leq x \leq 0.3$ for Ba and $0 \leq x \leq 1$ for Sr.

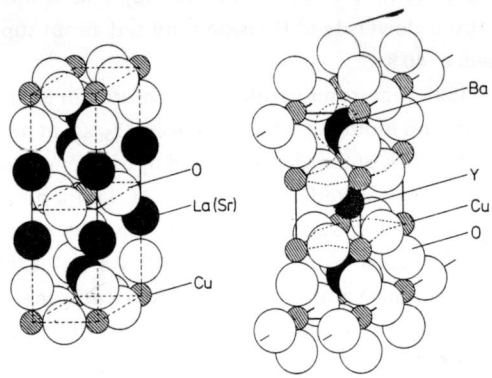

Fig. 2. Crystal structures of the two new high-temperature superconductors $La_{2-x}Sr_xCuO_4$ ($0.1 \leq x \leq 0.3$) and $YBa_2Cu_3O_7$. Both lattices have been represented on true scale in order to demonstrate the spacings and the occupation.

Figure 2 shows the unit cells of $La_{2-x}Sr_xCuO_4$ and $YBa_2Cu_3O_7$. The crystal symmetry in the case of $La_{2-x}Sr_xCuO_4$ is tetragonal (space group: I 4/mm). $YBa_2Cu_3O_7$ occurs in two modifications. The superconducting low-temperature phase has an orthorhombic structure (space group: P mm). It is now well known that above 620° C in air and 700° C in oxygen only the tetragonal modification (space group P 4/mmm) exists[36]. It is very impressive and fascinating to see how simple it is to demonstrate superconductivity in qualitative terms above the temperature of liquid nitrogen in the experiment with the "levitating superconductor:" A $REBa_2Cu_3O_7$ sample cooled down with liquid nitrogen to about 77 to 80 K using plastic tweezers. If one lowers now the cold sample and deposits it on the pole of a permanent magnet (e.g., a 30 mm ring magnet of 0.1 to 0.5 T) superconducting shielding currents are induced which, according to the Lenz rule, generate a repulsive force. In the state of equilibrium the sample floats above the magnet as long as the induced permanent currents (R = 0) flow in the sample, i.e., as long as the sample is superconducting. Depending on the sample size, a sample of 0.5 to 3 g weight with T_c = 90 to 100 K is capable of levitation on the pole of an 0.4 T ring magnet for 15 to 60 s (Fig. 3).

Fig. 3. "Levitating superconductor" made of $YBa_2Cu_3O_7$. Qualitative demonstration of superconductivity above the boiling temperature of liquid nitrogen. A ceramic sample in the superconducting state levitated above the pole of a permanent ring magnet.

The Meissner-Ochsenfeld effect can be demonstrated in a similar easy way by convenient means: The warm sample to be examined is placed on a ring magnet ($T > T_c$) which is placed in liquid nitrogen. While the sample is cooled down below $T = T_c$ field displacement occurs; the $REBa_2Cu_3O_7$ superconductor which has become diamagnetic is immediately repelled by the magnet and rises up to the height of equilibrium where it is levitating thereafter.

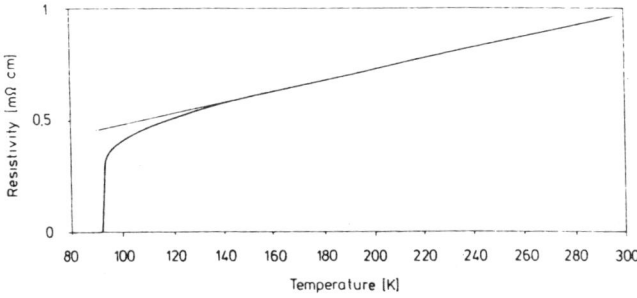

Fig. 4. Temperature dependence of the resistivity of a sample with the composition $YBa_2Cu_3O_7$. As a result of superconducting fluctuations a T_c onset of about 130 K is observed whereas the zero value occurs at 92 K. ($T_{c\ 90\%} = 104$ K, $T_{c\ 10\%} = 92.5$ K).

Figure 4 shows the temperature-dependent development of resistivity of a typical $YBa_2Cu_3O_7$ sample. The zero of resistivity is attained at T = 92 K. Here T_c onset, on account of superconducting fluctuations lies above 130 K. This superconducting behaviour with Tc values above 100 K has been observed frequently[34,37-39]. The total oxygen content of $YBa_2Cu_3O_{7-x}$ samples plays a major role. Samples with $0.5 < x < 1$ do not exhibit superconductivity. While stoichiometry is approached (x = 0) T_c increases above 100 K. During quenching (dT/dt $\geq$ 10^2 K s^{-1}) of $YBa_2Cu_3O_{7-x}$ samples from annealing temperatures between 400° C and 950° C development of resistivity similar to that of a semi-conductor is always observed[40]. A metal-type behaviour is displayed only if quenching takes place with annealing temperatures below 400° C. Also the superconducting state is observed to occur earlier for samples quenched or homogenized at lower temperatures. For many rare earth elements a coexistence is expected of superconductivity and magnetism. In fact, $GdBa_2Cu_3O_{7-x}$ exhibits a strong but normal paramagnetic state above the critical temperature[41].

There is no doubt that χ of $GdBa_2Cu_3O_7$ follows the Curie law with the Curie constant 1.017 x 10^{-2} emu K/g Oe. Magnetization at 300 K is likewise linear up to a maximum field of 10 kG with χ = 3.35 x 10^{-5} emu/g Oe. This data indicate that $GdBa_2Cu_3O_7$ is in a strong but normal paramagnetic state above T_c. Since the value of $YBa_2Cu_3O_7$ in the normal state is of the order of 10^{-7} emu/g Oe, the paramagnetic or diamagnetic contribution to χ of Cu^{2+}, Ba^{2+} and O^{2-} obviously cannot adopt higher values than corresponding to this order of magnitude. This means that the high paramagnetic value can with certainty be attributed solely to the magnetic ions of Gd^{3+}. The effective moment of Gd^{3+} was determined to be 7.77μ_B which comes very close to the theoretical value of 7.94μ_B. This means that the magnetic moments of Gd^{3+} ions are very well localized in $GdBa_2Cu_3O_7$ and totally decoupled from the conduction electrons. The magnetic susceptibility of $EuBa_2Cu_3O_7$ in the range T < 150 K during cooling down and heating exhibits the Meissner effect but also "flux pinning," a characteristic feature of the type II superconductors. For this $EuBa_2Cu_3O_7$ sample a 20 % diamagnetic state was achieved[42].

Copper is present in a number of compounds both in the monovalent and bivalent states. It is known that Cu occurs also in the trivalent state; however, trivalence has been confirmed so far only for complex compounds. Compounds in which Cu^{3+} is present have consequently commanded interest for a long time already. A substantial contribution to these mixed valent Cu^{2+}-Cu^{3+} oxides was made by C. Michel and B. Raveau[42] as early as in 1984 when they prepared by synthesis $La_3Ba_3Cu_6O_{14+\delta}$, $La_{2-x}A_{1+x}Cu_2O_{6x/2+\delta}$ and $La_{2-x}A_xCuO_{4-x/2+\delta}$ (A = Ca, Sr, Ba). In this basic work also the influence of oxygen stoichiometry on the electric properties was examined. Unfortunately, S. Michel und B. Raveau have not carried out their measurements at temperatures below 77 K (LN_2); therefore, no superconducting behaviour had been observed as a result of the abrupt vanishing of resistivity. Starting from these activities, J. G. Bednorz and K. A. Müller were able to pursue systematically the search for oxide superconductors.

An important question is why superconductivity does actually occur with these high T_c values. An essential feature of these oxide superconductors is the presence of Cu in a mixed valence state, namely as Cu^{2+} and Cu^{3+} which is a result of the substitution of some of La^{3+} by Ca^{2+} or Sr^{2+}, i.e., a situation similar to $BaPb_{1-x}Bi_xO_3$ (Bi^{3+}, Bi^{5+}) and $Li_{1+x}Ti_{2-x}O_4$ (Ti^{3+}, Ti^{4+}). It is crucial whether this mixed valence is steady-state, i.e., whether some Cu sites are bivalent and other trivalent or whether all Cu sites are equal in valence and a dynamic fluctuation of valencies prevails. Both the nature of mixed valence and the role it plays in superconductivity have not been unambiguously clarified to this date. It is still too early to make statements concerning the validity of one or the other theoretical models on high-temperature superconductivity. It should be briefly mentioned that more than a dozen of theoretical models are being discussed besides the model of electron phonon interaction. At present, the interest has been concentrated on experimental findings concerning the electronic structure, the bond, the stoichiometry and the structural processes taking place in these oxide superconductors. For this reason, a number of dedicated studies has been made using UPS, XPS and XANES[44-45]. In these studies the presence of Cu^{3+} in $REBa_2Cu_3O_7$ and a shorter Cu-O spacing than in CuO were detected although a clear assignment to high-temperature superconductivity is still missing.

Discussion and Outlook

High-temperature superconductivity above the nitrogen boiling temperature has been observed both in La-Ba(Sr)-Cu-O and in Y-Ba-Cu-O systems. La and Y can be substituted successfully by Sc, Nd, Sm, Eu, Gd, Dy, Ho, Er, Tm, Yb and Lu, i.e., all trivalent rare earth elements can be used instead of La and Y. Ce, Pr and Tb do not only occur in the trivalent state, but predominantly in the tetravalent state. It cannot yet be said with certainly whether the latter state implies that Y cannot yet be successfully substituted in $YBa_2Cu_3O_7$ by Ce, Pr and Tb. Whereas Ba and Sr can be substituted by Ca at a little degree only, successful substitution of O^{2-} has been made possible only by F. Also in this case it is still an open problem whether the F-atoms actually occur at the sites of O-atoms. The relevant investigations are under way. At present, the end of this development cannot yet be forecast. Always new and higher critical temperatures are reported; e.g., T_c's above 100 K are no longer rare[38,46,47]. After substituting F for O also a $T_{c\,o}$ of 155 K was measured[48,49]. Likewise, superconductivity was reported with an abrupt T_c onset at 238 K and a zero ($R \leq 10^{-4}\,\Omega$) at

228 K for a midpoint of 230 K in $EuBa_2Cu_3O_{6+\delta}$[50]. The recently discovered class of mixed-valence $Sr_2Bi_2Cu_2O_7$ compounds with T_c = 22 K by the Raveau-group[51] has opened the door for many new high-T_c compounds without any rare earths. By suitable substitution of Be, Mg and Ca but also Sb, Bi and In for Ba and Sr in this and related copper oxides there is hope for still higher T_c's as have been found until yet in $Sr_2Bi_2Cu_2O_7$.

Also the numerous press releases and information received by phone or at least temporarily observed critical temperatures above the room temperature indicate that it will be possible one day to demonstrate qualitatively in the laboratory room temperature superconductivity as a stable state.

References

1. B. T. Matthias, *Physics Today*, **24** (1971) 23
2. W. Buckel, *Supraleitung, Grundlagen und Anwendungen*, Physik Verlag 1983
3. W. Buckel, R. Hilsch, *Z. Phys.* **132** (1952) 420
4. R. E. Glover III, F. Baumann, S. Moser, Proc. *12th Int. Conf. Low Temp. Phys,.* (LT12) Kyoto 1970, S. 337, Academic Press of Japan
5. W. L. Johnson, S. J. Poon, P Duwez, *Phys. Rev.* **B11** (1975) 150
6. M. C. Krupka, A. L. Giorgi, N. H. Krikorian, E. G. Szklarz, J. Less, *Common Metals* **19** (1969) 113
7. G. F. Hardy, J. K. Hulm, *Phys. Rev.* **87** (1953) 884 und **93** (1954) 1004
8. B. T. Matthias, T. H. Geballe, S. Geller,E. Corenzwit, *Phys. Rev.* **95** (1954) 1435
9. G. Arrhenius, E. Corenzwit, R. Fitzgerald, G. W. Hull Jr., H. L. Luo, B. T. Matthias, W. H. Zachariasen, *Proc. Natl. Acad. Sei. US* **61** (1968) 621
10. J. R. Gavaler, Appl. Phys. Lett. 23, (1973) 480
11. D. A. Papaconstantopoulos, W. E. Pickett, B. M. Klein, L. L. Boyer, *Phys. Rev.* **B31** (1985) 752
12. Y. Zhao, S. He, *Solid State Commun.* **45** (1983) 281
13. C. Politis, W. Kaiser, unpublished results
14. D. Dew-Hughes, V. E. Rivlin, *Nature* **250** (1974) 723
15. V. M. Pan, V. P. Alekseevskij, A. G. Popov, Yu. I. Beletskij, L. M. Yupko, V. V. Yarosh, *JETP Lett.* **21** (1975) 228
16. B. Olinger, L. R. Newkirk, *Solid State Commun.* **37** (1981) 613
17. B. W. Roberts, *J. Phys. Chem. Ref. Data* **5** (1976) 581 und *NBS Technical Note* **983** (1978)
18. D. C. Johnston, H. Prakash, W. H. Zachariasen, R. Viswanathan, *Mat. Res. Bull.* **8** (1973) 777
19. A. W. Sleight, J. L. Gillson, P. E. Bierstedt, *Solid State Commun.* **17** (1975) 27
20. W. Buckel, *Phys. Bl* **43** (1987) 41
21. J. G. Bednorz, K. A. Müller, *Z. Phys. B* **64** (1986) 189
22. W. Buckel, *Phys. Bl* **43** (1987) 96
23. S. Uchida, H. Takagi, K. Kitazawa, S. Tanaka, *Jpn. J. Appl. Phys.* **26** (1987) L1
24. H. Takagi, S. Uchida, K. Kitazawa, S. Tanaka, *Jpn. J.. Appl. Phys.* **26** (1987) L123

25. C. W. Chu, Ph. H. Hor, R. L. Meng, L. Gao, Z. J. Huang, Y. Q. Wang, *Phys. Rev. Lett.* **58** (1987) 405
26. Z. X. Zhao, L. Q. Chen, C. G. Cui, Y. Z. Huang, J. X. Liu, G. H. Chen, S. L. Li, S. Q. Guo, Y. Y. He, *Kexue Tongao* **32** (1987) 522
27. K. Kishio, K. Kitazawa, N. Sugii, S. Kanbe, K. Fueki, H. Takagi, S. Tanaka, preprint, submitted to *Chemistry Letters*, Jan. 24, 1987
28. M. Sato, S. Hosoya, S. Shamoto, M. Onoda, K. Imaeda, H. Inokuchi, preprint, 08. Jan. 1987
29. R. J. Cava, R. B. van Dover, B. Batlogg, E. A. Rietman, *Phys. Rev. Letters* **58** (1987) 408
30. W. K. Kwok, G. W. Crabtree, D. G. Hinks, D. W. Capone, J. D. Jorgensen, K. Zhang, preprint, January 1987
31. C. Politis, J. Geerk, M. R. Dietrich, B. Obst, *Z. Physik B* **66** (1987) 141
32. M. K. Wu, J. R. Ashburn, C. J. Torng, P. H. Hor, R. L. Meng, L. Gao, Z. J. Huang, Y. O. Wang C. W. Chu *Physics Rev. Letters* **58** (1987) 908
33. J. M. Tarascon, L. H. Greene, W. R. McKinnon, G. W. Hull, *Phys. Rev. Lett.* (1987)
34. H. Takagi, S. Uchida, K. Kishio, K. Kitazawa, K. Fueki, S. Tanaka, *Jpn. J. Appl. Phys.* **26** (1987) L 320
35. C. Politis, J. Geerk, M. R. Dietrich, B. Obst, H. L. Luo, *Z. Phys. B* **66** (1987) 279
36. J. D. Jorgensen, M. A. Beno, D. G. Hinks, L. Soderholm, K. J. Volin, R. L. Hitterman, J. D. Grace, I. K. Schuller, C. U. Segre, K. Zhang, m. S. Kleefisch preprint, submitted to *Phys. Rev.* 01.06.87
37. Y. Zhongjin, Z. Naiping, J. Xiaoping, P. Dexing, Qi Hongbo, Su Guoyue, Z. Ze, Yu Huafeng, *J. Phys. C: Solid State Phys.* **20**, (1987) L 351
38. J. J. Caponi, C. Chailont, A. W. Hewat, P. Lejay, M. Merezio, N. Nguyen, B. Raveau, J. L. Sonbeypoux, J. L. Tholence, R. Tournier, *Europhys. Lett.* **3**, (1987) 1901
39. X. Cai, R. Yoynt, D. C. Larbalestier, *Phys. Rev. Letters* **58** (1987) 2798
40. Yu Mei, C. Jiang, S. M. Green, H. L. Luo, C. Politis, *Z. Phys. B* **69** (1987) 11
41. Yu Mei, S. M. Green, G. G. Reynolds, T. Wiczynski, H. L. Luo, C. Politis, *Z. Phys. B* **67** (1987) 303
42. C. Jiang, Yu Mei, S. M. Green, H. L. Luo, C. Politis, *Z. Phys. B* **68** (1987) 15
43. C. Michel, B. Raveau, *Rev. Chim. Meneral* **21** (1984) 407
44. A. Bianconi, A. Congiu Castellano, M. De Santis, C. Politis, A. Marcelli, S. Mobilio, A. Savoia, *Z. Phys. B* **67** (1987) 307
45. P. Steiner, S. Hüfner, V. Kinsinger, I. Sander, B. Siegwart, H. Schmitt, R. Schulz, S. Junk, G. Schwitzgebel, A. Gold, C. Politis, H. P. Müller, R. Hoppe, S. Kemmler-Sack, C. Kunz, *Z. Phys. B* **69** (1988) 449-458
46. C. Politis, M. R. Dietrich, G. M. Friedman, J. Geerk, S. M. Green, R. Hu, S. Hüfner, C. Jiang, W. Krauss, H. Küpfer, H. Leitz, G. Linker, H. L. Luo, Yu Mei, O. Meyer, B. Obst, P. Steiner, H. Wühl, Proc. of Symposium High Temperature Superconductors, D. U. Gubser, M. Schluter, (Eds.) 1987 *Spring Meeting of the MRS*; 23.24 April 87 Anaheim, USA
47. I. Kirschner, J. Bánkuti, M. Gál, K. Torkos, K. G. Sólymos, G. Hor, G. Horváth, *Europhysics Letters* **4** (1987) 37
48. S. R. Ovshinsky, R. T. Young, D. D. Allred, G. DeMaggio, G. A. Van der Leeden, *Phys. Rev. Letters* **58** (1987) 2579
49. R. N. Bhargava, S. P. Herko, W. N. Osborne, *Phys. Rev. Lett.* **59** (1987) 1468
50. C. Y. Huang, L. J. Dries, P. H. Hor, R. L. Meng, C. W. Chu, R. . Frankel, Preprint
51. C. Michel, M. Hervieu, M. M. Borel, A. Grandin, F. Deslandés, J. Provost, B. Raveau *Z. Phys. B* **68** (1987) 421

CERAMIC MATERIALS AND HIGH-Tc SUPERCONDUCTIVITY: THE 1:2:3 COMPOUND

R. ESCUDERO
Instituto de Investigaciones en Materiales, Universidad Nacional Autónoma de México, Apdo. Postal 70-360, 04510 México, D.F. México.

ABSTRACT.

Experimental results related to the substitution of different atomic species in the typical 1:2:3 compound ($Y_1Ba_2Cu_3O_{7-\delta}$) are presented and some characteristics of the recently discovered ceramic superconductors of high-Tc are described. This study is directed towards a better understanding of these new and fascinating systems.

INTRODUCTION.

The discovery of ceramic superconductors with critical temperatures of the order of 30 K has awaken great interest amongst the scientific and technological communities. But the wonder did not stop there. Within few months, new ceramic superconductors were discovered with the critical temperatures in the range of 90 K. Both groups of superconductors (30 K and 90 K) share in common a crystalline structure which is of the perovskite type.

In the first case (30 K materials) the crystalline structure is of the K_2NiF_4 type. The parent compound is La_2CuO_4 wich behaves as a semiconductor at low temperatures. When alkaline-earth impurities, such as Ba, Sr or La are introduced in this compound it becomes superconductive. One of the possible causes of this change of behaviour in the transport properties is the presence of $[Cu-O_2]_\infty$ chains, which produce a tendency towards a mixed valence behaviour in the copper atoms. Copper oscillates between valence states of Cu^{+2} and Cu^{+3}. Although it seems a fact that the valence oscillation is directly related with the transport properties in superconducting ceramics, the detailed relationship remains to be understood. We can put it in the following words: The experimental evidence indicates that the copper mixed-valence state and the low dimensionality are characteristics necessarily present in these superconducting materials [1] but there is no microscopic theory to explain the relationship between these evidences and the mechanism for their superconducting properties.

The above mentioned facts are also present in the 90 K superconducting ceramics. In this class of materials the best known and best studied one is the $Y_1Ba_2Cu_3O_{7-\delta}$ system, known as the 1:2:3 compound. It is precisely on this material that our research has been focussed. We have asked ourselves questions such as: Which would be the effect on the critical temperature if we substitute yttrium by other rare-earth elements?. How the compound would be affected if barium were substituted by other alkaline-earths in these compounds?. Which is the rôle of the oxygen content?. How can we enhance the critical temperature?. What will happen if we substitute yttrium by an element of similar atomic radius?. In order to answer some of these questions, we have performed different experiments. These results are discussed here together with other related reports from other people.

THE ROLE OF OXYGEN IN THE CERAMIC COMPOUNDS.

The 90 K compounds have crystalline structures of the perovskite type. They can be described as consisting of the stacking of three elemental perovskites, with oxygen defficiency in the surroundings of the so called Cu(II) and Cu(I) regions, yet greater in the Cu(II) regions. The slight oxygen defficiency in the Cu(I) sites could be altered depending on the preparation process: at temperatures of the order of 900°C the material presents the highest possible oxygen defficiency (close to $\delta = 1$) and a tetragonal structure due to the defficiency in the Cu(I) sites. As the temperature decreases, the oxygen content starts to increase, until, at temperatures around 750°C, there is a phase transition in the material from tetragonal to orthorhombic. The material reaches its optimum superconducting transition temperature (Tc) when the oxygen content has attained a value close to $\delta = 0$ [2].
The oxygen absorption and deabsorption in the Cu-O chains seems to be controlled by the large electric fields present along the c-axis [3, 4].
We have just mentioned that the 90 K compounds have three piled-up perovskites, while the 30 K compounds present only one. It is tempting to assume that increasing the number of perovskites in the stack, increases the Cu content and enhances the critical temperature.

YTTRIUM SUBSTITUTION FOR RARE-EARTHS.

It has been found that substituting yttrium by lanthanides, except Ce, Pr and Tb, gives superconducting materials with very similar characteristics to the $Y_1Ba_2Cu_3O_{7-\delta}$ compound. It has also been observed that when the ionic radius of the substituted rare earth increases, there is a slight increase in the transition temperature [5,6]. The case of the 1:2:3 compound with La, (in which, according to the last argument, Tc should be near 100 K) is particularly interesting because Tc is anomously low (between 50 K and 80 K) [7,8]. Trying to understand this contradiction would throw some light on the influence of the oxygenation process, and could even produce new ideas related to the superconductivity mechanisms. Apparently the La compound tends to present a tetragonal structure which is superconducting, contrary to the other 1:2:3 compounds in which only the orthorhombic structure is superconductive, although Tc is rather low as it can be seen in Figure 1 [8].
The fact that the compounds (RE) $Ba_2Cu_3O_{7-\delta}$ where RE is Y or a rare-earth have such a similar transition temperature, regardless of the rare-earth magnetic behaviour, seems to indicate that the fundamental rôle of the yttrium or the rare-earth element, in the 1:2:3 compounds is only to form the skeleton or support of the crystalline structure. On the other hand, the above mentioned dependence of Tc with the

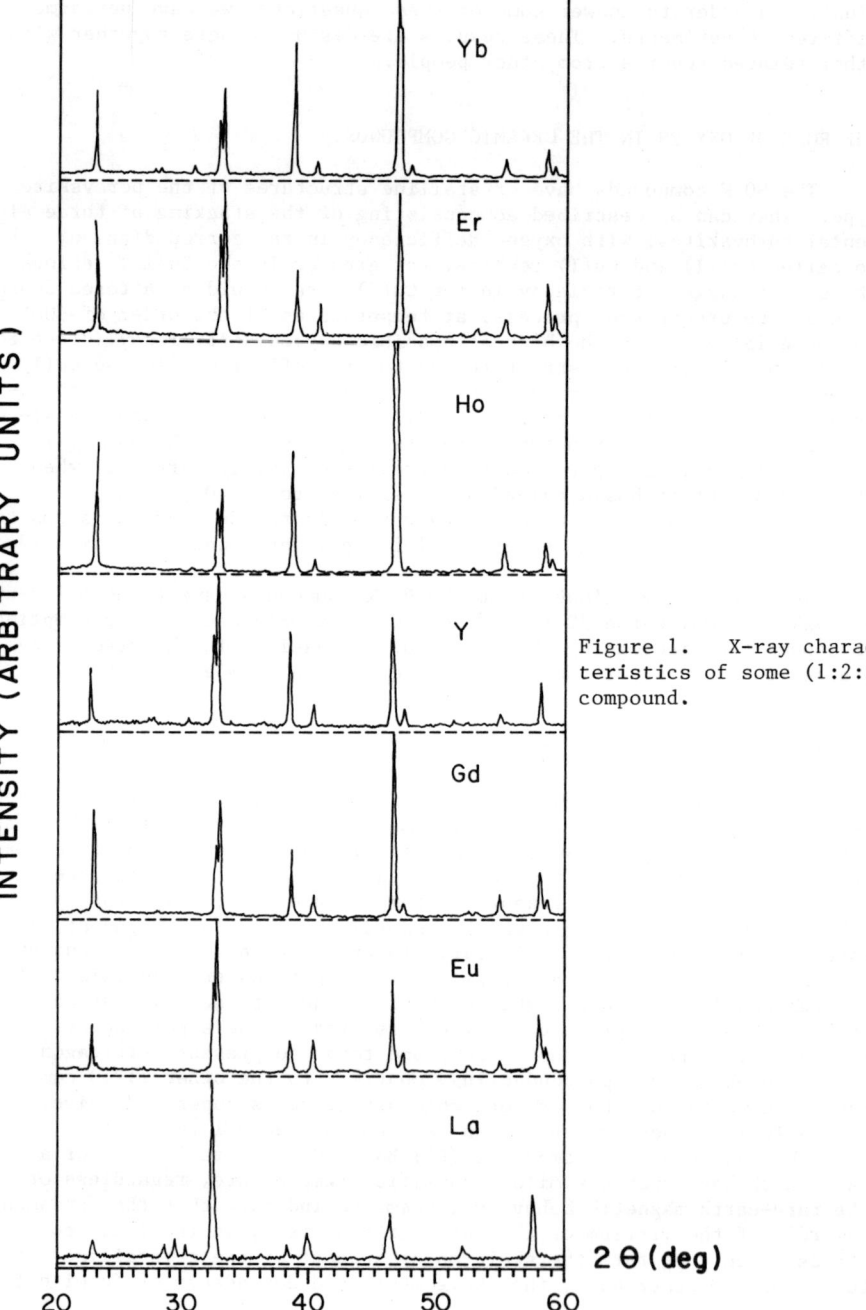

Figure 1. X-ray characteristics of some (1:2:3) compound.

ionic radii must be related to the way in which the oxygenation and deoxygenation takes place in the Cu(I) sites, as suggested by the electric fields along the c-axis [3,4].

The conclusion of this line of reasoning is that other superconducting ceramics could be designed without the presence of a rare-earth component. This suspicion has been supported by a recent discovery by the Caen group where the rare-earth element has been completely substituted by bismuth and presents a critical temperature of the order of 20 K.

THE ROLE OF BARIUM.

Let us discuss now the rôle of barium, or in general, the alkaline-earth elements in the 1:2:3 compounds.

In the 30 K compounds is well known that introducing alkaline-earth impurities (Sr, Ca, or Ba) affects dramatically the crystalline and electronic properties of the material. For instance, adding a small amount of Sr ($La_{2-x}Sr_xCuO_{4-\delta}$) with $x \leq 0.2$) produces a state of mixed valence in the copper and a change from semiconducting to superconducting behaviour. This change in the transport properties has been related to the fact that in the undoped material the Fermi surface is nested. When impurities are added, this nesting is supressed and therefore, there is no CDW or SDW transition [10,11].

To dope superconducting ceramics with alkaline-earth elements controls the copper oxidation but only to a certain limit. For instance, if in the 1:2:3 compounds Ba is partially substituted by Sr the transition temperature decreases slowly, until the compound is not superconductor any more. We have also observed that when substituting Ba by Ca, it is possible to obtain a metallic ceramic, but not a superconducting one. In these latter experiments it is observed that a metallic behaviour is obtained when annealing from 950°C to 77 K. On the other hand, in slow cooling to room temperature in the presence of air, the samples present a nearly constant R vs T curve.

It is observed that if the compound is oxygenated for several hours, the transport properties do not change, indicating that the material can not take more oxygen, thus mixed valence in the copper is possibily inhibited.

Clearly, the rôle of the alkaline-earths depends not only on their electronic properties but on their size. The strontium and the calcium might be too small in certain cases.

COPPER SUBSTITUTION IN THE 1:2:3 COMPOUND.

The last case to be analyzed is probably the most interesting: copper. Many experiments have been performed to this date [12,13,14,15]. Figure 2 shows the critical temperature for compounds where 10% of the total content of copper has been substituted. It can be noted that Ti, V, Cr and Mn have small effect in decreasing the transition temperature, while Fe suppresses the superconductivity rather quickly and, rather surprisingly substituting Zn which es quite similar to

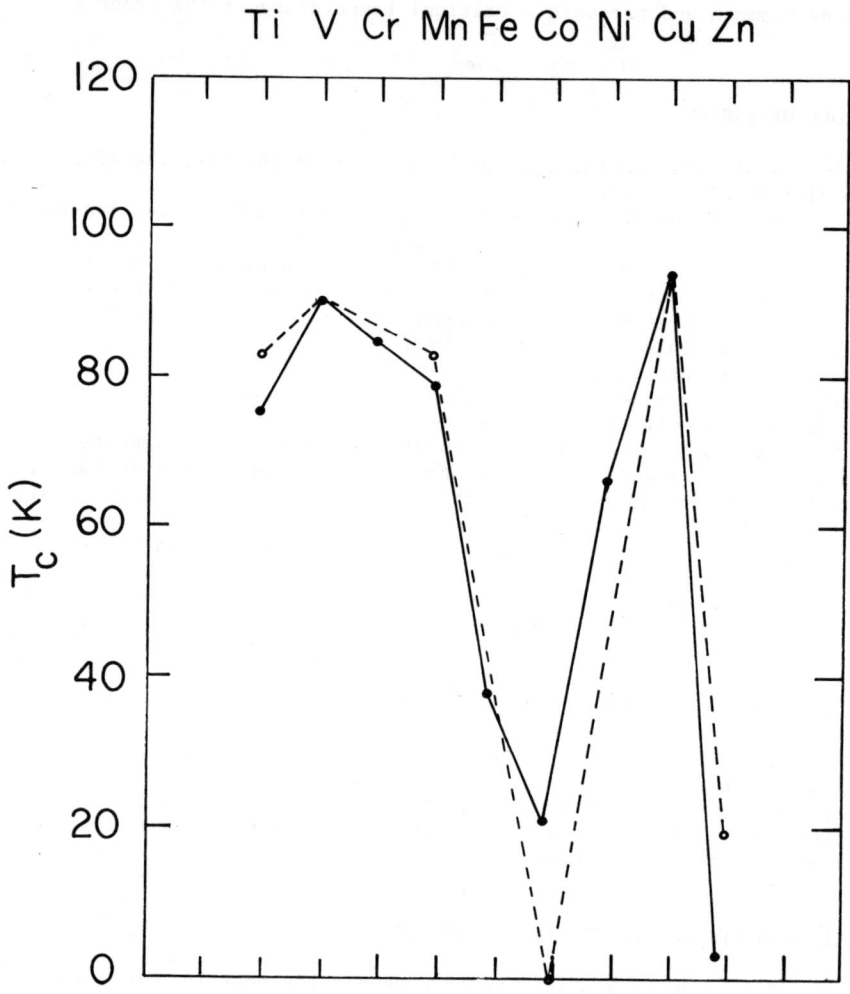

Figure 2. Transition metal substitutions made in $Y_1Ba_2Cu_{2.7}M_{.3}O_{7-\delta}$. The solid line is from reference [13]. The broken line is drawn through our data points.

copper, decreases the transition temperature notably. Some authors have reported [13] that the effect of zinc is very similar to the effect of iron. Our results indicate that it is iron the one that destroys superconductivity in a most efficient way. A possible explanation of this large effect of Fe can be taken from Mössbauer experiments [4]. It is seen there that Fe substitutes favorably copper in the Cu (I) sites and that it tends to be in a 3^+ state of oxidation. As the Cu(I) sites are the ones with the possibility of mixed valence and Fe goes to 3^+ Cu sites preferably, the disproportion of Cu^{3+} and Cu^{+2} decreases and consequently the number of holes decreases. The reaction that would explain this process is:

$$(Cu^{3+}-O^{2-}) \quad (Cu^{2+}-O^{-})$$

Other points to consider are: a) the magnetic effects of iron, which somehow could tend to break the electron pairs and thus inhibit the superconducting transition, b) the fact that zinc tends to be in a 2^+ oxidation state which inhibits, the creation of holes, although in a less effective way.

These latter results seem to indicate that the so called oxygen holes play a very important rôle in the superconductivity of high Tc materials.

ACKNOWLEDGMENTS.

I would like to thank J. Tagüeña and R.A. Barrio for their useful comments. Financial support from Programa Universitario de Superconductores de Alta Temperatura Crítica, UNAM and CONACYT is greatly acknowledged.

REFERENCES.

1. B. Raveau, C. Michel, A. Maignan, M. Hervieu and J. Provost. (This volume).

2. R.J. Cava, B. Batlogg, C.H. Chen, E.A. Rietman, S.M. Zahurak, D. Werder. Nature 329, 423 (1987); Phys. Rev. B36, 5719 (1987).

3. J.B. Goodenough and A. Manthiram. (This volume).

4. R. Gómez, S. Aburto, V. Marquina, M.L. Marquina, M. Jiménez, C. Quintanar, T. Akachi, R. Escudero, R.A. Barrio, and D. Ríos-Jara. Phys Rev. B36, 7226 (1987).

5. R. Escudero et al, Solid State Conmm. 64, No.2, 235 (1987).

6. L. Govea, R. Escudero, D. Ríos, C. Piña, C. Wang and R.A. Barrio. (This volume).

7. A. Maeda, T. Yabe, K. Uchinokura, M. Izumi and S. Tanaka, Jpn. J. Appl. Phys. 26, L1550 (1987).

8. S. Lee, J.P. Golben, S.Y. Lee, X. Chen, Y. Sang, T.W. Noh, R.D. Michael, J.R. Garnes, D.L. Cox and B.R. Patton, Submitted to Phys. Rev. Lett. (1987).

9. C. Michel, M. Hervieu, M.M. Borel, A. Grandin, F. Deslandes, J. Provost and B. Raveau, Z. Phys. B. 68, 421 (1987).

10. L.F. Mattheiss, Phys. Rev. Lett. 58, 1028 (1987).

11. M. Schluter. (This volume).

12. Y. Maeno et al. Nature 328, 512 (1987).

13. G. Xiad, F.H. Streitz, A. Gavrin, Y.W. Du, and C.L. Chien, Phys. Rev. B35 8782 (1987).

14. B.A. Richard and R.E. Allen, to be published in "Proceeding of the 34th National Symposium of the American Vacuum Society". Anaheim, Cal. (1987).

15. G. Gonzalez, D. Ríos-Jara, T. Akachi, R.A. Barrio and R. Escudero. Proc. of the MRS Fall Meeting, Boston Mass. (1987).

PHYSICAL PROPERTIES

BEFORE HIGH-T_c SUPERCONDUCTORS

NOW

PHYSICAL PROPERTIES OF HIGH Tc OXIDES

S. UCHIDA
Department of Applied Physics and Engineering Research Institute,
University of Tokyo, Hongo 7-3-1, Tokyo 113, JAPAN

ABSTRACT

A brief review is given on the current status of the experimental research for the high Tc superconductors in Japan. Based upon the recent experimental investigation, the variation of electronic states is discussed for three classes of oxides, $BaPb_{1-x}Bi_xO_3$, $(La_{1-x}Sr_x)_2CuO_4$ and $YBa_2Cu_3O_{7-y}$ with the change of compositions x and the oxygen vacancies y.

INTRODUCTION

It was been revealed that the discovered oxide superconductors potentially possess superior properties, high critical temperatures over 30K, even higher than the liquid nitrogen temperature, high critical fields and sufficient critical current at 77K(1). On the other hand, it is not so clear why Tc is so high in the Cu-oxides and which type of interactions is most relevant to the realization of high Tc. Many theoretical modes have so far been proposed based either on the pairing mechanisms mediated by phonons, excitons or plasmons or on the mechanisms which emphasize the interelectron Coulomb interaction (2,3). In order that the experiments give an answer to the question, which theoretical model best approximate the actual systems, more complete and systematic data must be accumulated on the well-characterized, if possible, single crystalline samples.

Here, I would like to report the current state of the experimental research on the high Tc superconductors being made by the researchers in Japan. First, fundamental superconducting parameters, the critical temperature (Tc), the critical current (Jc) and the critical magnetic field (Hc), are briefly summarized for the 90K-class materials. These three parameters are most important for the practical applications. In the second part, are summarized the experimental investigation of the electronic states in the oxide superconductors, referring also to another oxide superconductors $Ba(Pb,Bi)O_3$. Emphasis is made on the phase diagrams of three classes of oxide superconductors. Systematic studies have been made to see the change of the properties with varying composition of Bi and alkaline earth elements in $BaPbO_3$ and La_2CuO_4 respectivly and varying the oxygen content in $YBa_2Cu_3O_7$. It is shown that superconductivity competes or coexists with the charge-density-wave (CDW) instability in $Ba(Pb,Bi)O_3$, which it is realized just in contact with the antiferromagnetic (AF) order both in La_2CuO_4 based materials and in $YBa_2Cu_3O_{7-y}$. The latter seems to indicate strong interelectron correlation in cuprates which gives rise to quite distinct normal state properties and maybe to much higher Tc.

SUPERCONDUCTING PARAMETERS-Tc, Jc and Hc

Since the discovery of 90K-class 1-2-3 comounds, $YBa_2Cu_3O_7$ and related magnetic superconductors $LnBa_2Cu_3O_7$ with Ln= rare earth elements other than Pr and Ce, no fully new materials with Tc comparable to 90K or higher have appeared up to now. Although several reports of finding higher Tc have

been presented, they have not been confimed by other institutes.
Because the higher Tc date still awaits reliable experimental confirmation, the higherst Tc achieved so far is still 90K and something around it for the 1-2-3 compounds. Some researchers become aware that Tc may be slightly higher as the ionic radius of the lanthanide element becomes larger in $LnBa_2Cu_3O_7$. Such trends have been recognized for a series of well characterized samples. As an examples, single crystals with high quality have been grown in Nippon-Telegraph-Telephone (NTT)(4), University of Tokyo (5) and several other institutes using so called nonstoichiometric melt method. They demonstrated that Tc, defined as an end-point of the resistive transition, is 91K for Y-, 93K for Gd- and 95K for $Eu-Ba_2Cu_3O_7$. Similar data have been obtained also for polycrystalline samples after careful heat treatment (Fig.1).

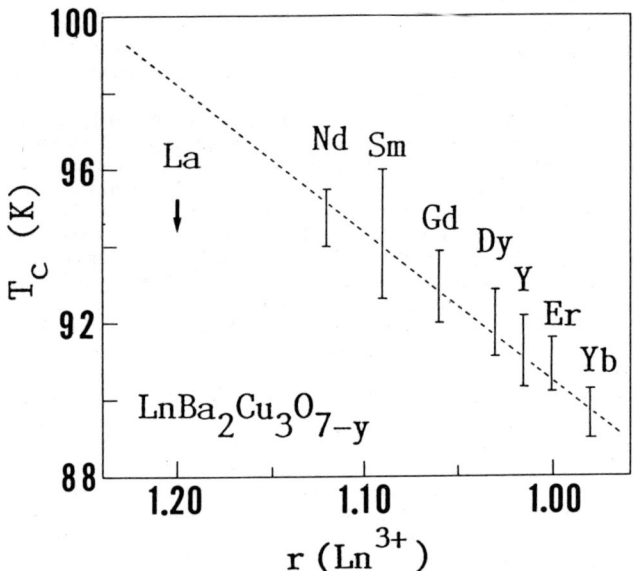

Fig. 1 Tc vs Ionic radius of rare earth element in $LnBa_2Cu_3O_7$ (unpublished data by T.Wada, Univ. of Tokyo).

If one extrapolates this trends to the lanthanides with larger ionic radius, than one may get Tc of about 100K for $LaBa_2Cu_3O_7$. Unfortunately, the ionic radius of La is almost identical to the Ba ionic radius, so that it is very difficult to synthesize pure, i. e. ordered, 1-2-3 compound of $LaBa_2Cu_3O_7$. The highest Tc, the end point, so far achieved in La-Ba-Cu-O is 81K with transition width more than 10K (6,7).
Concerning the critical current density, the value $Jc \simeq 2 \times 10^6 A\ cm^{-2}$ at 77K, was achieved by an NTT group for single crystalline films of $YBa_2Cu_3O_7$ synthesized on $SrTiO_3$ substrates(8). In contrast to thin films or single crystals, the critical current density of ceramic wires has only recently reached the value $10^4 A\ cm^{-2}$ at 77K (9).

The measurement of the upper critical magnetic field Hc_2 was made on single crystalline $YBa_2Cu_3O_7$ by one of the groups in the Institute for Solid State Physics (ISSP) of University of Tokyo (10). Magnetic fields up to 40 Tesla were applied in a pulsed form and superconducting state, zero-resistence state, at 77K was not destroyed even at 40 Tesla when the field was applied parallel to the CuO planes of the crystal. This is practically the highest Hc data so far reported for high Tc cuprates.

ELECTRONIC STATES IN OXIDE SUPERCONDUCTORS

In most 3d- transition-metal oxides, the O 2p band is well separated from the d block-band. This stabilizes O^{2-} and the positively charged metal ionic states. The band calculations (8,9) indicate that the energies of the Cu 3d block band almost coincides with the energy of O 2p band in La_2CuO_4 as well as in $YBa_2Cu_3O_7$. The coincidence seems accidental, but the same situation occurs also in another oxide superconductor $BaPb_{1-x}Bi_xO_3$ (BPB) where Pb(Bi) 6s band is almost degenerate with the O 2p band. It is also noteworthy that in the end (starting) materials $BaBiO_3$ and La_2CuO_4 there is exactly one electron per active B-site cation (both BPB and cuprates have the crystal structure basically of perovskite type) and as a result the Bi 6s (Cu 3d)- O 2p antibonding conduction band is just half-filled. In these respects it is worthwhile to review what kind of electronic state is realized in $BaBiO_3$ and the solid solutions with $BaPbO_3$. Then we argue the ground state of La_2CuO_4 and its analogue $YBa_2Cu_3O_6$ and the phase variation by doping with alkaline earth elements or by adding oxygen atoms.

$Ba(Pb,Bi)O_3$

According to the band calculations(11,12), the Bi 6s-2p antibonding orbitals form a wide (5 eV or more) conduction band in $BaBiO_3$ which is half-filled as stated above. In the perovskite structure, Bi atoms arrange themselves to form a simple cubic lattice if orthorhombic distortions of the lattice are neglected which are due to the rigid rotations of the BiO_6 octahedra and thus does not affect the electronic states in the conduction band. The Fermi surface of the half-filled band in a simple cubic lattice nests perfectly thereby leading to the instability of the electronic states, either CDW or spin-density-wave (SDW) or antiferromagnetic order when the Coulomb repulsion U between two electrons on one site is sufficiently strong.

The result of the band calculations favors the CDW instability, because U should not be large in such a wide band material and also because the strongly hybridized Bi 6s- O 2p states near the Fermi level lead to a strong coupling between the electronic states and the oxygen breathing-mode vibrations(11). The breathing-mode distortions if the oxygen octahedra surrounding Bi atoms are accompanied with the charge fluctuation around the Bi site as $-Bi^{4+}-Bi^{4+} \rightarrow -Bi^{4+\delta}-Bi^{4-\delta}-$. The formation of CDW opens a gap at the Fermi level and will make $BaBiO_3$ a semiconductor.

Various experiments seem to support this viewpoint. The neutron diffraction(13) as well as infrared and Raman spectroscopic studies(14,15) indicate the presence of the breathig-mode distortions $BaBiO_3$, and the material is actually a semiconductor up to about 1000K.

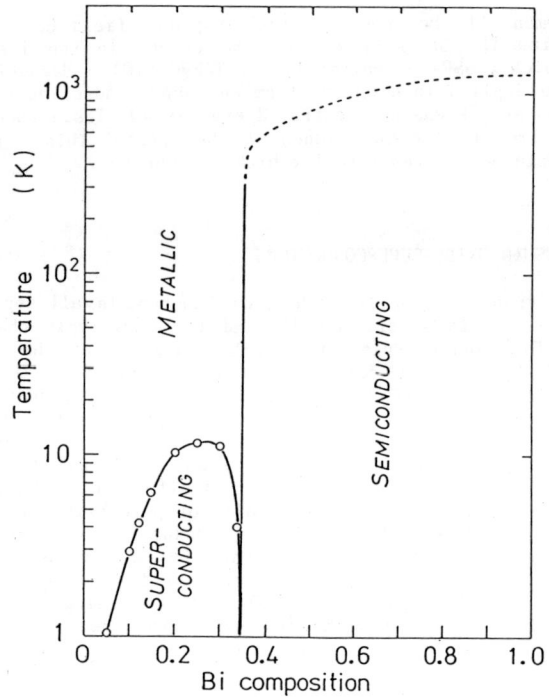

Fig. 2 Phase diagram of $BaPb_{1-x}Bi_xO_3$. The dotted line which separates the metallic and semiconducting phase is a speculation from the resistivity measurements.

Substitution of Pb with one less valence electron than Bi might shrink the Fermi surface and thus violate the nesting condition, making metallic state more stable. However, the semiconductivity persists up to Pb- substitution of 65 % (Fig.2). This fact can be understood as a CDW stabilized locally on Bi sites due to extremely strong coupling of electrons with the breathing-mode phonons (16.17). The relatively high Tc superconductivity of 10K range is observed in the Bi composition range $0.2 \leq x \leq 0.35$ and Tc decreases rapidly with decrease of x for $x < 0.20$. The Hall effect(18) and the plasma reflectivity edge(19) indicate that the conduction in the metallic phase $x \leq 0.35$ is due to the electrons donated by Bi atoms substituted for Pb in $BaPbO_3$.

$(La_{1-x}A_x)_2CuO_4$

La_2CuO_4 is a mysterious material. Its property is quite sensitive to the oxygen nonstoichiometry and to the local fluctuation of the La Cu cation ratio. Now it is widely believed that the ground state is nonmetallic and exhibiting a three-dimensional antiferromagnetic (AF) order. The AF order was observed by the neutron diffraction(20,21) and the onset (T_N) is indicated by a cusp in the temperature dependence of the static magnetic susceptibility. The recent joint US-Japan neutron scattering experiment(22) has revealed that two-dimensional AF correlation within CuO planes persists well above T_N (∼ 250K).

Similarly to BPB, the band calculations for La_2CuO_4 show that the conduction band is composed of the antibonding Cu $3d^{2-}$ O $4 2p$ orbitals and is half-filled with the nesting Fermi surface(23). This is due to the fact that just one electron resides on each $3d_{x2-y2}$ orbital of Cu atom which forms a square planar lattice. However, the experiments do not support CDW in La_2CuO_4 — the orthorhombic to tetragonal structural transition at around 500K has nothing to do at least with the CDW formation. The ground state different from $BaBiO_3$ would arise either from lower dimensionality or larger onsite Coulomb repulsion U or both.

A striking difference is seen in the doping effect. The semiconductivity La_2CuO_4 disappears by only 2.5 % substitution of alkaline earth element (Ba, Sr or Ca) for La. For x > 0.025, the material is a metal exhibiting a superconducting transition. Different from the case of $BaBiO_3$,

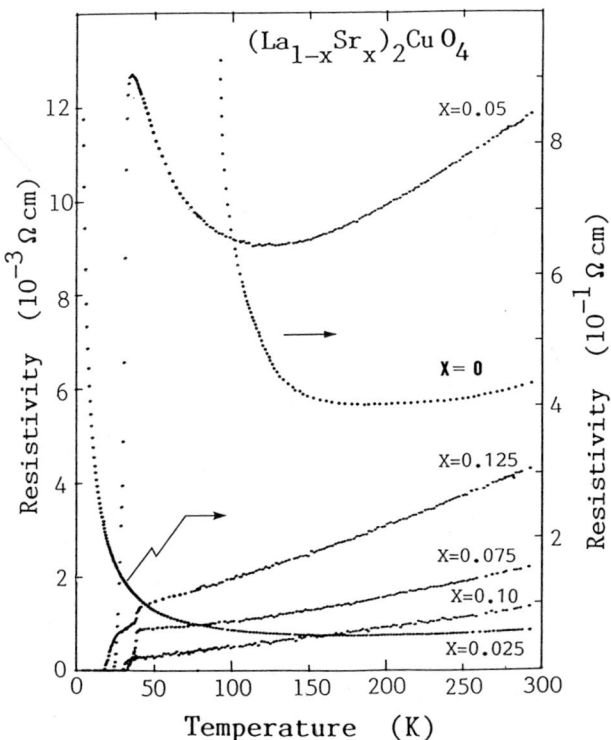

Fig. 3(a) Temperature dependence of resistivity for $(La_{1-x}Sr_x)_2CuO_4$.

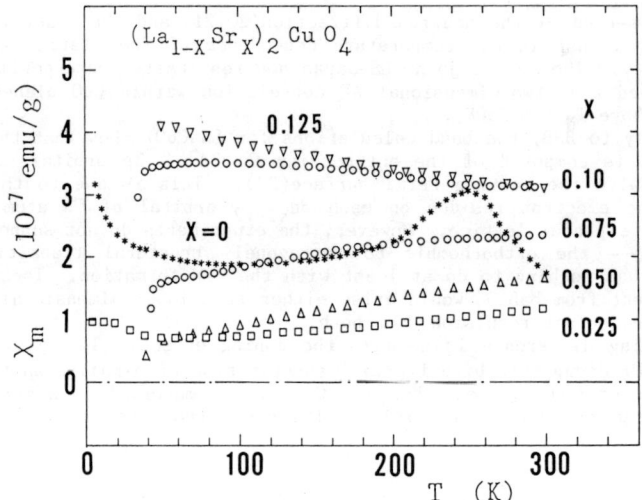

Fig. 3(b) Temperature dependence of magnetic susceptibility for $(La_{1-x}Sr_x)_2CuO_4$.

substitution of trivalent La with divalent element corresponds to doping of mobile holes in the CuO planes. The 3D AF order is not seen in this compositional range. Figures 3(a) and 3(b) show the resistivity and the magnetic susceptibility data for $(La_{1-x}Sr_x)_2CuO_4$ measured by the University of Tokyo group.

More detailed phase diagram has been studied using ^{139}La nuclear quadrupole response (NQR) technique by the Osaka University group(24). NQR gives local information around La sites and the spectrum is determined by the combination of electric quadruple interaction and the Zeeman energy arising from the internal magnetic field produced by magnetic order. The La-NQR detected the internal field produced by the AF order of the Cu magnetic moments.

Thus obtained magnetic phase diagram is schematically reproduced in Fig.4 The temperature T_N below which the internal magnetic field sets in is found to decrease steeply to the Ba composition around 0.01 in agreement with the result of the magnetic susceptibility measurement(25). For x > 0.01 T_N decreases rather gradually toward 0K at about x=0.025 and then superconductivity sets in. T_N in this region coincides with the temperature where the spin-spin relaxation rate $1/T_2$ diverges as found by the Hokkaido Univ. group(26). The Osaka Univ. group speculates that a two-dimensional magnetic order may persist within each CuO plane but the coherence along the c-axis (perpendicular to the CuO planes) may be very short.

These findings indicate that the magnetic coupling in between layers in La_2CuO_4, i.e. 3D AF order, is weak enough to be destroyed by a few percent doping or by raising temperature up to about 200K. In fact, as seen in Fig.3, the magnetic susceptibility exhibits only a weak cusp and one cannot observe any appreciable anomaly in various transport properties at T_N (27).

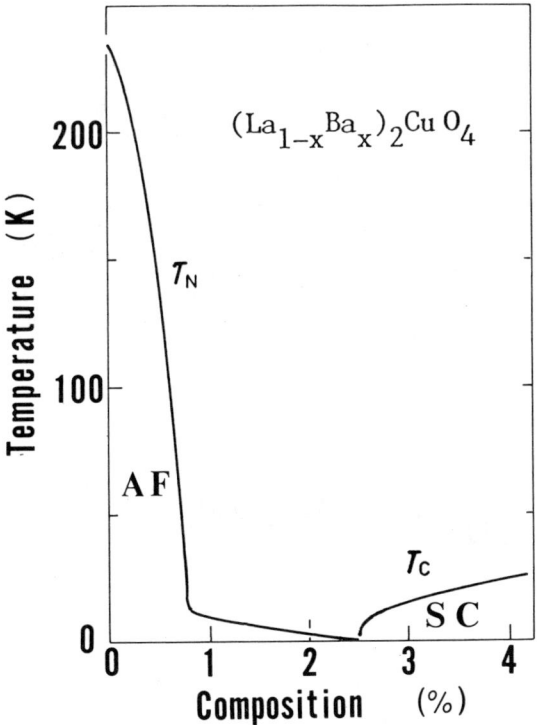

Fig. 4 Schematic phase diagram of $(La_{1-x}Ba_x)_2CuO_4$. The T_N-line is based on the NQR measurement(24).

On the other hand, the AF correlation within CuO plane seems to be strong as suggested by the above NQR result and the neutron scattering experiments. The optical spectrum of La_2CuO_4 also gives a support for this view. The observed spectrum below T_N is typical of a semiconductor with an energy gap of eV order(28), the most likely origin of which would be a Mott-Hubbard type gap derived from electron correlation. It is found that the spectrum does not show any appreciable change when the sample temperature is raised above T_N.

$YBa_2Cu_3O_{7-y}$

The oxygen content in this compound can easily be varied from 7 to 6 (29). With increasing oxygen deficiency y, the superconducting Tc decreases and at some value of y, which is dependent on the heat treatment of samples — quenching temperature and oxygen pressure, the compound becomes semiconducting associated also with the orthorhomic to tetragonal structural transformation.

From various experiments it is inferred that the CuO planes play a dominant role in the conduction of carriers and in superconductivity as in the case of La_2CuO_4-based materals. It is also speculated that the base material corresponding to undoped La_2CuO_4 would be $YBa_2Cu_3O_6$. The Hall effect measurements(30,31) indicate that holes are supplied in the CuO planes by adding oxygen atoms to linear Cu chains in $YBa_2Cu_3O_6$.

Recent important finding is that an AF order is identified for the compound with y close to 1 by muon spin rotation (μSR)(32), neutron diffraction(33) and Cu- NQR(34). The Cu- NQR measurement has been made by the joint Osaka-Tokyo Univ. research group. From the set-in of the internal magnetic field T_N is estimated to be 20K for $YBa_2Cu_3O_{6.3}$ which was quenched from 950 °C under an atomspheric oxygen pressure. It is also found that the nuclear-spin relaxation rate diverges at ~20K as was observed for $(La_{1-x}Ba_x)_2CuO_4$. The experiment for various oxygen contents is now underway.

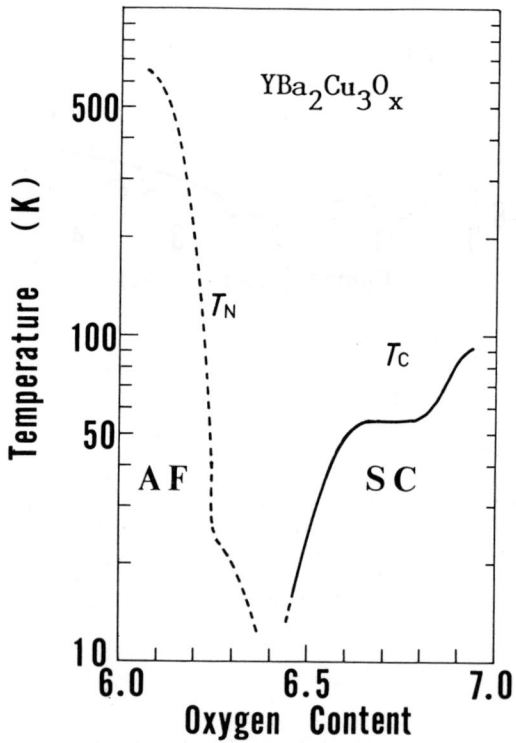

Fig. 5 Schematic phase diagram of $YBa_2Cu_3O_x$ with x in 6 x 7. Tc vs x line is derived from the data in ref. (31) and T_N is from ref.s (32-34).

In view of the result of the neutron diffraction which indicates Tn 400K for $YBa_2Cu_3O_{6.15}$ and Tn≅500K for $YBa_2Cu_3O_{6.0}$, One can imagine a similar phase diagram to that for $(La_{1-x}Ba_x)_2CuO_4$ with $YBa_2Cu_3O_6$ as a base material. This is schematically depicted in Fig.5. Close similarity between two classes of cuprates indicates that the electronic states and the electronic interaction within CuO plane play a primary role in determining the properties of these cuprates.

SUMMARY

I have shown the phase diagrams of three classes oxide superconcuctors, $BaPb_{1-x}Bi_xO_3$, $(La_{1-x}A_x)_2CuO_4$ and $YBa_2Cu_3O_{7-y}$, based on the recent experimental investigation of the electronic states in these oxides. The base materials of these three oxides, $BaBiO_3$, La_2CuO_4 and perhaps $YBa_2Cu_3O_6$, have similar crystal structure and electronic configuration. The active Bi and Cu atoms form a dimensional cubic and a 2 dimensional square lattice in the respective basically perovskite-type structure and there is one electron per site or the conduction band is half-full in both $BaBiO_3$ and Cu oxides.

In spite of these similarities, the ground state of Cu oxides is antiferromagnetic or resonating-valence-band state(35) in contrast to the CDW ground state of $BaBiO_3$. The difference maybe ascribed either to lower dimentionality or rather to U in the Cu oxides. An indication of the strong correlation was obtained by photoemission spectroscopy(36), various transport measurement(37) and the temperature dependence of NQR spin-lattice relaxation rate(38). Thus, the strong interelectron correlation apparently gives rise to quite distinct mormal-state properties and maybe to much higher Tc of Cu oxides, although conclusive evidence have ot been obatained yet.

References

1) See the Proceedings of 18th Int. Conf. on Low Temperature Physics, Kyoto,1987.
2) T.M. Rice, Z.Phys. B 67, 141 (1987).
3) P.W. Anderson and E. Abrahams, Nature 327, 363 (1987); V.J. Emery, Nature 328, 756 (1987).
4) Y. Tajima, M. Hikita, A. Katsui, Y. Hidaka, T. Iwata and S. Tsurumi, to be published in J. Cryst. Growth.
5) H. Takei, H. Takeya, Y. Iye, T. Tamegai and F. Sakai, Jpn. J. Appl. Phys. 26, L1425 (1987).
6) A. Maeda, T. Yabe, K. Uchinokura and S. Tanaka, Jpn. J. Appl. Phys. 26, L1368 (1987).
7) R. Yoshizaki, H. Sawada, T. Iwazumi, Y. Saito, Y. Abe, H. Ikeda, K. Imai and I. Nakai, Jpn. J. Appl. Phys. 26, L1703 (1987).
8) Y. Enomoto, T. Murakami, M. Suzuki and K. Moriwaki, Jpn. J. Appl. Phys. 26, L1248 (1987).
9) Fujikura-Densen Press Release, November 1987.
10) T. Sakakibara, T. Goto, Y. Iye, N. Miura, H. Takeya and H. Takei, Jpn. J. Appl. Phys. 26, L1892 (1987).
11) L. F. Mattheiss and D. R. Hamann, Phys. Rev. B28, 4227 (1983).
12) K. Takegahara and T. Kasuya, J. Phys. Soc. Jpn. 56, 1478 (1987).
13) C. Chaillout, Doctor Thesis, L Universite Scientifique et Medicale de Grenoble, 1986.
14) S. Uchida, S. Tajima, A. Masaki, S. Sugai, K. Kitazawa and S. Tanaka, J. Phys. Soc. Jpn. 54, 4395 (1985).
15) S. Sugai, Phys. Rev. B35, 3621 (1987).

16) D. Yoshioka and H. Fukuyama, J. Phys. Soc. Jpn. 54, 2996 (1985).
17) E. Jurczek and T. M. Rice, Europhys. Lett. 1, 225 (1986);
    E. Jurczek, Phys. Rev. B35, 6997 (1987).
18) T. D. Thanh, A. Koma and S. Tanaka, Appl. Phys. 22, 205 (1980).
19) S. Tajima, K. Kitazawa and S.Tanaka, Solid State Commun. 47, 659 (1983)
20) Y. Yamaguchi, H.Yamauchi, M.Ohashi, H. Yamamoto, N. Shimoda, M. Kikuchi
    and Y. shono, Jpn. J. Appl. 26, L447 (1987).
21) D. Vaknin, S. K. Sinha, D. E. Moncton, D. C. Johnston, J. M. Newsam,
    C. R. Safinya and H. E. King, Jr., Phys. Rev. Lett. 58, 2802 (1987).
22) G. Shirane, Y. Endoh, R. J. Birgeneau, M. A. Kastner, Y. Hidaka,
    M. Oda, M. Suzuki and T. Murakami, Phys. Rev. Lett. 59, 1613 (1987).
23) e.g. K. Takegahara, H. Harima and A. Yanase, Jpn. J. Appl. Phys. 26,
    L352 (1987); L.F.Mattheiss, Phys. Rev. Lett. 58, 1028 (1987).
24) Y. Kitaoka, K. Ishida, S. Hiramatsu and K. Asayama, preprint.
25) T. Fujita, Y. Aoki, Y. Maeno, J. Sakurai, H. Fukuba and
    H. Fujii, Jpn. J. Appl. Phys. 26, L368 (1987).
26) I. Watanabe, K.Kumagai, Y. Nakamura, T. Kimura, Y. Nakamichi
    and H. Nakajima, J. Phys. Soc. Jpn. 56, 3028 (1987).
27) e.g. S. Uchida, H. Takagi, H. Yanagisawa, K. Kishio, K. Kitazawa,
    K. Fueki and S. Tanaka, Jpn. J. Appl. Phys. 26, L445 (1987).
28) S. Tajima, S. Uchida, H. Ishii, H. Takagi, S. Tanaka, U. Kawabe,
    H. Hasegawa, T. Aita and T. Ishiba, to be published in
    Modern Physics Letters B.
29) K. Kishio, J. Shimoyana, T. Hasegawa, K. Kitazawa and K. Fueki,
    Jpn. J. Appl. Phys. 26, L1228 (1987).
30) N. P. Ong, Z. Z. Wang and J. Clayhold, Proc. Int. Workshop on
    Novel mechanisms of Superconductivity, Edited by S. A. Wolf and
    V. Z. Kresin, 1987 (Plenum, New York) P.1061.
31) H. Takagi, S. Uchida, H. Iwabuchi, H. Eisaki, k. Kishio, K. Kitazawa,
    K. Fueki and S. Tanaka, to be published in Physica B.
32) N. Nishida et al. Jpn. J. Appl. Phys. 26, L1856 (1987).
33) J. M. Tranquada, D. E. Cox, W. Kunnmann, H. Moudden,
    G. Shirane, M. Suenaga, P. Zolliker, D. Vaknin, S. K. Sinha,
    M. S. Alvarez, A. J. Jacobson and D. C. Johnston, preprint.
34) Y. Kitaoka, S. Hiramatsu, K. Ishida, K. Asayama, H. Takagi,
    H. Iwabuchi, S. Uchida and S. Tanaka, preprint.
35) P. W. Anderson, Science 235, 1196 (1987).
36) A. Fujimori, E. Takayama-Muromachi, Y. Uchida and B. Okai.
    Phys. Rev. B35, 8814 (1987).
37) H. Ishii, H. Sato, N. Kanazawa, H. Takagi, S. Uchida, K. Kitazawa,
    K. Kishio, K. Fueki and S. Tanaka, to be published.
38) Y. Kitaoka, S. Hiramatsu, T. kondo and k. Asayama, preprint.

HOLE CONCENTRATION COMPENSATION EFFECT IN $Ln_{1+x}Ba_{2-x}Cu_3O_{7-\delta}$

AND ANISOTROPIC UPPER CRITICAL FIELD OF $LnBa_2Cu_3O_{7-\delta}$

KÔKI TAKITA
Institute of Materials Science, University of Tsukuba, Tsukuba city, 305 Japan.

ABSTRACT

Hole concentration compensation effect and a simple correlation of the superconducting transition temperature T_c with the hole concentration have been found in solid solution systems $Ln_{1+x}Ba_{2-x}Cu_3O_{7-\delta}$ (Ln=La,Nd,Sm and Eu) by changing x. Anisotropic upper critical field studied by using preferentially oriented pellets are also reported for $LnBa_2Cu_3O_{7-\delta}$ (Ln=Y,Eu,Gd,Dy,Ho and Er).

§.1. INTRODUCTION

The discovery of the unprecedented new materials of high T_c superconducting oxides has excited extraordinary interest and research activity in physics as well as in the applications research. We report here some recent results of our experimental investigation for understanding the mechanism of superconductivity in these compounds and for improving the materials properties such as the upper critical field.

§.2. HOLE CONCENTRATION COMPENSATION EFFECT IN $Ln_{1+x}Ba_{2-x}Cu_3O_{7-\delta}$

Recently, it has been suggested by a work on $La_{2-y}Sr_yCuO_{4-\delta}$ whose copper-oxygen interactions are systematically varied by changing y, that the superconductivity is related to the hole concentration in the copper oxide layer [1]. Variation of superconductivity with carrier concentration is also investigated recently in $YBa_2Cu_3O_{7-\delta}$ by changing the oxygen vacancy concentration [2]. The work we report here is an extension of these works. Recent studies on Nd-compound revealed that solid solution is formed with a formula $Nd_{1+x}Ba_{2-x}Cu_3O_{7-\delta}$ [3]. According to a structural analysis of $Nd_{1.2}Ba_{1.8}Cu_3O_{7-\delta}$ based on a neutron powder diffraction data, it has been revealed that the Ba ions are substituted partially by the excess Nd ions and the compound is represented as $Nd(Nd_{0.2}Ba_{1.8})Cu_3O_{7-\delta}$ [4]. If it is possible that the trivalent Nd ions substituted for the divalent Ba ions compensate the hole concentration in $Nd_{1+x}Ba_{2-x}Cu_3O_{7-\delta}$, we can investigate the hole concentration dependence of the superconductivity by changing x. In fact, a systematic dependence of T_c on the value of x has been found in $Ln_{1+x}Ba_{2-x}Cu_3O_{7-\delta}$ where Ln stands for La, Nd, Sm and Eu [5]. It can be interpreted as a hole concentration compensation effect and consistent with the hole concentration dependence of T_c found in $La_{2-y}Sr_yCuO_{4-\delta}$.

Samples were prepared by the solid-state reaction method under identical conditions for the same compound system. After the sintering, it was furnace-cooled to about 350°C during 7 h in the oxygen atmosphere, except for La-compound. These samples were examined by X-ray powder diffraction combined with the Rietveld analysis. Preparation of the compounds was attempted for x ranging from -0.10 to +0.40 with the interval of 0.05 and single phase compounds were obtained for $0 \leq x \leq 0.40$.

Figure 1 shows some typical temperature dependence curves of the electrical resistivity for $Nd_{1+x}Ba_{2-x}Cu_3O_{7-\delta}$ with various x ranging from 0 to 0.40. Rather sharp superconducting transition was observed, except the sample with x = 0.40, as seen typically in the figure.

Figure 2 shows T_c as a function of x together with the lattice parameters, a, b and c/3 for $Nd_{1+x}Ba_{2-x}Cu_3O_{7-\delta}$ with $x = 0 \sim 0.40$. T_c shows a monotonic decrease with x except for the ranges $0 \sim 0.05$ (90K-range T_c) and $0.25 \sim 0.30$ (35K-range T_c) where it shows a plateau as seen in the figure. Increase of Nd-substitution for Ba induces an orthorhombic-tetragonal phase transition at x = 0.20. It is an interesting fact that the variation of T_c with x is insensitive to the structural phase transition.

Figure 3 shows results of the AC susceptibility measurement for the samples with x = 0, 0.2 and 0.3, which indicate that the volume fraction showing the Meissner effect is almost the same for these samples.

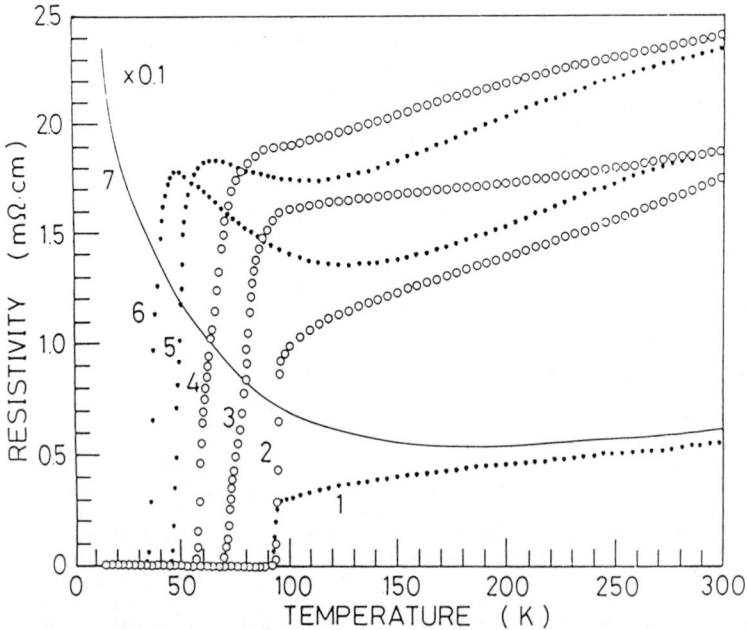

Fig. 1. Temperature dependence of the resistivity for $Nd_{1+x}Ba_{2-x}Cu_3O_{7-\delta}$ with various x values. 1; x=0. 2; x=0.05. 3; x=0.10. 4; x=0.15. 5; x=0.20. 6; x=0.25. 7; x=0.40. [From Ref.5]

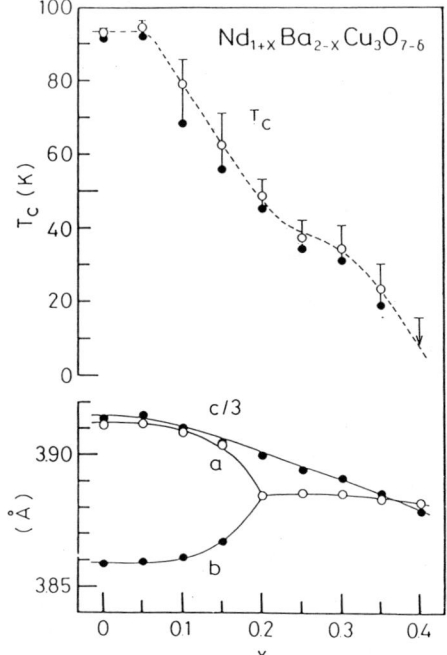

Fig. 2. x-dependence of T_c; Open and closed circles represent the midpoint and zero-resistance points, respectively, and horizontal bars are the onset temperature determined at 90% points. In the lower portion, x-dependence of the lattice parameters are shown.

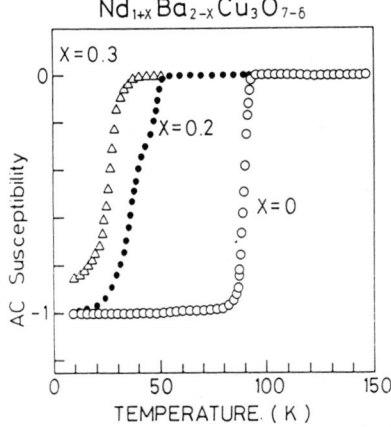

Fig. 3. Temperature dependence of AC susceptibility for $Nd_{1+x}Ba_{2-x}Cu_3O_{7-\delta}$. The scale of the ordinate is the self inductance change of the same coil normalized by sample size.

It has been shown by Shafer et al. [1] that the superconducting transition temperature T_c of $La_{2-y}Sr_yCuO_{4-\delta}$ has a direct correlation with the hole concentration $[Cu-O]^+$ determined by a chemical method. They also suggested that the same relation might apply to $YBa_2Cu_3O_{7-\delta}$. In Fig.4 we show an example of the plot of T_c vs $[Cu-O]^+$ concentration for the present experiment of $Nd_{1+x}Ba_{2-x}Cu_3O_{7-\delta}$ by closed triangles, together with the same plot for $La_{2-y}Sr_yCuO_{4-\delta}$ by Shafer et al.[1]. Here, the hole concentration $[Cu-O]^+$ is calculated based on a simple hole concentration compensation model, where it is assumed that the hole concentration of $Nd_1Ba_2Cu_3O_{7-\delta}$ is compensated by the partial substitution of trivalent Nd for divalent Ba in $Nd_{1+x}Ba_{2-x}Cu_3O_{7-\delta}$. The substitution has been confirmed by neutron powder diffraction for $Nd_{1.2}Ba_{1.8}Cu_3O_{7-\delta}$ [4]. In the $Nd_1Ba_2Cu_3O_{7-\delta}$ structure there are two different copper layers, one between the barium planes and two between neodymium and barium planes. Therefore, the ratio of $[Cu-O]^+$ to active Cu depends on whether only one, two layers or all three layers are active. The data are plotted under the assumption that two layers between neodymium and barium planes are active, because the plot shows excellent agreement with the $La_{2-y}Sr_yCuO_{4-\delta}$ under this assumption as seen in Fig.4.

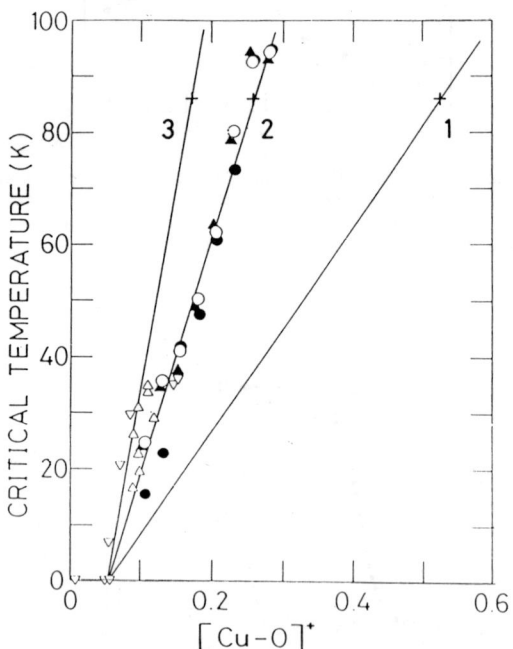

Fig. 4. T_c vs hole concentration $[Cu-O]^+$, as a fraction of total copper of $Ln_{1+x}Ba_{2-x}Cu_3O_{7-\delta}$; closed triangles; Ln=Nd: open circles; Ln=Eu: closed circles; Ln=Sm. The plots and the line 2 are based on a simple compensation model in the two active layers. The cases of one active layer and three active layers are indicated by line 1 and line 3, respectively. Down and up open triangles are for $La_{2-y}Sr_yCuO_{4-\delta}$ and the crosses are for $YBa_2Cu_3O_{6.6}$, which are replotted from Ref.1.

Thus, the relation of line 2 in Fig.4 is assumed for $Nd_{1+x}Ba_{2-x}Cu_3O_{7-\delta}$, which is expressed by eq.(1) where $[Cu-O]^+$ is expressed as a fraction of total copper in the two active layers;

$$[Cu-O]^+ = 0.278 - x/2 \tag{1}$$

Here, the hole concentration of $Nd_1Ba_2Cu_3O_{7-\delta}$, 0.278, is estimated from the value of $YBa_2Cu_3O_{6.6}$ of Ref.1. If only one layer or all three layers are active, the correlation is expressed by line 1 or line 3 in Fig.4. Fig.4 shows further that the Sm-compounds and Eu-compounds have the same properties. Preliminary investigation of La-compounds also shows similar result.

Fig.4 shows a surprisingly simple correlation between T_c and the hole concentration estimated by the above mentioned assumption. In order to confirm such a correlation and the hole concentration compensation effect in these systems, the variation of Hall coefficient R_H has been investigated. R_H is all hole-like and has a strong temperature dependence with decrease of x. The effective hole concentration n_H ($n_H = 1/R_H e$) is shown in Fig.5 for $Nd_{1+x}Ba_{2-x}Cu_3O_{7-\delta}$ with x = 0 ∼ 0.3. n_H is replotted against x in Fig.6 and compared to the hole concentration (broken line) estimated from the simple compensation model. The hole concentration compensation effect is clearly shown here, although the quantitative comparison is difficult because of the unknown origin of the strong temperature dependence of R_H.

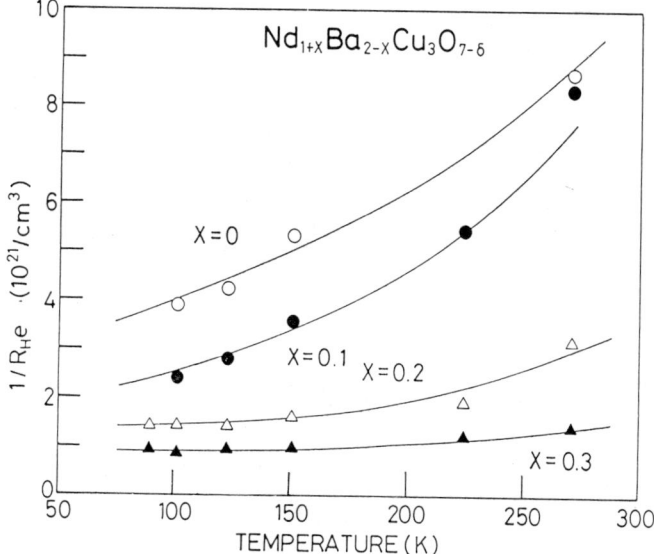

Fig. 5. Temperature dependence of the effective carrier concentration $n_H = 1/R_H e$ for $Nd_{1+x}Ba_{2-x}Cu_3O_{7-\delta}$.

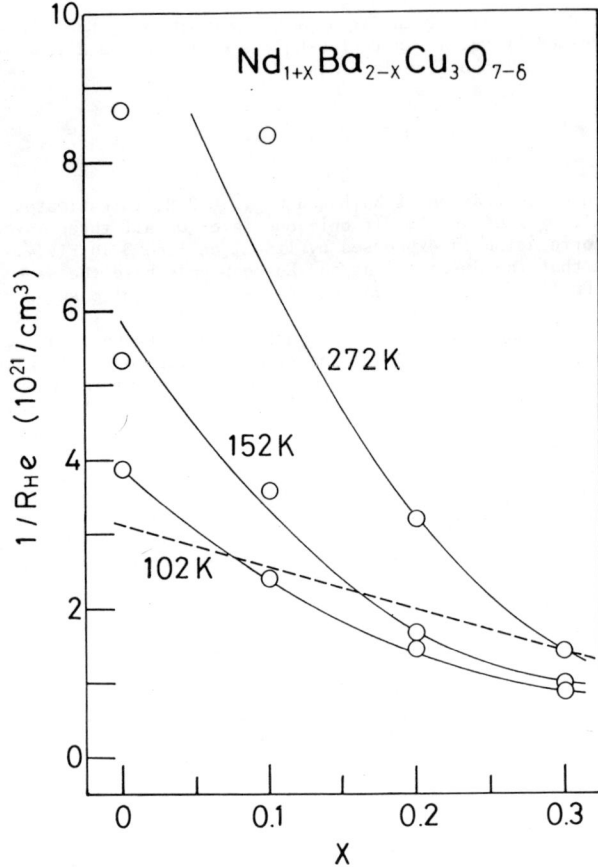

Fig. 6. $n_H = 1/R_H e$ plotted against the composition x for various temperatures (solid lines are guide of the eye). The broken line indicates x-dependence of the hole concentration in the compensation model shown in Fig.4.

In summary, we have studied how the superconducting properties change with the hole concentration compensation by changing x in $Ln_{1+x}Ba_{2-x}Cu_3O_{7-\delta}$, where Ln stands for La, Nd, Sm and Eu. The correlation of T_c with hole concentration is rather direct and almost independent of the element of Ln. These results may give an insight into the mechanism of the high T_c superconductivity in copper-oxide-type superconductors.

§.3. ANISOTROPIC UPPER CRITICAL FIELD OF $LnBa_2Cu_3O_{7-\delta}$

The structure of high T_c superconductors $YBa_2Cu_3O_{7-\delta}$ and $LnBa_2Cu_3O_{7-\delta}$ (in this section Ln stands for the lanthanoid elements except for Ce, Pr, Pm and Tb) is an oxygen-deficient triperovskite with orthorhombic symmetry. This structure indicates that the electric and magnetic properties could be highly anisotropic for all three axes a, b and c. The anisotropy of the upper critical field H_{c2} in these compounds has been investigated using highly oriented pellets by the present author's group [6,7]. In the investigations, an analysis based on a model of quasi-two-dimensional anisotropic superconductor has been proposed. It is reported that the anisotropy factor ε between the c-axis and the a-b plane is about 0.2 for these compounds. These results are consistent with the anisotropy observed in $YBa_2Cu_3O_{7-\delta}$ single crystal by Iye et al. [8].

Figure 7 shows an example of the anisotropic behavior of the resistive transition under magnetic field for highly oriented sintered pellet of $YBa_2Cu_3O_{7-\delta}$. The magnetic field direction is parallel ($H_{//}$) and perpendicular ($H_\perp$) to the disk plane of the pellet. The calculated curves based on a model are shown by broken lines.

It should be pointed out here that so-called midpoint of the resistance in the magnetic field generally depends not only on the material properties but also on the percentage of the preferred orientation of the sample pellets and, therefore, the upper critical field determined at midpoint has not clear meaning unless the preferred orientation is specified, which is demonstrated by a series of studies on the anisotropic upper critical fields of various high-T_c compounds using sintered pellets.

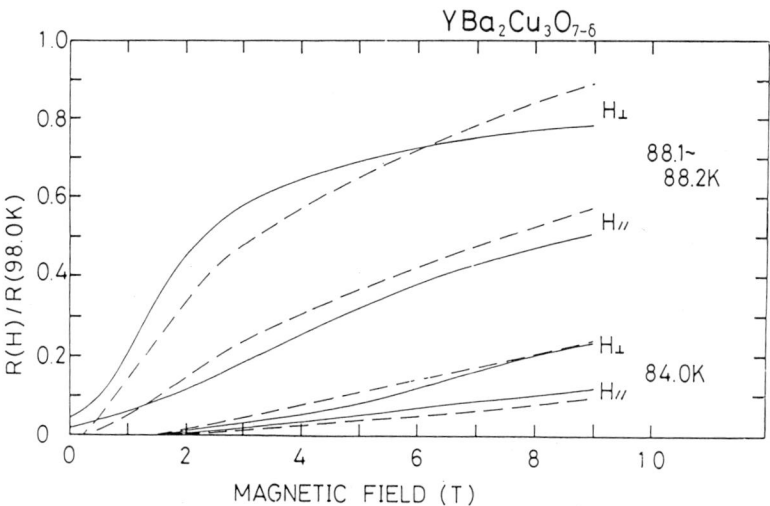

Fig. 7. Comparison of calculated curves (broken lines) with the experimental ones (solid lines) of magnetic field dependence of the resistance of the highly oriented $YBa_2Cu_3O_{7-\delta}$ pellet. $H_{//}$ and $H_\perp$ indicate the magnetic field parallel and perpendicular to the disk plane, respectively.

On the other hand, the upper critical field determined at offset point is material parameter even in the sintered sample and it is hardly affected by the preferred orientation. It should be noted here that the initial increase of the resistance in sintered sample may be caused by the appearance of the resistance in the grains whose c-axis is almost parallel to the applied magnetic field. If this is the case, it is expected that the H_{c2}^{off} of the sintered pellet should be close to that of the single crystal in $H/\!/c$ case (the magnetic field parallel to the c-axis). The comparison of H_{c2}^{off} with the single crystal values is summarized in Fig.8.

In Fig.8, the temperature dependence of the upper critical field for the offset points obtained in the present experiment are shown. They are the results of $LnBa_2Cu_3O_{7-\delta}$ (Ln = Y, Er, Ho, Dy, Gd and Eu) obtained for the sintered samples which showed the highest H_{c2} among the same compounds. For comparison, the results of $YBa_2Cu_3O_{7-\delta}$ and $GdBa_2Cu_3O_{7-\delta}$ single crystals for the magnetic field parallel ($H/\!/c$) and perpendicular ($H \perp c$) to the c-axis are shown by the broken lines in the figure [8]. The single crystal values in the magnetic field parallel to the c-axis are almost comparable to the values of the upper critical fields of sintered pellets for both materials.

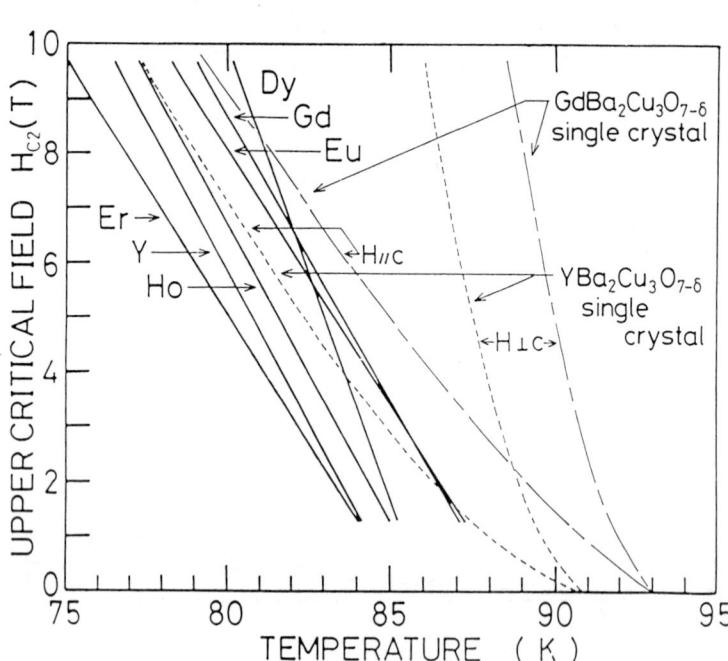

Fig. 8. A comparison of the upper critical fields determined at offset point H_{c2}^{off} of $LnBa_2Cu_3O_{7-\delta}$ pellets, where Ln stands for Y, Er, Ho, Dy, Gd and Eu. H_{c2}^{off} of single crystals $YBa_2Cu_3O_{7-\delta}$ and $GdBa_2Cu_3O_{7-\delta}$ are also shown by broken lines and dot-dashed lines, respectively [From Ref.8].

ACKNOWLEDGMENT

This work has been performed in collaboration with Prof. H. Asano and Prof. K. Masuda. The author is also indebted to H. Akinaga and H. Katoh for their assistance in the experiment.

REFERENCES

[1] M. W. Shafer, T. Penney and B. L. Olson, Phys. Rev. **B36** 4047 (1987).
[2] Z. Z. Wang, J. Clayhold, N. P. Ong, J. M. Tarascon, L. H. Greene, W. R. McKinnon and G. W. Hull, Phys. Rev. **B36** 7222 (1987).
[3] Y. Matsui, S. Takekawa and N. Iyi, Jpn. J. Appl. Phys. **26** L1693 (1987).
[4] F. Izumi, S. Takekawa, Y. Matsui, N. Iyi, H. Asano, T. Ishigaki and N. Watanabe, Jpn. J. Appl. Phys. **26** L1616 (1987).
[5] K. Takita, H. Akinaga, H. Katoh, H. Asano and K. Masuda, Jpn. J. Appl. Phys. **27** No.1 (1988).
[6] K. Takita, H. Akinaga, H. Katoh, T. Uchino, T. Ishigaki and H. Asano, Jpn. J. Appl. Phys. **26** L1323 (1987).
[7] K. Takita, H. Akinaga, H. Katoh and K. Masuda, Jpn. J. Appl. Phys. **26** L1552 (1987).
[8] Y. Iye, T. Tamegai, H. Takeya and H. Takei, Jpn. J. Appl. Phys. **26** L1057 (1987).

TUNNELING SPECTROSCOPY
OF HIGH TEMPERATURE SUPERCONDUCTORS

ALEX L. DE LOZANNE
University of Texas, Austin, Texas 78712-1081, USA

ABSTRACT

A review of our results on tunneling into high temperature superconductors is given. Bulk samples of $La_{1.85}Sr_{0.15}CuO_{4-y}$ and $YBa_2Cu_3O_{7-y}$ have been measured with a low temperature scanning tunneling microscope using tips made out of tungsten, aluminum and indium. A wide variety of I-V curves are obtained ranging from nearly ideal S-I-N and S-I-S characteristics, to strongly asymmetric and multiple-peaked characteristics. Large values of $2\Delta/kT_c$ are obtained. We discuss possible explanations for these large values.

INTRODUCTION

The recent discoveries by Bednorz and Muller[1] of superconductivity in the 30 K range and by Wu et al.[2] in the 90 K range have generated tremendous interest because of their scientific and technological importance. We have performed spectroscopic measurements on both classes of materials with a newly developed scanning tunneling microscope (STM). This instrument operates at low temperature in an ultra-high-vacuum chamber and has a number of unique capabilities:
* Temperature range: 10K to 400K
* Topographic imaging and I-V characteristics
* In-situ Auger analysis, capable of scanning
* In-situ LEED
* In-situ ion milling, annealing and thermal evaporation
* Capability to cleave samples in-situ at low temperature
* Load-lock exchange of samples and tips

The last feature is very important in this experiment since it has allowed us to measure a large number of combinations of tips and samples in a short time. In-situ cleaving is also essential because the native surfaces of these materials have poor properties.

RESULTS

Samples were prepared by standard sintering methods. The best surface preparation for these materials is to first cool them to 77K or lower, and then break them in situ. This gives a surface with properties as close as possible to those of the bulk and minimizes the loss of oxygen. Ion milling should not be used in this case because the complex stoichiometry of the surface can be easily altered by preferential sputtering. Furthermore, damage caused by ion milling cannot be annealed in a vacuum system because of oxygen loss. In-situ breaking of sintered samples is still not ideal, however, because the sample is most likely to break along grain boundaries. All the data shown here, except figure 1, was obtained by breaking the sample in-situ.

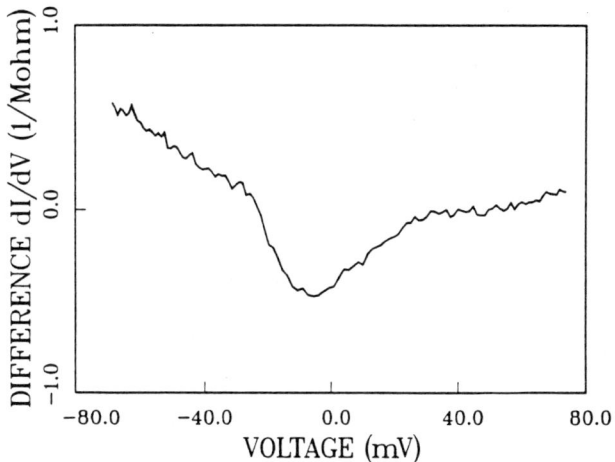

Figure 1. Differential conductance of a $La_{1.85}Sr_{0.15}CuO_{4-y}$ sample with a tungsten tip

Figure 1 shows the differential conductance (dI/dV) of a $La_{1.85}Sr_{0.15}CuO_{4-y}$ bulk sample with a tungsten tip[3]. A gap-like structure is clearly seen with a value of approximately =12 mV . The superconducting transition temperature, T_c, was measured resistively on a sample from the same batch, giving an onset of 40 K , a midpoint of 38.5 K and a width (10%-90%) of 1.1 K . Therefore the ratio $2\Delta/kT_c=7$ is twice as large as the BCS prediction.

Figure 2 shows similar data for a $YBa_2Cu_3O_{7-y}$ bulk sintered sample[4]. The three different curves illustrate the improvement achieved by using softer tips. A tungsten tip was used for the top curve, while an aluminum tip was used for the lower two. Sharper characteristics and lower subgap conductance are obtained with the softer aluminum tip. Our motivation for using softer tips was to increase the effective tunneling area, since a soft tip will deform to match the surface contours[4]. A larger area reduces the current density and the concomitant depairing within the superconductor. This depairing causes smearing in the I-V curve.

We have also obtained data in the vacuum tunneling mode (figure 4 in ref.3). In this case the tunneling current proceeds from the last few atoms on the tip into the sample, with current densities in excess of 10^6 A/cm^2. This causes strong depairing and smearing, but, on the other hand, improves the ideality of the conductance at high bias. Therefore the nonideal rise of the conductance at high bias seen in all the data shown here is probably due to a contaminated or damaged surface layer. All of our data is currently taken with the tip gently touching the sample in order to avoid smearing of the gap features.

The most common value of $2\Delta/kT_c$ for $YBa_2Cu_3O_{7-y}$ is approximately 11, with values even twice as large sometimes observed. The peaks in dI/dV usually associated with large gaps

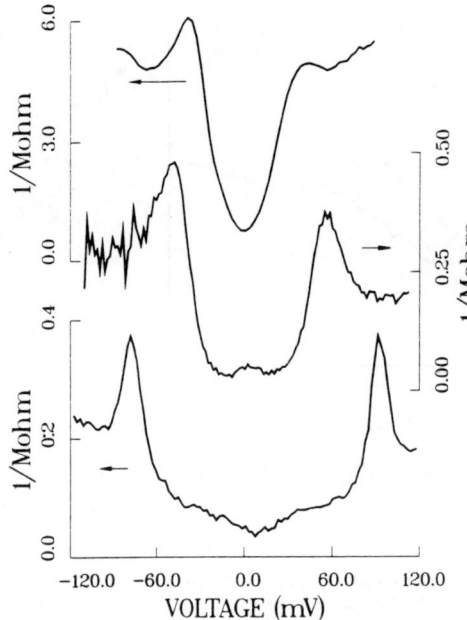

Figure 2. Differential conductance of a $YBa_2Cu_3O_{7-y}$ bulk sample and a tungsten tip (top) or aluminum tip (center and bottom)

are sharper than expected for S-I-N tunneling, and may be an indication of S-I-S tunneling (as in figure 2c). This may explain an extra factor of two.

Examples of less ideal results taken on a similar sample of $YBa_2Cu_3O_{7-y}$ are shown in figures 3 to 5. The temperature was 14K and the tip was made of aluminum in all cases. Figure 3 shows more asymmetry, a parabolic background in dI/dV, and a small voltage shift of unknown origin. Both figures 4 and 5 show multiple peaks in dI/dV but are qualitatively different. Figure 4 shows peaks at $\pm V_o$ and $\pm 3V_o$, with $V_o \sim 25mV$, while figure 5 shows peaks at $\pm V_o$ and $\pm 2V_o$, with $V_o \sim 50mV$. Small voltage shifts of about 5mV with opposite polarities are also observed.

DISCUSSION

One must ask whether these large gaps are real and whether they are related to superconductivity. The experimental setup has been calibrated with resistors to check the voltage scale. The value of the resistors has been chosen to be similar to our typical tunneling resistances, about 1 Mohm. Smaller resistors also yield the expected voltages. Furthermore, we do not believe that there is a series voltage drop anywhere in the circuit, because this would reduce the peak-to-valley ratios in the dI/dV curves.

A strong possibility is that at least some of the data may be explained by assuming that the tunneling involves a small particle, and that the structure we observe is due to the charging energy of this particle. This model has been used

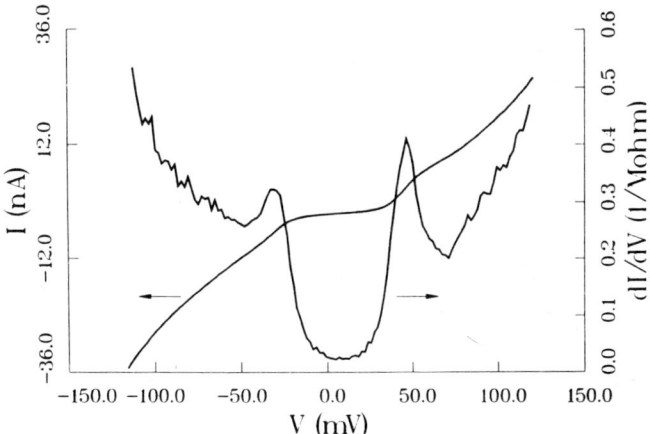

Figure 3. Differential conductance of a $YBa_2Cu_3O_{7-y}$ sample with an aluminum tip

recently by Barner and Ruggiero[5] to explain the behavior of a macroscopic tunnel junction with particles deposited in the barrier. This model predicts peaks in dI/dV at voltages of e/C, ±3e/C, ±5e/C,....Here C is the capacitance of the small particle and e is the electronic charge. This behavior is seen in figure 4 but not in figure 5, where the peak spacing is different. Unfortunately we cannot always sweep the voltage to three times the voltage of the first peak since this sometimes burns out the junction.

The effects of charging on the tunneling characteristics of small tunnel junctions have been studied recently by van Bentum et al. [6]. At low voltages they observe a linear dependence of

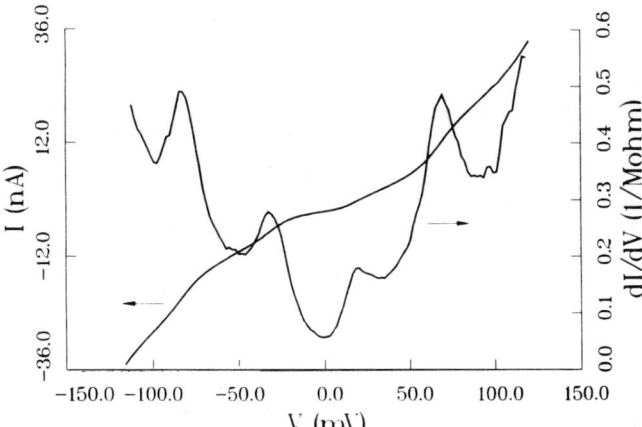

Figure 4. Data similar to that of figure 3, obtained on a different location on the same sample.

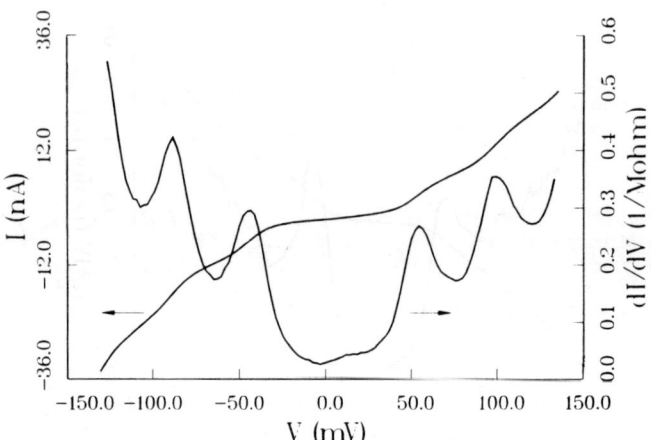

Figure 5. Data similar to that of figure 3, obtained on a different location on the same sample.

dI/dV on voltage as predicted by Averin and Likharev [7]. Their data is qualitatively different from ours, therefore we do not expect that our data can be explained solely in terms of a charging model. Similar results were obtained by Fulton and Dolan[8] with microfabricated tunnel junctions.

One last possibility is that the large gaps are due to the addition of several junctions in series. These junctions may be inside the material due to its granular and twinned structure, or may be between the sample and a superconducting particle stuck to the tip. Finally it should be pointed out that given the current state of the theoretical understanding of these materials, the BCS value for $2\Delta/kT_c=3.5$ need not be appropriate.

ACKNOWLEDGEMENTS

This work is supported by the Texas Advanced Technology Project and by NSF (DMR-8553305) through a Presidential Young Investigator Award (A.L. de Lozanne) with matching funds from IBM, Bell Communications Research (Bellcore), Texas Instruments, Kodak, and the Microelectronic and Computer Technology Corporation (MCC). LSCO samples were prepared by J.M. Tarascon and L.H. Greene (Bellcore). YBCO samples were prepared by A.J. Panson and J. Talvacchio (Westinghouse). Measurements at the University of Texas were performed by S. Pan and K.W. Ng.

REFERENCES

1. J.G. Bednorz and K.A. Muller, Z. Phys. B **64**, 189 (1986)

2. M.K. Wu, J.R. Ashburn, C.T. Torng, P.H. Hor, R.L. Meng, L. Gao, Z.J. Huang, Y.Q. Wang, and C.W. Chu, Phys. Rev. Lett. **58**, 908 (1987)
3. S. Pan, K.W. Ng, A.L. de Lozanne, J.M. Tarascon, and L.H. Greene, Phys. Rev. B **35**, 7220 (1987)
4. K. W. Ng, S. Pan, and A.L. de Lozanne: Jap. J. Appl. Phys. **26-3**, 993 (1987)
5. J.B. Barner and S.T. Ruggiero, Phys. Rev. Lett. **59**, 807 (1987)
6. P.J.M. van Bentum, H. van Kempen, L.E.C. van de Leemput, and P.A.A. Teunissen, Phys. Rev. Lett. **60**, 369 (1988)
7. D.V. Averin and K.K. Likharev, J. Low Temp. Phys. 62, 345 (1986)
8. T.A. Fulton and G.J. Dolan, Phys. Rev. Lett. 59, 109 (1987)

Microstructure and Properties of Ceramic $Ba_2YCu_3O_7$

G. J. Fisanick†, S. Nakahara†, S. Jin†, T. F. Tiefel†, R. C. Sherwood†, M. Yan†,
R. Moore‡, and R. B. van Dover†
†AT&T Bell Labs, Murray Hill, NJ 07974 USA
‡Perkin Elmer Physical Electronics, Edison, NJ 08822 USA

A. Introduction

It has been widely found[1] that the critical current of $Ba_2YCu_3O_7$ prepared by solid state ceramic processing is low (j_c(77 K,H=0)~10–10^3 A–cm^{-2}) despite an intense and multifaceted effort at improvement. Indeed, the initial paper which reported identification and isolation of $Ba_2YCu_3O_7$ also reported[2] a measurement of j_c (j_c(77 K,H=0)>1100 A–cm^{-2}) which has scarcely been exceeded since. Other preparation techniques have been found to give similarly discouraging results, with the notable exception of thin films, for which j_c(77 K,H=0)>10^6 A–cm^{-2} has been reported by at least two groups.[3] Recent results for melt-processed material[4] are substantially superior to those for conventionally sintered material, yet the basic cause improvement has not been been conclusively established. The intent of this paper is to present a consistent description of both normal-state (*i. e.*, resistivity) and superconducting transport in ceramic $Ba_2YCu_3O_7$, based on observations of the microstructure and properties of a variety of ceramic samples, and to relate this model to recent developments.

B. Grain Boundary Defect Structure

The structure of grain boundaries in bulk sintered samples of $Ba_2YCu_3O_7$ was investigated using transmission electron microscopy (TEM).[5,6] These samples were prepared using standard[7] processing: $Ba(OH)_2$, Y_2O_3, and CuO were mixed in stoichiometric proportions and reacted at 900°C/6h/O_2. The use of $Ba(OH)_2$ rather than $BaCO_3$ minimized contamination by C (but see below). The powder was ground, compacted, and then sintered at 975°C/40h/O_2. A final oxygen anneal was performed at 400-600°C. The resulting material is dense (6.1 g-cm^{-3}) and large-grained (1-40 μm), and exhibited critical currents in the range 40-100 A-cm^{-2} in zero field at 77 K, and a resistivity ρ(300 K)=790 μohm–cm.

Samples for TEM analysis were cut from regions adjacent to those used for j_c measurement, and were mechanically polished and further thinned by ion milling. Precautions were taken to avoid artifacts caused by exposure to humidity and to the ion milling itself[5]. TEM micrographs and electron diffraction patterns were taken with a Phillips 420T microscope operated at 120

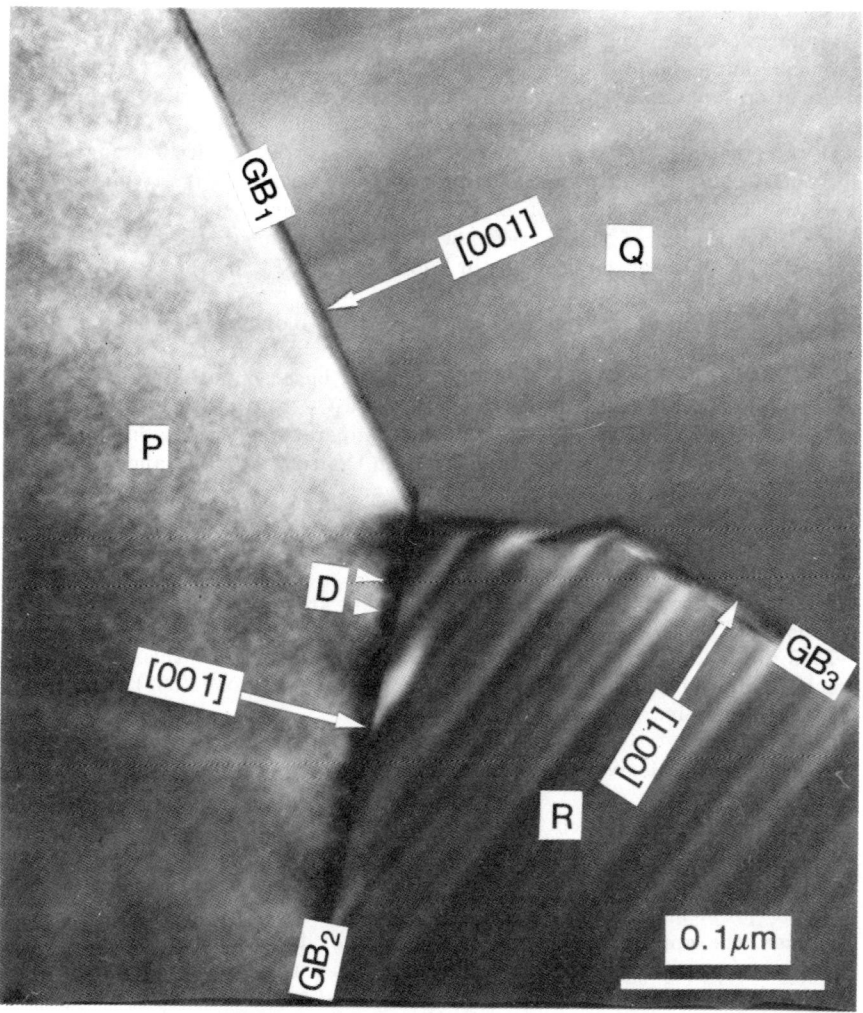

1. Transmission electron micrograph of the boundary of three grains of $Ba_2YCu_3O_7$: P, Q, and R. The [001] directions are indicated for each grain. The boundaries between P and Q, and Q and R are coherent (type 1) and the (001) planes lattice image to within a lattice spacing of the boundary. GB_2 contains extensive dislocation loops; it is the only grain boundary with an exact (001) face. It is classed as a type 2 grain boundary.

keV. Three classes of grain boundaries were observed[5]: 1) coherent grain boundaries, containing only the inevitable intrinsic sheet of dislocations, 2) semicoherent grain boundaries, with an extensive region (5-150 nm) of dislocation loops and 3) completely incoherent grain boundaries, in which there are cracks and voids within a layer of disordered material. The latter two classes of grain boundary defects are found to correlate with boundaries faced by a (001) plane, $i.\ e.$, a basal plane, and their formation is attributed[5] to the anisotropic thermal expansion of $Ba_2YCu_3O_{7-x}$ which generates stresses above the yield strength of the material when cooled from above about 700°C. A typical TEM micrograph is shown in Figure 1.

Electrical transport across such defective grain boundaries must be strongly impeded. In particular it is likely that the extensively defective regions of Type 2 boundaries are semiconducting or insulating, based on studies of the effect of ion-implantation-induced disorder[8] in $Ba_2YCu_3O_7$ and other measurements on amorphous $Ba_2YCu_3O_7$. At least at low temperatures the regions introduce a series resistance which is much greater than that of the grain, and will not pass a supercurrent. Type 3 boundaries, with voids as large as 1 nm or wider would pass a negligible current also, since the decay length for vacuum tunneling is ~0.14 nm (assuming the simplest model, with a WKB square barrier of ~5 eV).

Optical microscopy reveals that many grains grow as platelets, whose broad face is (001), with an average face-diameter to thickness ratio of ~6 to 1. Thus ~75% of the surface area of the grain is an (001) face, so this is roughly the fraction of grain boundaries which comprise at least one (001) face and are therefore defective.

C. Effect on Resistivity

The effect of having such a large fraction of disordered and cracked grain boundaries is easy to appreciate qualitatively but would be difficult to estimate quantitatively even if very detailed statistics were available, as this represents a difficult problem in effective medium theory. Clearly the result will be to decrease the apparent conductivity, by roughly the ratio $(1-f_{disorder}-f_{crack})$, $i.\ e.$, $\sigma_{obs} \simeq \sigma_{grain} \cdot (1-f_{disorder}-f_{crack})$, where $f_{disorder}$ is the fraction of the total area of type 2 grain boundaries, and f_{crack} is the areal fraction which is of type 3. So by the above estimate $f_{crack}+f_{disorder} \simeq 0.75$ and $\sigma_{obs} \simeq \sigma_{grain} \cdot 0.25$. Equivalently we might write $\sigma_{obs} = \sigma_{grain} \cdot f_{tight}$ and consider f_{tight} a parameter of order 0.25. This model is shown schematically in Figure 2. Intrinsic conductivity anisotropy[9] due to the crystal structure of $Ba_2YCu_3O_7$ presents an additional complication and might affect the measured resistivity by roughly another factor of two in a random polycrystalline sample, even though the conductivity anisotropy is quite large.

The observed resistivity of these samples at 300 K is 790 μohm—cm, which

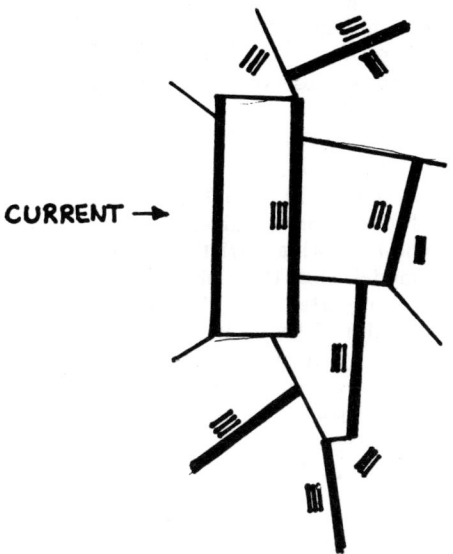

2. Model for grain-boundary connections in ceramic $Ba_2YCu_3O_7$. Schematic (001) planes are shown for each grain. Blocked transport is indicated by heavily shaded faces. The effective cross-sectional area for current transport is much less than the geometrical one.

implies a resistivity ρ_{grain}=200 μohm−cm by this estimate. This is remarkably close to the scaled resistivity inferred by Beasley[10] in order to explain the fluctuation conductivity in thin films of $Ba_2YCu_3O_7$, namely $\rho(300\ K)$=150−190 μohm−cm. In that case also, the discrepancy between the observed and scaled resistivity was attributed to sub-macroscopic inhomogeneities, such as cracks, poor grain boundaries, or uneven oxidation. However, those films are highly oriented ($\hat{c} \| \hat{n}$) so the close agreement may be fortuitous. Furthermore, fine-grained samples typically do not show such large shape anisotropy of the grains, so that one would not *a priori* expect the same value of f_{tight}, yet the lowest observed bulk resistivities are quite similar (500-800 μohm−cm).

The temperature dependence of the normal resistivity of $Ba_2YCu_3O_7$ bulk samples has been observed to fit into two classes: for 500$<\rho(300\ K)<$5000 μohm-cm, $\rho(T)/\rho(300\ K)$ falls on a universal curve, while samples with $\rho(300\ K)>$5000 μohm-cm exhibit a wide variety of temperature dependences. The essential point of this model, that the resistivity is modified by a multiplicative geometrical factor, is supported by the universality of the temperature dependence of the normal resistivity in the first class. Since the measured resistivity is determined by a geometrical factor, no added temperature dependence is introduced in our model as f_{tight} is varied from sample to sample. If, on the other hand, a source of additional series resistance had been indicated, it would be necessary to also postulate an identical temperature dependence for that source as for the resistivity within a grain—an unlikely coincidence.

Bulk samples where additional features associated with grain boundary wetting phases are present can have $f_{tight} \rightarrow 0$. Then the cracked, defective, and chemically inhomogeneous grain boundaries, with a low but nonzero conductivity, will form a parallel channel contributing to the observed resistivity, so that the temperature dependence will no longer be universal. This is indeed observed in samples with $\rho(300\ K)>$5000 μohm-cm.

D. Critical Current and Weak Coupling

The observation that a substantial fraction of the grain boundary area in large-grained ceramic samples are extremely defective is sufficient to explain why the resistivity of these samples is so high and why the temperature dependence $\rho(T)$ is identical even for widely varying values of $\rho(300\ K)$. But while it can account for the fivefold increase in apparent resistivity, it cannot explain the thousandfold decrease in j_c from values expected on the basis of intrinsic energy considerations (depairing) or by analogy with observed behavior in metal alloys and compounds (dominated by flux pinning). Neither can it explain the strong dependence of j_c on an external magnetic field H. For this it is necessary to invoke a second concept, namely weak coupling between strongly superconducting regions. We will argue that the dominant effect is at grain boundaries. The source of weak coupling could be intrinsic, e. g., due to anisotropy in the conductivity of $Ba_2YCu_3O_7$ leading to inefficient coupling at high-angle grain boundaries[11] or certain twin boundaries,[12] or it could be extrinsic, due to chemical inhomogeneity, e. g., in oxygen stoichiometry, precipitation of a second phase, or inadvertent reaction with contaminants.

TEM observations imply an upper bound of ~1 nm on the thickness of any intergranular inhomogeneity at tight type 1 boundaries in our large-grained sample. We note that neither TEM nor windowless EDX are sufficiently sensitive to detect the small (~4 %) changes in O content which would be sufficient to decrease T_c of $Ba_2YCu_3O_7$ by at least 30 K. This upper limit makes some possibilities unlikely, such as amorphous decoration of grain boundaries, or the presence of merely semiconducting material, as such materials should present a relatively low barrier height, so that the scale length for tunneling would be correspondingly long. But it does not rule out the possibility of a large-gap insulating barrier, for which even a thickness of 1 nm can drastically attenuate the probability of electron tunneling.

E. Carbon Segregation at Grain Boundaries

To examine the intergranular region with greater sensitivity we used Scanning Auger Microscopy to map a surface which was fractured *in situ* in a UHV system.[13] Figure 3 shows and SEM image of a typical fracture surface, along with maps of C, O, Ba, Y, and Cu Auger intensities. The SEM image reveals that a substantial portion of the fracture surface involves transgranular fracture, a feature also observed by Cook, et al..[14] In Figure 3 it is seen that many grains are decorated by the fine step structure characterisically associated with transgranular fracture. It is obvious by inspection that this decoration is correlated with the absence of carbon and with increased Ba, Y, and Cu signals, as would be expected for a surface segregated C layer. Depth profiling confirms that the C does not extend into the bulk of the grain. The thickness of the C layer was estimated to be 1-3 monolayers[13] by measuring the O, Ba and Cu Auger intensities and comparing values with and without the intervening C layer (*i. e.*, on inter- and transgranular fractures) using estimated Auger electron escape depths.

From a chemical point of view it is plausible that Ba-based carbonates would form on the surface of $Ba_2YCu_3O_7$ by reaction with trace CO_2 in the atmosphere. (It is noteworthy that this carbon contamination is found in material which was not prepared using $BaCO_3$ reagent, as it is therefore likely that all similar preperations regardless of reagent will result in carbonate formation. Indeed, even a rigorous attempt to eliminate exposure to CO_2 during preparation failed to increase j_c.) Surface reaction with CO_2 has been widely reported[15] and the bulk chemistry of CO_2-$Ba_2YCu_3O_7$ has been

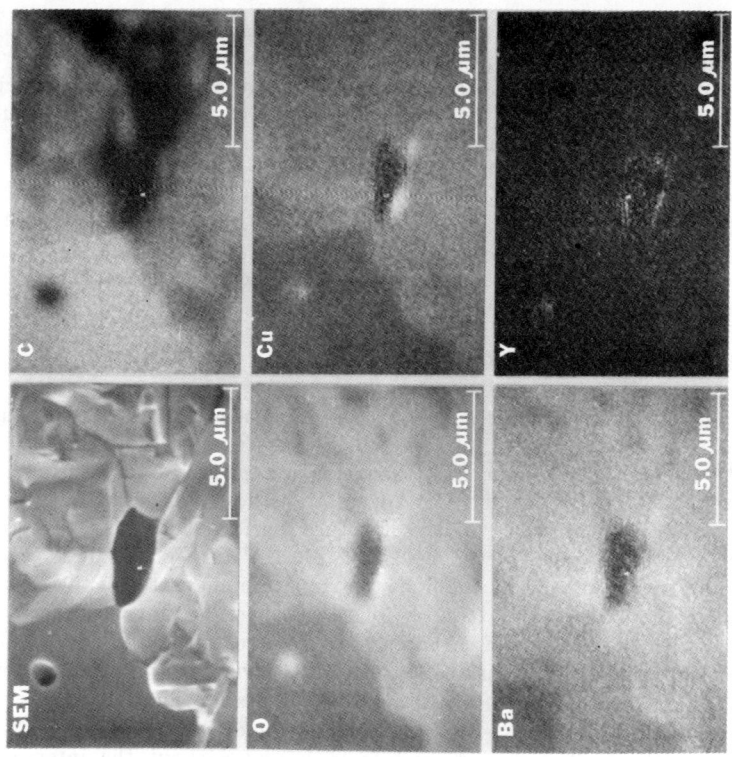

3. SEM and Scanning Auger maps of an *in situ* fractured surface of ceramic $Ba_2YCu_3O_7$. The high C signal is observed only for the intergranular fracture.

investigated by thermogravimetry,[16] explicitly revealing the strong tendency to form barium carbonates.

The universal presence of such an insulating layer at all grain or particle boundaries is sufficient to account for the low j_c obtained in bulk ceramic material. At 77 K this critical current density is $10-10^3$ A-cm^{-2}, dropping exponentially in an applied magnetic field with a characterisic field of 10-50 Oe. This is far lower than the upper limit for j_c set by the depairing current ($j_d = H_c(T)/3\sqrt{6\pi}\lambda(T) \sim 5\times10^6$ A-cm^{-2}) and the field dependence is far stronger than expected for a pinning-limited critical current by analogy with A15 and bcc superconductors at comparable values of T/T_c. Such behavior can be explained by the presence of weak coupling between superconducting regions,[17] and we have herein identified a plausible candidate for the weak links, as even a 1 nm carbonate tunneling barrier would have a low transmission probability: the WKB scaling length is $k^{-1} \sim (2mE/\hbar^2)^{-\frac{1}{2}} \sim 0.14$ nm assuming a barrier height E=5 eV, so the transmission probability for a 1 nm barrier would be reduced by $\sim e^{-1nm/0.14nm} \sim 10^{-3}$. As a result, the observed critical current is decreased by the same factor.

One can then ask if these junctions exhibit the ideal behavior associated with well-behaved granular superconductors, such as NbN[18,19]. In simplest form this model involves a cubic array of josephson-coupled grains, for which, in the low-temperature limit, and neglecting depairing, $j_c = \pi\Delta(0)/2e\rho_{eff}L$ where $\Delta(0)$ is the energy gap at T=0, L is the grain diameter, and ρ_{eff} is the effective normal state resistivity $\rho_{eff} = R_j L$, where R_j is the normal-state resistance of a single tunnel junction. It is assumed that $R_j \gg R_g$, the internal resistance of the grain.

To compare this result to the behavior of bulk ceramic Ba$_2$YCu$_3$O$_7$ we must first estimate R_j/R_g. For this we consider a typical grain as a rectangular prism and infer the tunneling resistance of the presumed junction at an end face, R_j, from the critical current using the full temperature-dependent Ambegaokar-Baratoff expression:

$$I_c R_j = (\pi\Delta(T)/2e)\tanh[\Delta(T)/2k_B T] \qquad (1)$$

The normal state resistance of the grain itself is simply $R_g = \rho L/A$. Then

$$\frac{R_j}{R_g} = \frac{\pi\Delta(T)\tanh[\Delta(T)/2k_B T]}{2eI_c} \cdot \frac{A}{\rho L} = \frac{\pi\Delta(T)\tanh[\Delta(T)/2k_B T]}{2ej_c \rho L} \qquad (2)$$

To estimate $\Delta(77)$ we assume[20] $2\Delta/k_B T_c \sim 4.5$, so $\Delta(0)=18$ meV and $\Delta(77\ K) \sim 7.2$ meV. Taking $j_c = 400$ A-cm^{-2}, $\rho(300K) = 200$ μohm-cm and L=40 μm from our measurements on the large-grained sample, we find $R_j/R_g \simeq 17$ even at 300 K. That is, the resistivity should be dominated by the series resistance of the junction, as is necessary in the model. This limit is even more valid for fine-grained material. Second we verify that self-shielding does not dominate in the posited junctions: the Josephson penetration depth $\lambda_J=(h/2e\mu_0\lambda j_c)^{1/2}=18\mu m$ (for $j_c=400$ A-cm^{-2} and $\lambda(77\ K)=200$ nm), so even the largest grains are only barely out of the small-junction limit required by the ideal-junction model. For ideal junctions, it is found experimentally that j_c is inversely proportional to the observed resistivity, as is found in well-behaved microparticle conductors.[18] This is not found for $Ba_2YCu_3O_7$ where we have observed j_c=10-1000 A-cm^{-2} in samples with $\rho(300\ K)$=800μohm−cm, and $j_c\sim$1000 A-cm^{-2} for samples with 500<ρ<1000 μohm-cm. Additionally, for ideal junctions, ρ_{eff} is inversely proportional to grain size, and again this is not observed in our bulk $Ba_2YCu_3O_7$. It is necessary, therefore, to infer that most of these weak links have a small I_cR_j product. Such a result is not unprecedented—in materials which are sensitive to disorder and which also have a short coherence length, such as the A15 compounds ($\xi_s(0)\sim$2nm), the product I_cR_j of a tunnel junction is often orders of magnitude less than $\pi\Delta/2e$. Thus, even with the 1 nm upper bound on the junction depth, arrived at from TEM observation of tight grain boundaries, we must infer the presence of low quality tunnel junctions. The correlation of these electrical, TEM and Auger measurements suggests a picture for the (non-basal-plane-faced) tight boundaries, namely a core barium carbonate insulating layer 1-3 ML thick, sandwiched between disordered regions of $Ba_2YCu_3O_7$ roughly a coherence length thick. These observations may in part explain the difficulty of obtaining high-quality junctions for tunneling studies of the superconductor.

F. Enhanced j_c in Melt-Textured Growth

A significant improvement in the critical current of ceramic $Ba_2YCu_3O_7$ has been obtained by a radical change in the processing of this material: instead of conventional solid-phase sintering, the technique involves partial or complete melting and controlled solidification.[4] The resulting material is extremely dense and develops a microstructure which consists of locally aligned, high-aspect-ratio grains with short dimensions of 2-5 μm. The short grain axis is parallel to the **c** axis. A typical region is shown in Figure 4. Adjacent grains are only slightly misoriented with respect to each other, so low angle grain boundaries predominate..

Growth anisotropy of $Ba_2YCu_3O_7$ under quasiequilibrium conditions has been observed, as mentioned above, with the **a-b**-axis rate many times that of the **c** axis. Under those conditions (solid state growth or slow cooling from an off-stoichiometric melt[21]) a platelet morphology obtains. The quantative details of the growth of crystallites in these $Ba_2YCu_3O_{7-x}$ supercooled melts are not yet understood.

Enhanced critical currents have been achieved by melt-textured growth following two substantially different protocols. In the first, the compound is heated only into the two-phase (liquid+solid) region, *i.e.*, to below about

1100°C. This results in decomposition into a liquid phase which is Ba- and Cu-rich, and solid-phase BaY_2CuO_5, as indicated by the approximate phase diagram of Figure 5. The second protocol avoids the problem of decomposition by heating into the near fully melted region (e. g., 1320°C) and cooling quickly through the peritectic range. This protocol yields the highest critical currents.[4] Figure 6 shows the transport j_c at 77 K as a function of applied magnetic field for the best melt-processed material, and compares it to the range of values obtained in conventionally sintered $Ba_2YCu_3O_7$.

Clearly the melt-processing technique reduces the effect of weak-link connections between grains. Indeed, the signature property of those weak links, namely the extreme magnetic field dependence of j_c, is no longer evident. The field dependence seen in Figure 7 is low enough that it may in part be due to other phemonena, such as flux flow (although it should be noted that the field dependence is still stronger than that observed for single crystals.)

It is not possible at this time to definitively infer the mechanism responsible for these improved properties. The grains in this material are at least locally highly oriented so that there are many low-angle grain boundaries—avoiding the potential problem of intrinsic anisotropy in the crystal structure. This may be crucial, but it is also true that the high-temperature processing is likely to decompose any carbonates on the surface of grains, and that renucleation within the melt ensures that newly-formed grain boundaries will be made in the absence of CO_2 or any other source of C. Thus the success of the melt-processing technique does not argue exclusively for either mechanism. It is indeed fortunate that this technique yields at least two beneficial changes simultaneously, allowing the formation of clean grain boundaries in dense, highly oriented material.

It is a pleasure to acknowledge useful discussions with C. E. Rice and S. B. DiCenzo, and the expert technical assistance of T. Boone, M. E. Davis, G. W. Kamlott, and R. E. Fastnacht.

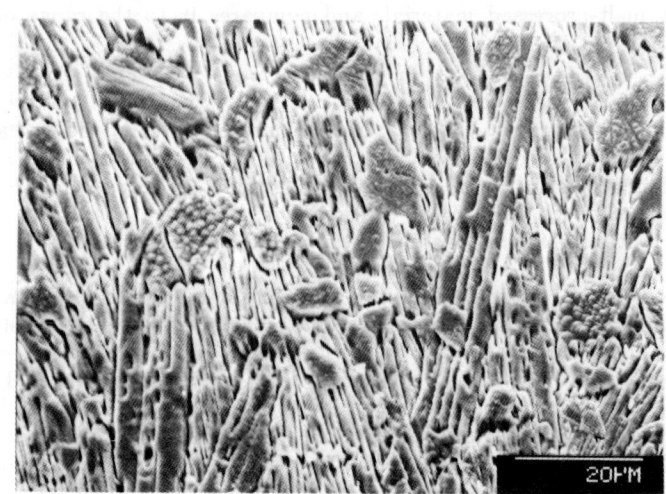

4. SEM photograph of a polished and etched melt-processed $Ba_2YCu_3O_7$ sample showing locally aligned grains.

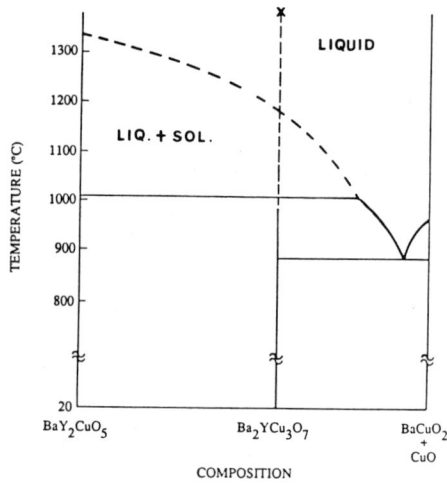

5. Tentative sectional phase diagram of $Ba_2YCu_3O_7$.

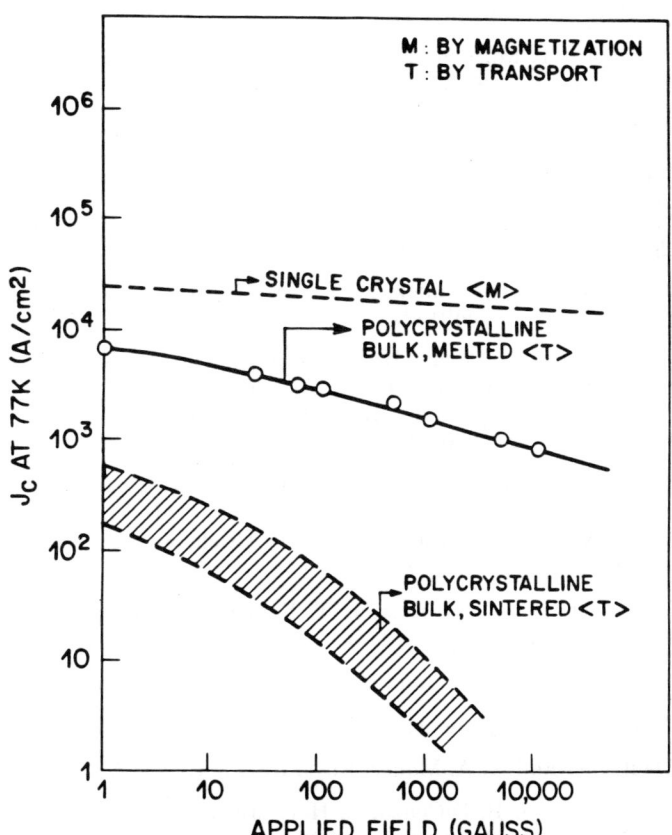

6. Critical current of various forms of $Ba_2YCu_3O_7$ as measured at 77 K, as a function of applied magnetic field.

References

1. See, *e. g.*, Proceedings of the Fall 1987 Materials Research Society Symposium on High-T_c Superconductors (Boston, 1987).
2. R. J. Cava, B. Batlogg, R. B. van Dover, D. W. Murphy, S. Sunshine, T. Siegrist, J. P. Remeika, E. A. Rietman, S. Zahurak, and G. P. Espinosa, Phys. Rev. Lett., **58**, 1676 (1987).
3. Y. Enomoto, T. Murakami, M. Suzuki and K. Moriwaki (preprint); P. M. Mankiewich, J. H. Scofield, W. J. Skocpol, R. E. Howard, A. H. Dayem and E. Good, Appl. Phys. Lett. **51**, 1753 (1987).
4. S. Jin, T. H. Tiefel, R. C. Sherwood, R. B. van Dover, M. E. Davis, G. W. Kamlott, R. A. Fastnacht, submitted.
5. S. Nakahara, G. J. Fisanick, M. F. Yan, R. B. van Dover, and T. Boone, J. Cryst. Growth **85**, 639 (1987).
6. S. Nakahara, G. J. Fisanick, M. F. Yan, R. B. van Dover, and T. Boone, Appl. Phys. Lett. submitted.
7. M. F. Yan, R. L. Barns, H. M. O'Bryan, Jr., P. K. Gallagher, R. C. Sherwood and S. Jin, Appl. Phys. Lett. **51**, 532 (1987); D. W. Johnson, E. M. Gyorgy, W. W. Rhodes, R. J. Cava, L. C. Feldman and R. B. van Dover, Adv. Ceram. Mat. **2**, 364 (1987).
8. A. E. White, K. T. Short, D. C. Jacobson, J. M. Poate, R. C. Dynes, P. M. Mankiewich, W. J. Skocpol, R. E. Howard, M. Anzlowar, K. W. Baldwin, A. F. J. Levi, J. R. Kwo, T. Hsieh, and M. Hong, Phys. Rev. Lett., submitted, submitted.
9. S. W. Tozer, A. W. Kleinsasser, T. Penney, D. Kaiser, and F. Holzberg, Phys. Rev. Lett., **59**, 1768 (1987).
10. M. R. Beasley, Proc. Yamada Conf. XVIII (Sendai, Japan, 1987).
11. J. W. Ekin, Adv. Ceram. Mat. **2**, 586 (1987).
12. G. Deutscher and K. A. Muller, Phys. Rev. Lett. **59**, 1745 (1987).
13. G. J. Fisanick, S. B. DiCenzo, E. Hartford, R. B. van Dover, M. F. Yan, C. E. Rice and R. Moore, to be submitted.
14. R. F. Cook, T. M. Shaw and P. R. Duncombe, Adv. Ceram. Mat **2**, 606 (1987).
15. See, *e. g.*, Proceedings 1987 Am. Vac. Soc. Meeting (Anaheim, CA).
16. P. Gallagher (private communication).
17. T. Y. Hsiang and D. K. Finnemore, Phys. Rev **B22**, 154 (1980).

18. M. Ashkin and M. R. Beasley, IEEE Trans, Mag. **MAG-23**, 1367 (1986).
19. R. T. Kampwirth and K. E. Gray, IEEE Trans. Mag. **MAG-17**, 565 (1981); J. R. Clem, B. Bumble, S. I. Raider, W. J. Gallagher, and Y. C. Shih, Phys. Rev. **B35**, 6637 (1987).
20. S. Pan, K. W. Ng, and A. L. de Lozanne, Proceedings of the Intnl. Workshop on Novel Mech. Supercon., (Plenum, New York, 1987), p. 1029.
21. L. F. Schneemeyer, J. V. Waszczak, T. Siegrist, R. B. van Dover, L. W. Rupp, B. Batlogg, R. J. Cava, D. W. Murphy, Nature **328**, 601 (1987).

MACROSCOPIC VIEW OF THE HIGH Tc SUPERCONDUCTIVITY

F. de la Cruz, L. Civale, E. Osquiguil, H. Safar and R. Decca
Basic Research Department, Centro Atómico Bariloche,
8400 Bariloche, R. N. Argentina

D. A. Esparza and C. D'Ovidio
Applied Research Department, Centro Atómico Bariloche,
8400 Bariloche, R. N. Argentina

ABSTRACT

In this paper we discuss some of the well established properties of the high Tc ceramic superconductors, in order to provide an adequate phenomenological superconducting frame for the analysis of experimental results. Problems arising when applying the accepted macroscopic theories are pointed out. Results related to electrical transport properties are discussed together with the important role played by the oxygen concentration, either in the normal or superconducting states. The work is based on experimental data taken in the La-Sr-Cu-O system.

INTRODUCTION

It has been experimentally established that superconductivity in high critical temperature ceramic materials [1] is characteristic of inhomogeneous systems. The granular superconducting picture has provided an adequate frame to describe [2-8] the magnetization behavior as well as the low critical currents typically found in these materials. It is also known [9] that magnetic impurities weakly modify the superconducting transition, suggesting that the order parameter, ψ, might be independent of the spatial distribution of the usually taken as strong pair breaking centers. In principle, these two results can be reconciled if the inhomogeneous character is associated to Josephson-coupling among the grains of the sintered material, while the magnetic impurities response is associated to some intrinsic intragrain superconducting property. Nevertheless, experimental data indicate that the previous picture is not adequate to describe the available results [10]. It has been shown that oxygen concentration determines the inhomogeneous characteristic behavior of these materials. It was also suggested [11] that the high density of twin boundaries might induce Josephson junctions in a scale much smaller than that of ceramic grains. These results point toward a behavior determined by a superconducting order parameter easily modulated in space. It will be shown that the superconducting response to the presence of magnetic moments can also be understood within the same picture. The suggestion is supported by the measured high critical field, H_{c2}, characteristic of a very short coherence length, $\xi(0)$. From this point of view the granular character of the ceramic superconductors differs from that of superconducting metallic grains surrounded by a non-metallic system. In this last case the superconducting material has a coherence length much longer than the size of the superconducting grain. As a consequence, Ψ can be considered constant within each grain and the effective coherence length is determined by the coupling between grains. In the high T_c oxides the small $\xi(0)$ seems to be an intrinsic property of the material [12] and the boundary conditions in each grain should determine the macroscopic response of the system.

The inhomogeneous character is also evident when studying the superconducting transition temperature as measured by the electrical resistivity. Although the onset of the superconducting transition, T_{co}, remains constant, the transition width, ΔT_c, is very sensitive [12] to oxygen concentration, changing from a reasonable sharp transition in the fully oxidized sample to a non-percolative superconducting regime when 1% of the total amount of oxygen is removed. Since the electrical resistivity is also sensitive [10, 12, 13] to oxygen concentration, it is important to determine the correlation between the normal transport properties and the non-homogeneous superconducting character.

The study of the electronic transport behavior of these materials is important in itself [13]. The fully oxidized samples show a metallic like behavior well represented [12-14] by

$$\rho = \rho(0) + \alpha T \qquad (1)$$

where $\rho(0)$ is the residual resistivity at T=0. The temperature dependence in (1) is characteristic of the electron-phonon contribution in metallic samples in the limit $T > \theta$, where θ is the Debye temperature. Concerning this point, it is remarkable that in the oxidized $La_{1.8}Sr_{0.2}CuO_{4-\delta}$ samples the linear temperature dependence is followed down to $T_c \cong 40K$, that is to $T/\theta \cong 10^{-1}$. The high resistivity value ($\rho(300\ K) \cong 1000\ \mu\Omega cm$) has induced to believe [15] that ceramic superconductors should be treated in the dirty limit, i.e. $\ell < \xi(0)$, where ℓ is the elastic electron mean free path. Nevertheless the lack of saturation [12, 14] at temperatures as high as 1000 K has led to suggest [12] that these materials are superconductors in the clean limit.

Oxygen deficiency induces a low temperature semiconducting like behavior of unknown microscopic origin. Very recently, the low temperature resistivity data of the superconducting samples were shown [13] to follow the Variable Range Hopping behavior (VRH) [16], as previously suggested [17] for the non superconducting La based oxides. It is important to remark that the analysis of the electrical resistivity is made [13] on the assumption that the material is homogeneous when discussing transport properties, although a non-homogeneous picture is used when interpreting the superconducting results. The controversy is not solved in this paper but we provide arguments based on experimental results supporting that dualism.

The results discussed in this work are obtained in samples of the La-Sr-Cu-O system although the conclusions might be extended to the Yttrium based compounds. Most of the experimental results have already been published.

SUPERCONDUCTING CHARACTERISTICS

The inhomogeneous character of the order parameter was pointed out by Müller, Takashige and Bednorz [2] from the beginning of the investigation of the magnetic properties of the high T_c superconductors. The dynamic response to magnetic field variations [18] as well as the magnetic flux behavior in short time scales [6] are found to be typical of systems that can remain in thermodynamic metastable states. Nevertheless, the model or combination of models representing the magnetic flux penetration and its distribution within the sample is still under discussion [8]. In this paper we address our attention to the response of the ceramic material to the presence of very small fields, applied after the sample has been cooled in a remnant field smaller than 1 mOe. As a result of the field shielding currents flowing at the sample surface. These currents are determined by the topology of the superconducting material in the sample. Since the superconductivity can be characterized by an order parameter, the topology of the superconducting state is determined by its spatial configuration. If $\Psi(x)$ is finite everywhere the superconducting state is represented by a well defined ground state (the Meissner state). The diamagnetic currents flow in a Meissner penetration depth, λ, which might be space dependent. On the contrary, if the order parameter iz zero in regions surrounded by superconducting material, the intensity and current distribution are determined by the fluxoid quantization applied around the superconducting circuit. This condition does not necessarily represent the minimum Gibbs free energy of the field sample system. The stability of the resulting metastable state depends on the spatial distribution of the order parameter and the geometry of the superconducting circuit. If the sample has a finite $\Psi(x)$ everywhere there will be a field, H_{c_1}, above which the penetration of magnetic vortices will be favoured. The vortices will penetrate through the sample surface until pinned by sample defects. It is important to realize that a vortex in this sense represents a solution of superconducting equations defined in the whole sample. In a multiple connected superconducting system the magnetic flux penetration has to be accomplished by depressing to zero the order parameter at some point of the superconducting path. Once the flux is locked into the sample the fluxoid quantization applies on the superconducting circuit. The intrinsically reversible state is the Meissner state; the other two cases discussed previously will generally show irreversible magnetic behavior and, eventually, time dependent effects when evolving toward lower free energy states.

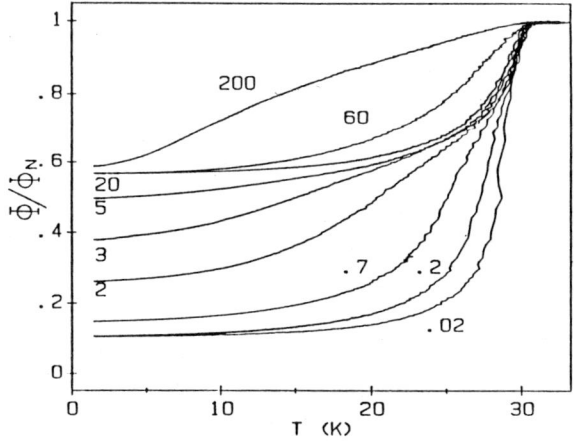

Figure 1
Magnetic flux through the sample normalized by the flux in the normal state, as a function of temperature for zero field cooling experiments. The number on each curve indicates the magnetic field in Oe.

Figure 1 shows the magnetic flux through a ceramic sample of $La_{1.8}Sr_{0.2}CuO_{4-\delta}$ as a function of temperature, normalized by the flux in the normal state. The sample has been cooled in zero field and then the field indicated in the figure was applied. Experimental details can be found in ref.6. At the lowest temperatures there are two field ranges where the flux through the sample seems to be normalized by the applied field. If the low field region is found to be reversible it could represent the Meissner state in the whole sample. However, the results of fig. 2 show the expanded low temperature region of fig. 1 for a field as

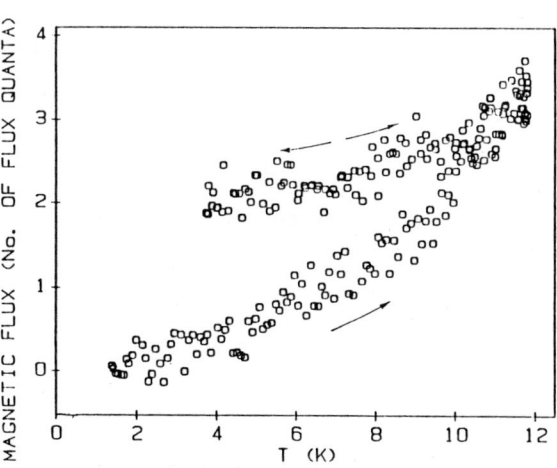

Figure 2
Expanded temperature dependence of the magnetic flux through the sample. The arrows indicate the sequence of the temperature sweeps.

low as 10 mOe, indicating that the flux penetration is irreversible even when only few flux quanta penetrate. This result, together with the magnetic flux dependence on the applied magnetic field [6] points toward a multiple connected superconducting model, where critical currents are already reached when applying fields of 10 mOe. The second regime where the flux is proportional to the applied field, has been interpreted [6] as the Meissner state of the superconducting regions connected through weak links. These results and those of references [2] to [10] are clear evidence of the non-homogeneous behavior of $\Psi(x)$. Although it is quite possible that this characteristic is not an intrinsic property of the new superconductivity, it is interesting to ask why superconductivity in these ceramics nucleates in a state of higher Gibbs free energy.

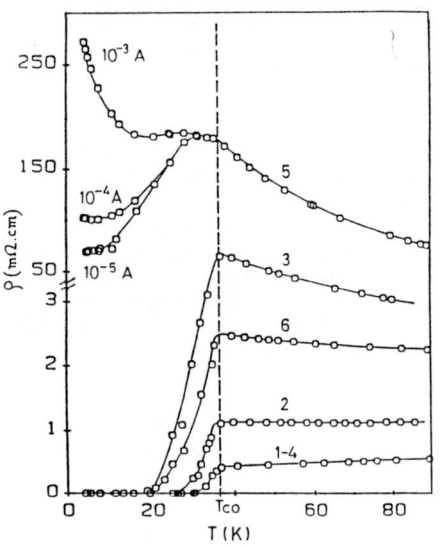

Figure 3
Superconducting transitions for a sample with different oxygen contents.
(1) fully oxidized sample,
(2) 10 hr in vacuum,
(3) 10 hr in vacuum,
(4) 10 hr in 1 atm-oxygen
(5) 160 hr in vacuum,
(6) 20 min in 1 atm-oxygen
All heat treatments were done at 770 K and the periods indicated for each run have to be added to the previous thermal history.

The non-homogeneous nucleation is also evident when studying the superconducting transition width as determined from resistive measurements [10, 12]. Figure 3 shows the superconducting transition of a sample with different amounts of oxygen content. The sample was heat treated at 700 K either in vacuum or in 1 atm-oxygen. The number on each curve indicates the heat treatment sequence as shown in the figure caption. The highest resistivity corresponds to the sample after removing 1% of the total amount of oxygen. It should be pointed out that the onset of the superconducting transition remains practically constant [10, 12] while ΔT_c changes drastically when the resistivity increases. It is also evident from fig. 3 that while ρ in the normal state is independent of the current density used in the measurements, the transition width is not. Although a detailed analysis of these results is beyond the scope of this paper we notice that the behavior of T_{co} indicates the presence of superconducting regions not affected by the reduction of oxygen content. The increase of the transition width is consistent with a grain bond model [19] where the oxygen depletion decreases the order parameter in the bonds. Preliminary results [20] suggest that the magnetization associated to the grains is diminished when removing oxygen, pointing out that the topology of the superconducting circuits is affected by oxygen concentration.

Taking into account the results already discussed and those found in the literature the following characteristics of the ceramic superconductivity in the La-Sr-Cu-O system are worthy of attention:

- The order parameter is easily modulated in space, indicating a short coherence length in agreement with critical field measurements.
- The coherence length, $\xi(0) \simeq 10\text{-}20$ Å is smaller than that of amorphous metals [21] ($\xi(0) \simeq 100$ Å), where the electron mean free path is of the order of interatomic distances.
- The onset of the superconducting transition does not change [9, 12] with magnetic or other impurities, up to concentration of the order of a percent.
- The transition width is sensitive to impurity concentration well below one percent.
- High resistivity values, $\rho \simeq 1000$ $\mu\Omega$cm, at room temperature with $d\rho/dT > 0$ in contradiction with Mooij's criterium [22].
- Low carrier concentration, $n \simeq 10^{21}$ cm^{-3}, as indicated from Hall effect measurements [23].
- Upper critical fields independent of the value of the resistivity [4].

It is interesting to notice that most of the properties indicated above are well understood within a picture provided by a superconductivity with an intrinsically small coherence length. Disregarding anisotropic effects it is possible to estimate [12] an order of magnitude of $\xi(0)$, following the rather general argument introduced by Pippard based on the uncertainty principle. Assuming an electron density $n \simeq 10^{21}$ cm^{-3}, an effective mass five times the electron mass [19] and the free electron model we obtain $\xi(0) \simeq 20$ Å, in close agreement with the values obtained from H_{c2}. Within the same rough picture we estimate the density of superconducting carriers n_p, assuming that $n_p \simeq n\, T_c/T_F$, where T_F is the corresponding Fermi temperature. In this way it is found that the distance between pairs is of the same order of magnitude as the size of the pairs This is usual compared to traditional superconductivity, where the pairs overlap strongly. Within this picture it is easy to understand why T_{co} remains constant as far as some regions in the sample keep their proper chemical stoichiometry. It is also easy to understand that magnetic or other impurities will influence T_{co} only at concentrations where impurity spacing is of the order of $\xi(0)$. In agreement with this suggestion, thermogravimetric data [24] indicate that the weight loss induced to obtain curve 5 in fig. 3 correspond to an average distance between oxygen vacancies of the order of $\xi(0)$. The increase in ΔT_c induced by the reduction of oxygen concentration makes clear the importance of the local oxygen stoichiometry in the determination of the magnitude of the space dependent order parameter. The behavior of ΔT_c and its current density sensitivity is in agreement with a picture where the order parameter is depressed locally and recovered in the superconducting grains. Preliminary magnetization results [20] indicate that oxygen concentration controls the granular behavior of the material, not only by a local depression of the order parameter in specific sites, but also by changing the total amount of superconducting material. These results show the sensitivity of the superconducting properties to material perturbations and rule out the possibility that the superconducting grains have their single origin in the ceramic grains.

ELECTRICAL TRANSPORT PROPERTIES

As mentioned in the introduction, the oxygen content reversibly determines [10, 12] the behavior of the superconducting transition as well as the resistivity in the normal state.

The characteristic time for oxygen removal in the La-Sr-Cu-O system [12] is orders of magnitude longer than in [25] Y-Ba-Cu-O. The measurements of the electrical resistivity, in the former compound, can be considered [10, 12] to be made at constant oxygen concentration up to a temperature of 1000 K.

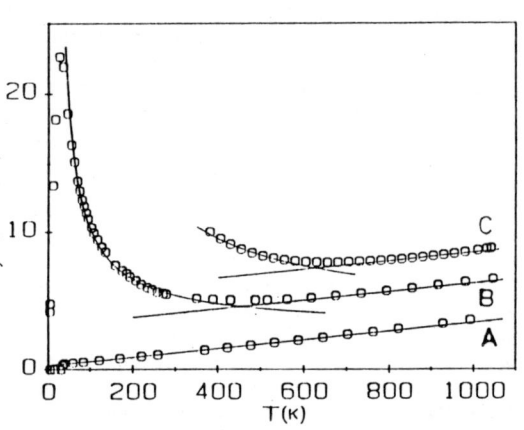

Figure 4
Temperature dependence of the resistivity
A fully oxidized sample,
B 2 hr in vacuum at 1045 K
C 3 hr in vacuum at 1045 K
The periods indicated for each run have to be added to the previous thermal - history. Solid lines are theoretical fits by VRH and metallic expressions, see text.

Figure 4 shows the electrical resistivity as a function of temperature for a sample with different oxygen contents. The procedure used to remove oxygen is describe [12] elsewhere. The resistivity follows the metallic behavior indicated by (1) in the whole temperature range only in the fully oxidized sample. The reduction of oxygen content increases the resistivity and changes [10, 12, 13] its temperature dependence at low temperatures, indicating the presence of other transport mechanisms. The low temperature semiconducting like behavior is very well fitted [13] by an expression of the form

$$\rho(T) = R \, (T/T_o)^{1/2} \, \exp(T_o/T)^{1/4} \qquad (2)$$

where R and T_o are constants.

Expression (2) represents the VRH regime [16] that has been used by Kastner et al.[17] to fit the resistivity of the non-superconducting $La_{2-y}Sr_yCu_{1-x}Li_xO_{4-\delta}$ system. The excellent fit obtained by the VRH at low temperatures and by expression (1) at higher temperatures is shown in fig. 4. It is important to remark that the curves representing the VRH fit and the metallic one are not obtained adding expressions (1) and (2). Adjusting the experimental data as described above, physically means that the material behaves homogeneously as far as the transport properties are concerned. In principle this result seems to be in contradiction with the experimental behavior of the superconducting state. Nevertheless, it should be noticed that the increase of the resistivity in the metallic regime is only due to an increase of $\rho(0)$. As a result a fit obtained adding expressions (1) and (2) is rather artificial, since for each set of parameters R, T_o and $\rho(0)$ it would be necessary to use a different α to obtain the constant experimental slope found at high enough temperatures. The previous argument, the fit of the low temperature data with only two parameters for each curve and the fact that the coefficient α is expected to be the same as that of single crystals [26], strongly suggest that the transport properties are typical of homogeneous systems.

Figure 5 shows the resistivity of a sample of $La_{1.95}Sr_{0.05}CuO_{4-\delta}$. The data follow the same behavior as that of the other compound under study (see fig. 4), and are equally well fitted by expressions (1) and (2) in the corresponding temperature regimes (full lines).

The lack of saturation in the resistivity [10, 12, 14] at high temperatures has been used to suggest [12] that these superconductors are materials in the clean limit. This suggestion is made assuming that the electrical conduction in these oxides behaves as expected for the

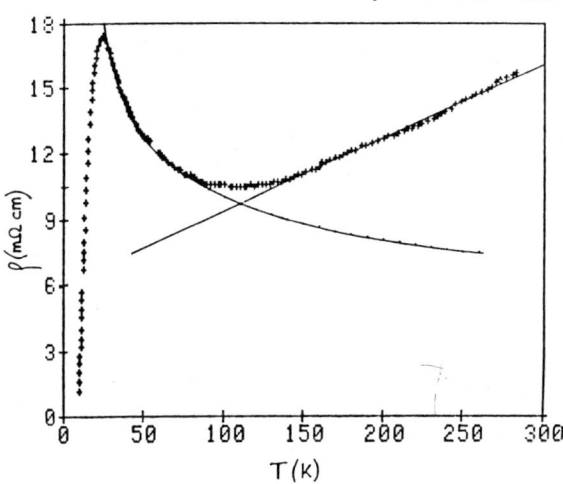

Figure 5
Resistivity vs. Temperature for a sample of $La_{1.95}Sr_{0.05}CuO_{4-\delta}$. Solid lines represent theoretical fits with VRH (low temperature regime) and metallic behavior (high temperature regime), see text.

usual electron transport in metals. The linear temperature dependence down to 0.1θ might suggest an anomalous behavior. However, recent work [27, 28] has pointed out that these results can be expected when the number of carriers is low enough and the phonon-electron interaction is dominant.

CONCLUSIONS

The study of the superconducting properties of ceramics has led to a picture for the new superconductivity, based on the macroscopic behavior of the order parameter. From this point of view the small coherence length (of the order of the size of the crystallographic cell) is its fundamental characteristic. Although the picture is based on the concepts introduced by the Ginzburg-Landau theory, the resulting small $\xi(0)$ suggests a critical review of the applicability criteria of the theory in its traditional form: several authors have indicated [29, 30] the importance of studying the fluctuations of $\Psi(x)$ near T_c. The small $\xi(0)$ can make experimentally accessible the critical region. Recent experimental results [31] show that the contribution of the fluctuations to the electrical conductivity, well above T_c, is properly described by the mean field three dimensional theory. This result is not surprising since there is no reason to expect geometrical limitation confining the variation of the order parameter in the samples investigated [31]. As a consequence of the small size of the superconducting pairs, the usual averaging effects due to pair overlapping will be diminished in the superconducting oxides. The macroscopic boundary conditions on $\Psi(x)$ should be revised [11], in particular the boundary conditions between a superconductor and vacuum are of experimental relevance. The effect of the free surface on the order parameter was pointed out by De Gennes [32] and might be important when the coherence length is small, as recently demonstrated by Simonin [33].

It is suggested that the inhomogeneous character of the superconductivity is associated to the intrinsically small coherence length found in these materials, rather than to the specific method used to prepare, the samples. We do not imply that sample preparation is not important. On the contrary, it is emphasized that almost any crystalline defect might introduce local perturbations in the order parameter.

The study of the electronic transport properties in the normal state is important and the existing results point toward [12] a superconductivity where the typical length is characterized by intrinsic properties. This is an issue that deserves further investigation: if the superconducting parameters respond to the clean limit [12] the experimental T_c will result from the strongest electron-electron interaction; on the contrary if the dirty limit applies the experimental T_c might be considerably reduced and determined by some average interaction induced by mean free path effects [34].

The study of the electrical conductivity shows [17] that in the non-superconducting $La_{2-y}Sr_yCu_{1-x}Li_xO_{4-\delta}$ the electrical transport is better described by a hopping mechanism [16] than by a semiconducting gap in the density of states. In this work we provide evidence showing that the hopping mechanism can be induced by decreasing oxygen content. The transition from the localized regime to a metallic like conduction occurs in a narrow range of temperatures and oxygen concentration. The mechanism inducing this transition is unknown. The temperature dependence of the resistivity in the metallic regime is not modified by the change in oxygen content, the whole variation is induced [10, 12, 14] in the temperature independent term, $\rho(0)$. Since oxygen concentration is expected to control both, the number of carriers and the order of the crystalline structure, it has been argued [14] that a change in carrier density does not affect the temperature dependence of ρ, when due to electron-phonon scattering. This argument is useful to explain the constancy of α with the amount of oxygen in the sample, but it leaves unclear why α depends on Sr concentration [35]. Considering this discussion, Mooij's criterium, based on the behavior of metals with $n \simeq 10^{23}$ cm^{-3} and band structures with the Fermi level far away from the band edge, does not necessarily apply to these oxides.

Recent experimental results [36, 37] show that electron localization induced by oxygen disorder is also observed in Y-Ba-Cu-O. The sensitivity of the resistivity to small changes in oxygen content, in the two families of superconducting oxides, is another indication of the important role played by the Cu-O configurations either in the superconducting or in the normal properties.

ACKNOWLEDGEMENTS

One of us, F. C., thanks the hospitality of Los Alamos National Lab., where part of this paper was written. Discussions with A. Malozemoff, J. Jose, C. Balseiro, B. Alascio, A. Rojo, J. Lorenzana and the Low Temperature Group of Bariloche are acknowledged. E. O. acknowledges financial support from the Consejo de Investigaciones Científicas of Argentina (CONICET). This work was partially supported by CONICET.

REFERENCES

1) J. G. Bednorz and K. A. Müller, Z. Phys. B64, 189 (1986); C. W. Chu, P. H. Hor, R. L. Meng, L. Gao, Z. J. Haung and Y. O. Wang, Phys. Rev. Lett. 58, 405 (1987).
2) K. A. Müller, M. Takashige and J. G. Bednorz, Phys. Rev. Lett. 58, 1143 (1987).
3) H. Maletta, A. P. Malozemoff, D. C. Cronemeyer, C. C. Tsuei, R. L. Greene, J. G. Bednorz and K. A. Müller, Solid State Commun. 62, 323 (1987).
4) D. A. Esparza, C. A. D'Ovidio, J. Guimpel, E. Osquiguil, L. Civale and F. de la Cruz, Solid State Commun. 63, 137 (1987).
5) S. Senoussi, M. Dussena, M. Ribault and G. Collin, Phys. Rev. B 36, 4003 (1987).
6) L. Civale, H. Safar, F. de la Cruz, D. A. Esparza and C. A. D'Ovidio, Solid State Commun. 65, 129 (1987).

7) A. Raboutou, P. Peyral, J. Rosenblat, C. Lebeau, O. Peña, A. Perrin and M. Sergent, Europhysics Lett., to be published.
8) D. C. Cronemeyer, A. P. Malozemoff and T. R. McGuire, Proc. of the Material Research Society, Fall Meeting, Boston (USA) 1987.
9) M. T. Causa, S. M. Dutrús, C. Fainstein, H. R. Salva, L. B. Steren, M. Tovar and R. Zysler, 32nd Annual Conf. on Magnetism and Magnetic Materials, Chicago (USA) 1987 and references therein.
10) E. Osquiguil, L. Civale, R. Decca, H. Safar, E. N. Martínez, G. Nieva and F. de la Cruz, Proc. of the Material Research Society, Fall Meeting, Boston (USA) 1987.
11) G. Deutscher and K. A. Müller, Phys. Rev. Lett. $\underline{59}$, 1745 (1987).
12) E. Osquiguil, R. Decca, G. Nieva, L. Civale and F. de la Cruz, Solid State Commun. $\underline{65}$, 491 (1987).
13) E. Osquiguil, L. Civale, R. Decca and F. de la Cruz, to be published.
14) M. Gurvitch and A. T. Fiory, Phys. Rev. Lett., to be published.
15) T. P. Orlando, K. A. Delin, S. Foner, E. J. McNiff, J. M. Tarascon, L. H. Green, W. R. McKinnon and G. W. Hull, Phys. Rev. B $\underline{35}$, 5347 (1987).
16) N. F. Mott and E. A Davis, Electronic Processes in non Crystalline Materials, 2nd.ed. (Clarendon Press, Oxford 1979).
17) M. A. Kastner, R. J. Birgeneau, C. Y. Chen, Y. M. Chiang, D. R. Gabbe, M. P. Jenssen, T. Junk, C. J. Peters, P. J. Picone, T. Thio, T. R. Thurstone and H. L. Tuller, to be published.
18) M. Tuominen, A. M. Goldman and M. L. Mecartney, Proc. of the Material Research Society, Fall Meeting, Boston (USA) 1987 and references therein.
19) G. Nieva, E. N. Martínez, F. de la Cruz, D. A. Esparza and C. A. D'Ovidio, Phys. Rev. B, to be published.
20) L. Civale, private communication.
21) W. L. Johnson: in "Glassy Metals I", ed. by H. J. Günterodt and H. Beck, (Springer Verlag, Berlin Heidelberg, N. Y., 1981) chap. 9.
22) J. H. Mooij, Phys. Status Solid A $\underline{17}$, 521 (1973).
23) M. W. Shafer, T. Penney and B. L. Olson, Phys. Rev. B $\underline{36}$, 4047 (1987).
24) G. Polla, private communication.
25) I. Haller, M. W. Shafer, R. Figat and D. B. Goland, Pure and App. Chem., to be published.
26) S. W. Tozer, A.W. Kleinsasser, T. Penney, D. Kaiser and F. Holtzberg, Phys. Rev. Lett. $\underline{59}$, 1768 (1987).
27) R. Micnas and J. Ranninger, Phys. Rev. B Rapid Commun. $\underline{36}$, 4051 (1987).
28) L. Civale and E. N. Martínez, unpublished.
29) C. J. Lobb, preprint
30) Aharon Kapitulnik and M. R. Beasley, to be published
31) P. P. Freitas, C. C. Tsuei and T. S. Plaskett, Phys. Rev. B $\underline{36}$, 833 (1987).
32) De Gennes in "Superconductivity of Metals and Alloys", ed. W. A. Benjamin INC., N. Y., Amsterdam (1966).
33) J. Simonin, Phys. Rev. B $\underline{33}$, 7830 (1986).
34) D. Markowitz and L.P. Kadanoff, Phys. Rev. $\underline{131}$, 536 (1963).
35) E. Osquiguil, L. Civale, R. Decca and F. de la Cruz, (this volume)
36) S. I. Park, . C. Tsuei and K. N. Tu, preprint.
37) Yu Mei, C. Jiang, S. M. Green, H. L. Luo and C. Politis, Z. Phys. B $\underline{69}$, 11 (1987).

Mössbauer Spectroscopy in High T_c Superconductors.

Raúl W. Gómez.
Facultad de Ciencias, UNAM.
Ciudad Universitaria, México D.F., 04510.

INTRODUCTION.

The recent development of high critical temperature superconductors [1,2] has generated world-wide interest in these new materials and much work is being done in order to characterize them [3]. Among the different analytical techniques available, Mössbauer spectroscopy seems to be particularly suitable to help disentangling some of the intriguing aspects of their behaviour, such as possible antiferromagnetic or Jahn-Teller transitions that could be associated with electron-pairing mechanisms. In this work, I will first give a brief and qualitative description of this effect and of the experimental information that can be obtained from it. Then I will proceed to review what has been done so far in high T_c superconductors.

THE MÖSSBAUER EFFECT.

Mössbauer effect [4] is just the nuclear resonant absorption of radiation emited by another nucleus in such conditions that during the emission and absorption events, the nuclei involved do not recoil. If these conditions are fullfiled the emission and absorption lines, which have a Lorentzian profile (Fig. 1), do not shift and have a natural width Γ, wich is related to the life-time of the excited state through Heisenberg's uncertain relations. To avoid the recoil of the nuclei it is necessary to anchor them in some way; this is accomplish by placing the Mössbauer atoms in crystal lattice-sites. When this is done, the recoil energy is transfered to the crystal. However, the crystal is a quantized system and if the recoil energy E_R is less than the minimum phonon excitation energy, the crystal lattice can not accept it and the γ-ray is then emmited with its natural width Γ, which is in the range of 10^{-9} to 10^{-6} eV. The recoil energy in normal Mössbauer atoms ranges from 10^{-4} to 10^{-1} eV when the γ-ray energies ϵ_0 are between 10^4 to 10^5 eV. The intrinsic tunning power Γ/ϵ_0 is then around 10^{-10} to 10^{-14} (compared to 10^{-8} in atomic line spectra), giving Mössbauer spectroscopy an enormous potential for measuring small fractional energy changes in the systems under study. Actually, the interaction of the charge distribution that surrounds the Mössbauer atoms will produce small shifts and splittings on the nuclear levels, among which the nuclear transitions take place. The small energy differences involved are sufficient to break the resonant condition between a source and an absorber. However, these energy differences can be compensated by moving the source (or the absorber) with a velocity such as to produce Doppler energy-shifts that account exactly for the differences. A Mössbauer spectrometer then consists of a controlled-motion servomechanism and a data-acquisition system (Fig. 2) that produces a velocity vs. intensity spectrum.

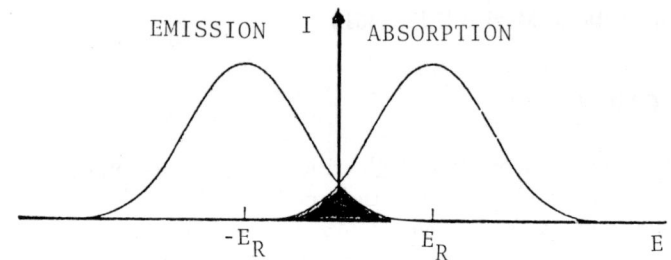

Figure 1. Emission and absorption lines.

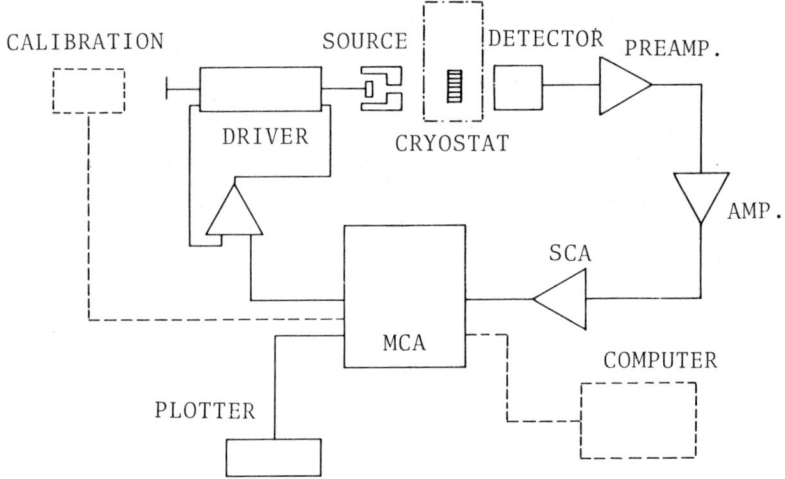

Figure 2. Block diagram of Mössbauer spectrometer.

THE MÖSSBAUER PARAMETERS.

a) Mössbauer fraction.

The possibility of recoil-free emission depends on temperature and on the nature of the crystal lattice in which the Mössbauer atom is embedded. Using a Debye model for the crystal, the following expresion for the so called Mössbauer fraction f can be obtained:

$$f = exp\left[\frac{-6E_R}{k\theta_D}\left(\frac{1}{4} + \frac{T^2}{\theta_D^2}\int_0^{\frac{\theta_D}{T}}\frac{xdx}{e^x - 1}\right)\right] = e^{-2W} \quad (1)$$

where θ_D is the Debye temperature, k is Boltzmann constant, E_R is the recoil energy of a free atom and W is the Debye-Waller factor.

b) Electrostatic interactions.

Considering that the nucleus is surrounded by a charge distribution, one can calculate the interaction energy of the nuclear charge Ze and that distribution as:

$$E = \int \rho(r)U(r)\,dr \quad (2)$$

where $\rho(r)$ is the nuclear charge distribution and $U(r)$ is the potential due to the charge distribution outside the nucleus. Making a Taylor expansion of $U(r)$ and separating the different contributions, the interaction energy can be expresed as:

$$E = eZU_0 + \sum_\alpha^3 \left(\frac{\partial U}{\partial x_\alpha}\right)_0 \int \rho(x_1, x_2, x_3)x_\alpha\,d\tau + \frac{1}{2}\sum_\alpha \left(\frac{\partial^2 U}{\partial x_\alpha^2}\right)_0 \int \rho(x_1, x_2, x_3)x_\alpha^2\,d\tau + \cdots \quad (3)$$

Experimentally, one is only interested in the energy differences between the excited and ground states. After some considerations, the energy difference can be writen as:

$$\Delta E = \frac{1}{2}\sum_\alpha U_{\alpha\alpha}\int \rho(r)\left[x_\alpha^2 - \frac{r^2}{3}\right]d\tau + \frac{1}{6}\sum_\alpha U_{\alpha\alpha}\int \rho(r)r^2\,d\tau \quad (4)$$

where $U_{\alpha\alpha} = (\delta^2 U/\delta x_\alpha^2)_0$; or, taking into account that $\Sigma U_{\alpha\alpha} = 4\pi e|\Psi(0)|^2$, as:

$$\Delta E = \frac{2}{3}\pi e|\Psi(0)|^2 \int \rho(r)r^2\,d\tau + \frac{1}{2}\sum_\alpha U_{\alpha\alpha}\int \rho(r)\left[x_\alpha^2 - \frac{r^2}{3}\right]d\tau \quad (5)$$

The first term represents the interaction of the (finite) spherical part of the nuclear charge and gives the so called isomer shift:

$$\delta = \frac{4\pi}{5}e^2 ZR^2 \frac{\Delta R}{R}\left[|\Psi(0)|_a^2 - |\Psi(0)|_s^2\right] \quad (6)$$

where $\Delta R = (<r>_e + <r>_g)/2$, $<r>_e$ and $<r>_g$ are the excited and ground state radii and $\Psi(0)_s$ and $\Psi(0)_a$ are the electron densities at the nuclear position of the source and absorber. Taking into account that not only the s states contribute to the electron density in the nuclear position, but also the $p_{\frac{1}{2}}$ states, the expression for the isomer shift can be writen as:

$$\delta = A\frac{\Delta R}{R}\left[|\Psi(0)|_a^2 - |\Psi(0)|_s^2\right] \quad (7)$$

where $A = 4Ze^2R^2S'(Z)/5$ and $S'(Z)$ is a relativistic correction factor when one uses nonrelativistic wave functions. The second term of Eq. (5) represents the interaction energy of the nuclear quadrupole and the electric field gradient and it can be expresed as:

$$\Delta Q = \frac{e^2qQ}{4I(2I-1)}[3I_z^2 - I(I+1) + \eta(I_x^2 - I_y^2)] \qquad (8)$$

where $\eta = (U_{xx} - U_{yy})/U_{zz}$ is called the asymmetry parameter, and I_i is the ith component of the nuclear spin I. In the usual case of half-integer values of the nuclear spin, this interaction energy partially removes the nuclear level degeneracy producing a splitting into Kramer's degenerated states.

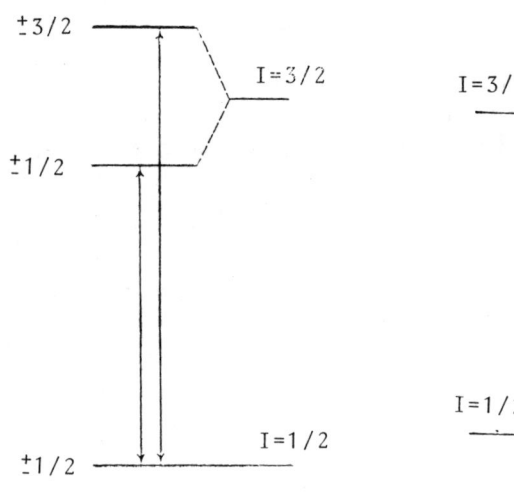

Figure 3. Quadrupole splitting of the energy levels.

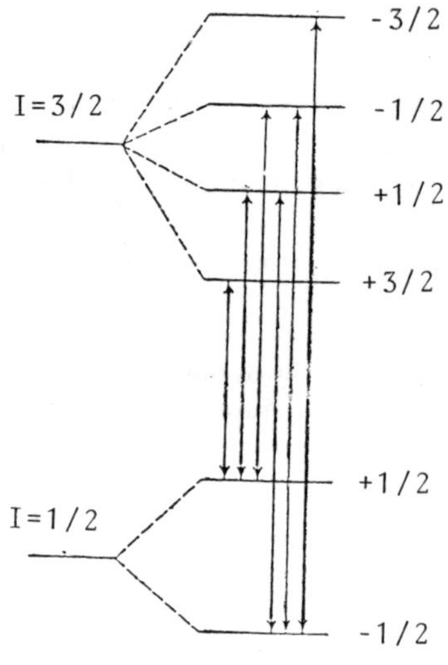

Figure 4. Magnetic splitting of the energy levels.

c) Magnetic interactions.

In the presence of a magnetic field (internal or external) there is an interaction between the nuclear magnetic dipole moment and the field. The interaction energy is given by:

$$E = -\mu \cdot \mathbf{H} \qquad (9)$$

where $\mathbf{H}$ is the magnetic field and μ is the nuclear magnetic moment. It is clear that this interaction completely removes the degeneracy and produces a splitting of the nuclear levels that manifests itself as a complicated spectrum. As an example, Fig

3 shows the splitting of the energy levels of iron-57; in the presence of a magnetic field, the spectrum consists of six Mössbauer lines.

APPLICATIONS.

The main objetives for using Mössbauer spectroscopy are: (a) to stablish the ionic and spin states of the Mössbauer atoms, whether as a part of the original crystaline structure or as substitutes of other atoms of the structure. (b) to look for anisotropies in the vibrational states, (c) to determine if there is some kind of structural change in the material when its temperature is changed and (d) to detect magnetic ordering in some temperature range.

As is well known [5], the 40 K superconductors have a tetragonal K_2NiF_4 layered structure with two cystalographic inequivalent oxygen sites. One layer consists of Cu bonded in a square network to O(1) which is linked to two adjacent planes consisting of the La and O(2) sites. On the other hand, the 90 K compounds have an oxygen-deficient perovskite like structure [6] with two inequivalent Cu sites, which are called Cu(1) and Cu(2). The later are in five-fold coordination sites that are quite insensitive to oxygen content; the former are situated in planar Cu-O squares (between the Ba-O planes) which present oxygen vacancies.

Several papers have been published since the discovery of high T_c superconductors for different systems. In the first published one, Giapintzakis et al [7] used Sn to probe the Cu sites in the 40 K superconductor $(La_{1.85}Sr_{0.15})Cu_{1-x}Sn_xO_{7-\delta}$ for $x = 0.05$, 0.10 and 0.15. They obtained an unresolved symmetrical quadrupole doublet and, from the measured parameters, they conclude that: (a) any anisotropy between vibrations in the Cu-O(1) plane and perpendicular to the plane in the direction of O(2) is small, (b) the valence state of Sn is $+4$, and (c) there is an electron change, characteristic of a phase transition or a phonon softening, near 75 K and 170 K.

All the other papers deal with 90 K type superconductors. Two papers [8,9] deal with the Eu-Ba-Cu-O system using ^{151}Eu as the Mössbauer atom. Unfortunately, in this case there is no (measurable) quadrupole moment nor magnetic splitting, giving a single line spectrum. The analysis of their spectra yields similar results and they conclude that the ionic state of the Eu atoms is $3+$, with no evidence of Eu^{2+}, and that the Debye temperature is about 285 K. They do not detect any anomalies with temperature of any of the Mössbauer parameters.

The other works deal with 1-2-3 superconductors doped with Fe, which is used as the Mössbauer atom to probe the local surroundings of the Cu sites, and it is with them where a systematization of results can be looked for. The following paragraphs describe the principal results obtained with these systems.

Coey et al [9] studied the $EuBa_2Cu_{3-x}Fe_xO_{7-\delta}$ system and obtain three quadrupole doublets with splittings of 1.14, 1.83 and 0.56 mm/s and isomer shifts characteristic of low spin iron states. They do not observe any anomalous behaviour with temperature nor magnetic structure. They associate these doublets with Fe^{3+} in the four and five-coordinated sites of Cu.

Tang et al [10] established, in a very convincing way, the coexistence of superconductivity and magnetism at 4.2 K for the $GdBa_2(Cu_{0.94}Fe_{0.06})_3O_{9-\delta}$ system and suggest that this coexistence is actually in the CuO layers or in the CuO-Ba-CuO structures. However, at temperatures above 10 K they observed only two quadrupole doublets with different splittings (2 mm/s and 1 mm/s) and small isomer shifts (-0.12 and -0.17 mm/s). They assigned the small doublet to Cu(2) sites, using an argument that assumes that the electric field gradient is due only to the lattice ions and not to the valence electrons of the proposed Fe^{3+} ion.

Gómez et al [11] showed that there is an anomalous behaviour of the Mössbauer parameters in the $YBa_2Cu_{3-x}Fe_xO_\delta$ system (with $x = 0.0625$) near 100 K, corresponding to a change of character of their sample from metallic to semiconductor. They atribute it to a possible Jahn-Teller transition based on the fact that the measured isomer shifts of their spectra imply the existence of high local internal fields. They also conclude that the Fe atoms have low spin states and that they occupy only Cu(1) sites.

Zhou et al [12], working with $x = 0.045$ and 0.3, obtained three quadrupole doublets with different quadrupole splittings. Their meassured isomer shifts are characteristic of low spin states, although they do not comment this fact. Actually, they use the same (wrong) argument of Tang et al to assign the Fe atoms, responsible of the doublet with larger splitting, to the Cu(2) sites. The spectrum with 4.5 % of iron is similar to the one obtained by Gómez et al, but presents a magnetic ordering at 1.8 K with relaxation effects and they speculate that this magnetic order can be due to long range effects of the Cu sublattices.

Takano and Takeda [13], working with $x = 0.03$, also obtain three doublets with isomer shifts characteristic of low spin state of Fe; however they assign a intermediate spin state of 3/2 for the assumed Fe^{3+}. Furthermore, they propose that the site preference of the Fe ion changes from Cu(2) to Cu(1) as the oxidation increases. They find magnetic ordering at 4 K, with an internal field of 49 T, and suggest the formation of magnetic islands around the Fe atoms.

Finally, Bauminger et al [14], working with $x = 0.03$, 0.06, and 0.3, obtain innumerable quadrupole doublets that are difficult to explain. However, there are two interesting aspects in this paper. One of them is that they obtain magnetic ordering at low temperatures, in agreement with the results of Tang and of Zhou; the other one is that the Mössbauer spectra for quenched samples consist of a prominent doublet which has a quadrupole splitting of about 2 mm/s. The possibility of oxygen vacancies in these conditions is high. Actually, one could think that the Fe atoms responsible for this doublet are only surrounded by the two vertex oxygen atoms of the square pyramids. In this sense, their results are in contradiction to the assigment made by Tang and by Zhou.

To conclude this paper, I will present our last results obtained with a 1-2-3 sample doped with 4.16 % of Fe ($x = 0.125$). Fig. 5 shows the Mossbauer spectra obtained for this sample in a temperature range from 12 to 300 K. The interesting things to note are that the external doublet (which has a quadrupole splitting of about 2 mm/s) has a temperature dependent asymmetry and that its absorption changes drastically with temperature. The value of the quadrupole splitting corresponds precisely to the one obtained by Bauminger for the quenched samples. This, together with the results of Tao et al [15] and with the strong temperature variation of the absorption of this peak, firmly support the idea that the Fe atoms responsible for this doublet are in Cu(1) sites surrounded only by the vertex oxygen atoms of the square pyramids. If such is the case, these atoms would have much more mobility than the ones surrounded by four or by six oxygen atoms, and this would explain the strong variation of it absorption. On the other hand, the asymmetric behavior of this doublet is rather intriguing, due to the fact that it reverses when temperature is lowered. We believe that this interesting behavior can be explained by the presence of a weak internal magnetic field.

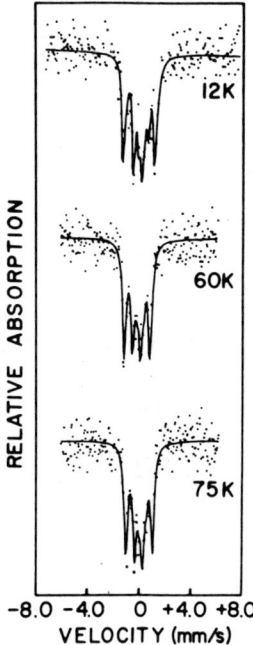
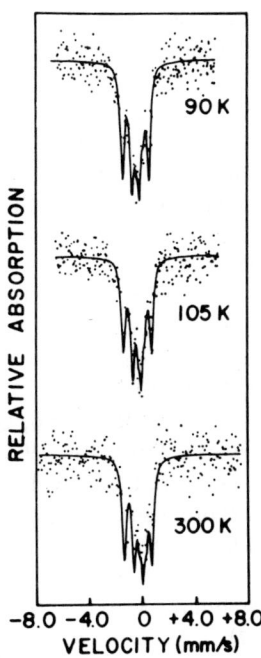

Figure 5. Mössbauer spectra from 12 to 300 K.

ACKNOWLEDGMENTS

I would like to thank all the members of the Atomic and Molecular Laboratory of the Faculty of Science of the National University of México for their invaluable suggestions and help in this work. I will also like to thank R. Gómez-Aiza for his help in preparing the manuscript.

REFERENCES

[1] J.G. Bednorz and K.A. Müller. Z. Phys.**B 64**, 189, (1986).
[2] C.W. Chu, P.H. Hor, R.L. Meng, L. Gao, Z.J. Huang and Y.Q. Wang. Phys. Rev. Lett. **58**, 405, (1987).
[3] See, for example "Novel Superconductivity" edited by Stuart A. Wolf and Vladimir Z. Kresin. Plenum, (1987).
[4] See, for example "An Introduction to Mössbauer Spectroscopy" edited by Leopold May, Plenum, (1971).
[5] R.J. Cava, R.S. van Dover, B. Batlogg and E.A. Rietman. Phys. Rev. Lett. **58**, 408, (1987).
[6] J.J. Capponi, C. Chaillout, A.W. Hewatt, P. Lejay, M. Marezio, N. Nguyen, B. Raveau, J.L. Soubeyroux, J.L. Tholence and R. Tournier. Europhysics Letters **3**, 1301, (1987).
[7] J. Giapintzakis, J.M. Matykiewicz, C.W. Kimball, A.E. Dwight, B.D. Dunlap, M. Slaski, and F.Y. Fradin. Phys. Lett. **A 121**, 307, (1987).
[8] P. Boolchand, R.N. Enzweiler, I. Zitkouvsky, R.L. Meng, P.H. Hor, C.W. Chu and C.Y. Huang. Solid State Commun. **63**, 521, (1987).
[9] J.M.D. Coey and K. Donnelly. Z. Phys. **B 67**, 513, (1987).
[10] H. Tang, Z.Q. Qiu, Y.W. Du, Gang Xiao, C.L. Chien, and J.C. Walker. Phys. Rev. **B 36**, 4018, (1987).
[11] R. Gómez, S. Aburto, M.L. Marquina, M. Jiménez, V. Marquina, C. Quintanar, T. Akachi, R. Escudero, R.A. Barrio, and D. Rios-Jara. Phys Rev. **B 36**, 7226, (1987).
[12] X.Z. Zhou, M. Raudsepp, Q.A. Pankhurst, A.H. Morrish, Y.L. Luo, and I. Maartense. Phys. Rev. **B 36**, 7230, (1987).
[13] M. Takano and Y. Takeda. Jpn. J. Appl. Phys. **26**, L1862, (1987).
[14] E.R. Bauminger, M. Kowitt, I. Felner, and I. Nowik. Solid State Comunn. **65**, 123, (1988).
[15] Y.K. Tao, J.S. Swinnea, A. Manthiram, J.S. Kim, J.B. Goodenough, and H. Steinfink. Material Research Society Fall Meeting. Boston, Mass, USA, (1987).

STRUCTURAL
PROPERTIES
AND
DEFECTS

STRUCTURAL CHANGES

DISTORTIONS

METASTABILITY PROPERTIES OF GAUGE GLASSES AND THEIR RELEVANCE TO HIGH TEMPERATURE SUPERCONDUCTORS

Jorge V. José

Department of Physics
Northeastern University, Boston, Massachusetts 02115

ABSTRACT

The metastable magnetic properties found in the oxide superconductors are discussed. A random lattice superconductor model is introduced to model the history dependent properties of these systems. A Monte Carlo simulation of the model leads to results which are qualitatively analogous to those found experimentally. A critical assessment of the model results against the experimental findings is discussed.

INTRODUCTION

Among the many unusual properties found in the oxide superconductors, I want to single out for discussion here their low temperature metastable magnetic properties. These were first reported by Muller et.al. [1] (MTB henceforth) in the $La_2CuO_4 : Ba$ superconductor, and have since been found in other $40\,K$ superconductors [2−4], as well as in the $1-2-3$ ($YBa_2Cu_3O_{7-x}$) compound [5,6]. They find that below the superconducting critical temperature these materials show striking *history dependent* magnetic properties.

Let me go through the measurement procedure followed in the experiments since in the Monte Carlo simulation, to be discussed below we try to follow a similar calculational route. The results presented here are preliminary results obtained in collaboration with J. Choi, and will appear in a complete form elsewhere[7]. In the experiment the sample is first cooled down to a final low temperature (in the MTB experiment $T \approx 2\,K$) in essentially zero magnetic field. A magnetic field is then applied ($H \sim 0.03$ Tesla), and the temperature is raised at a slow rate ($T < T_c(H)$). The magnetization and susceptibility are then measured with a $SQUID$ susceptometer. If the warming is stopped at, say $6\,K$, and the sample is cooled again, at the same rate, it is found that M does not retrace its path but follows a different one which slope is temperature dependent. Warming the sample again brings the susceptibility back from where it left, and the monotonic increase of M continues. The nonergodic behavior of M persists up to a magnetic field critical temperature, $T_c(H) \leq T_c(H=0)$, above of which warming and cooling the sample in the field gives a perfectly reversible, or ergodic, curve. As recognized by MTB this type of metastable, or history dependent, behavior for M is reminiscent of what is expected in a magnetic spin-glass (SG). As I will discuss below, although there are indeed similarities between a SG and the metastable properties of the oxide superconductors, there are also some important differences that need special study.

When the external magnetic field is switched off, the equilibrium magnetization should go to zero. How does $M(t)$ reach zero after the field is off will give us another hint as to the microscopic mechanism behind the metastability. MTB found that the the time dependence of the relaxation is non-exponential, which is typical of systems with a spectrum of relaxation times. This characteristic is also analogous to the magnetization relaxation in a SG. From their experimental results MTB surmised that these properties are similar to those found in a canonical SG, and that we may have a representative of a superconducting glass (SCG) in the oxide superconductors.

Some controversy has arisen with regard to this surmise in that there is an old mechanism well known in type II superconductors known as *flux creep*, that leads to similar metastability and non-exponential decay of supercurrents [3]. This mechanism was introduced by Anderson and consists or having thermally activated flux motion [8]. The change in the magnetic field due to this motion is logarithmic in time and with a coefficient with a linear temperature dependence. Flux creep in ordinary superconductors has been tested extensively and it is believed to be the correct explanation for supercurrent decays [9]. Since the idea of having a SCG in the oxide superconductors materials would have a similar qualitative behavior, one needs to ask other questions to ascertain which explanation is the correct one.

Let me mention that there are recent experimental results in the $1-2-3$ compounds that appear to disagree with the flux creep explanation [6]. In a detailed time and temperature dependent analysis of the magnetization as a function of applied magnetic field, for the zero-field-cooled (ZFC) and field cooled

(FC) experiments, it was found that the time decay was indeed logarithmic but with a coefficient with a different T dependence, to wit: for a field of 500 Gauss, the temperature dependent coefficient of the remanent magnetization has a maximum at around $30K$, and decreases for higher temperatures. This results indicates that the slowing down seen in these materials is largest at low and high temperatures. This behavior has no counterpart in the standard SG systems, where the metastability increases monotonically as one lowers the temperature. Hence, the oxide superconductors become more "sluggish" as one gets closer to T_C, a rather unusual result indeed. Above T_C, on the other hand, the system behaves as a regular metal, with no metastability properties.

We have both an experimental as well as a theoretical challenge to determine if indeed the metastable properties seen in the oxide superconductors correspond to those of a SCG or are just related to a modified version of the flux creep phenomena.

THEORETICAL MODELS

One could question any attempt at formulating a theoretical analysis aimed at explaining the metastability properties found in the high T_c superconductors, while we are still far from having a universally accepted microscopic theory to explain the superconductivity in these compounds. We should remember that, qualitatively, the essential properties of the superconducting state e.g. the Meissner effect, infinite conductivity, flux quantization, and the Josephson effect, only depend on having electromagnetic gauge invariance broken in the superconducting state [10]. The breaking of this symmetry implies the existence of a Goldstone or propagating mode described by a phase $\phi(\vec{x})$. It is in terms of this phase that we can write a coarse grained expression for the free energy of the system as,

$$F = \frac{1}{2}\int (\vec{E}^2 + \vec{B}^2) + f[\vec{\nabla} - \frac{e}{c}\vec{A}, \phi(\vec{x})]. \tag{1}$$

where $\vec{E}$ and $\vec{B}$ are the usual electric and magnetic fields with $\vec{B} = \vec{\nabla}\times\vec{A}$. The expression for F is gauge invariant with $\vec{A} \to \vec{A} + \vec{\nabla}\Delta(\vec{x},t)$, $V(\vec{x}) \to V(\vec{x}) + \frac{\partial\Delta(\vec{x},t)}{\partial t}$ while $\phi(\vec{x},t) \to \phi(\vec{x},t) + \frac{e}{\hbar}\Delta(\vec{x},t)$. An expression for F could be derived from an explicit microscopic Hamiltonian in an appropriate coarse grained limit. However, even in the case of ordinary BCS superconductors, the study of glassy behavior has only been done in terms of models, whose microscopic derivation is not given but where the essential ingredients of having a spectrum of relaxation times is included.

To get an idea as to how to model F for the oxide superconductors we go back to results from other experiments for further help. The ceramic compound are formed by grains with sizes ranging from 1μ to 10μ. In the early experiments that showed the existence of the Josephson effect with charge $2e$ by Esteve et.al. [11], they used a nonsuperconducting metal tip to induce the effect. They found that in order to see the Shapiro steps they had to apply some pressure to the tip or else the effect was not there. This result led them to suggest that the couplings were not made at the interface between grains but it was likely to occur inside the grains. Further mention of this possibility was put forward by MTB as well as in reference [12], where a possible justification for the existence of intragrain Josephson junctions was related to the extremely short coherence length found in these compounds ($\xi_0 \sim 10-20A$). They surmised that the junctions may form at the twin boundaries seen in the diffraction experiments, or at grain boundaries.

A possible model to describe the magnetic properties of these materials was suggested to be then[1],

$$\mathcal{H}_J = \sum_i E_J(i+1,i)\left[1 - \cos(\phi_{i+1} - \phi_i + 2\pi f_{i+1,i})\right] \qquad (2)$$

with,

$$f_{i+1,i} = \frac{1}{\Phi_0} \int_{i+1}^{i} \vec{A} \cdot d\vec{\ell}. \qquad (3)$$

Here $i+1$ and i give the location of the superconductors i and j. $E_J(i+1,i)$ denotes the Josephson coupling between superconducting regions. This is a quantity that will need to be derived explicitly from a microscopic theory. The frustration parameter $f = \sum_p f_{i+1,i}$ is given by $f = \frac{\Phi}{\Phi_0}$, where Φ is the magnetic flux through a closed loop formed by an array of junctions, and Φ_0 is the fundamental quantum of flux. Eq. (2) with E_J constant has been studied extensively [13 − 15]. Notice that the electromagnetic contribution has not been considered in Eq(2). This seems to be justified by the fact that the London penetration length is much larger than ξ_0 and, therefore, the screening currents can be neglected.

A very important question is, what is the appropriate dimensionality at which we should study Eq(2). There is a significant amount of evidence that the experimental properties of the oxide superconductors are highly anisotropic, with the conduction mechanism mainly taking place in the CuO_2 planes. Thus, one can begin by considering a two dimensional model and later on include the weaker interplane coupling. If we consider a two dimensional model we have to have in mind the special critical properties of the model without and with a field, to wit: The periodic model ($f = 0$) has a Berezinskii-Kosterlitz-Thouless (BKT)

type transition with vortex pairs generated thermally [16].. These excitations are responsible for the occurrence of the transition. In the periodic case with a field, the model has been studied extensively recently[13 − 15]. The thermodynamics of the model is rather complicated depending intrinsically on the lattice geometry and the rationality of f. In the periodic case the critical temperature as a function of the magnetic field shows a nonmonotonic behavior that is related to the lower edge of the spectrum of a lattice Schrödinger equation with an incommensurate potential. Theoretically, there is an important difference between an irrational and a rational f [17]. This problem has been studied extensively and there are still a number of unanswered questions.

An essential property of the "intragrains" that may form the junctions within the grains is that they are located at random. This means that the $f_{i,j}$ are themselves random variables. We could also consider the $E_J(i,j)$ as random, but this may in some sense be included in the randomness of the $f_{i,j}$. The lattice model given in Eq(2), with the $f_{i,j}$ random, has been studied within the context of granular BCS superconductors by Ebner and Stroud[18] They considered a three dimensional random network of superconducting grains. A continuous version of Eq(2) has been studied by John and Lubensky, with dilution randomness [19]. None of these prior calculations considered the remanent properties that relate directly to the oxide superconductors. A first specific study aimed at understanding the MTB results has been carried out by Morgensten et.al. [20]. They considered a positional disordered model where each site is allowed to be at a maximum radius r from its regular lattice position. This radius is chosen at random. We leave a discussion and specific comparison of their results after I discuss our own results.

Part of the richness in the properties of the model given in Eq(2) comes from an intrinsic *lattice gauge invariance* i.e.: H_J is invariant under the transformation,

$$\phi_i \to \phi_i + n_i, \qquad (3.a)$$

$$f_{ij} \to f_{ij} + (n_i - n_j), \qquad (3.b)$$

where the $\{n_i\}$ are integers. This transformation allows to *gauge away* any irrelevant disorder while keeping the relevant one. To be more specific, consider four intragrains with with a supercurrent that circulates around them in a closed loop. The supercurrent between the $i - th$ and $(i + 1) - th$ grains is,

$$I_{i,i+1} = J sin(\phi_i - \phi_{i+1} - 2\pi f_{i,i+1}).$$

The total current about this loop is the sum $I_T = I_{1,2} + I_{2,3} + I_{3,4} + I_{4,1}$. Since we have the freedom to choose a gauge, we can choose the Landau gauge, $\vec{A} =$

$(0, Hx, 0)$. For a square loop say, only the y component of the I's contribute to the sum, i.e. $I_{2,3}$ and $I_{4,1}$. Take, for example $f_{2,3} = 0$ and $f_{4,3} = \frac{1}{2}$, thus we have a conflict with regard to the direction of the supercurrent: the direction can be either clockwise or counterclockwise, but the system does not which one to choose since both have the same energy while being topologically disconnected in energy space. The fact that one can not find a set of n_i's in Eq(3) to connect one circulation state to the other is typical of a SG situation. This random *frustration* effect is believed to be the essential element that leads to metastability in this model. A thorough discussion of this concept as applied to the model in Eq(2) can be found in ref. [21]

It should be mentioned that the type of circulating current mentioned here, often called vortices [22], do not share the same properties as the Abrikosov vortices. The latter type are defined in continuous superconductors while the ones considered here have voids in the center of the current loops. For this reason it is convenient to call the model defined by Eq(2) a *gauge glass model* and leave the term superconducting glass for a continuous superconductor. It is for the Abrikosov vortices that the flux creep phenomena was tailored for. There is recent experimental evidence for the formation of a vortex lattice at low temperatures in the $1 - 2 - 3$ compound [23]. At low temperature they form a vortex lattice in just the same way as ordinary superconductors. However, at higher temperatures their properties are different. It is likely that further studies of these vortices will shed light into their intrinsic properties.

THE MODEL

We have introduced a model to study the effects of topological disorder in the Hamiltonian given in Eq(2) with the aim of mimicking what is seen in the oxide superconductors [7]. Motivated by the fact that the Josephson junctions are surmised to form at the twin boundaries [12], with the boundaries being metallic, the model consists of an array of Josephson junctions equidistant along the y-axis but with random separations along the x-axis. This model has the advantage that it emphasizes the trapping of a particular amount of flux along the y-axis. The size of the lattice is fixed and equal to $L_x \times L_y$. The lattice sites along the x axis are given by the points $x_m = ma + \delta$, with m an integer, a the periodic lattice spacing, that we shall take equal to one, and δ is a random variable defined in the interval $[a - \delta, a + \delta]$ and with uniform probability distribution $P(\delta) = 1/2\delta$.

Given the Hamiltonian and $P(\delta)$ we can calculate the thermodynamic properties associated with the model. The *quenched* averaged free energy is given as,

$$\{F\}_c = -k_B T \{\ell n Z[f, \delta]\}_c. \qquad (4)$$

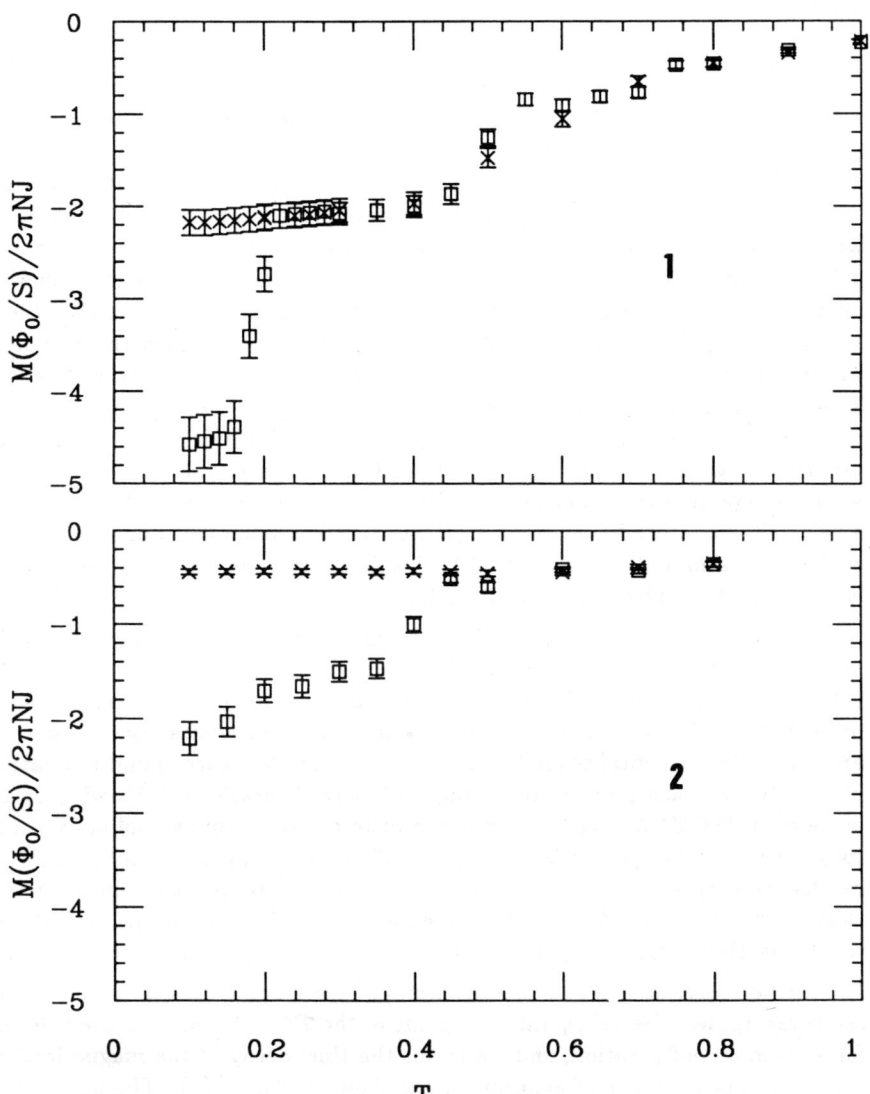

Here Z is the partition function associated to H for a given realization of δ's: the brackets $\{,\}_c$ stand for configurational average with respect to $P(\delta)$. Needless to say the calculation of $\{F\}_c$ analytically is nontrivial, but some progress can be made by transforming the partition function to its dual representation [7]. Here I will instead present some preliminary results for the calculation of the magnetization,

$$M(t) = -\frac{\partial \{F\}_c}{\partial H}, \qquad (5)$$

using the Metropolis or Monte Carlo (MC) algorithm. Other physical quantities are calculated at the same time but here I will concentrate on the Magnetization and discuss the other properties elsewhere [7]. One follows essentially the same procedure as described in the introduction i.e. for a given configuration of δ's we equilibrate the system at low temperatures in the absence of a magnetic field. A field is then turned on and the system is warmed up slowly, allowing for the system to reach a minimum energy state. In the periodic case, without a field, the model has a BKT critical temperature ~ 0.9, thus the system is warmed up passed this temperature, always with the constant field on. At about $T \sim 1.2$, the system is cooled down back to the initial temperature, and back to the high temperature again. We observe (figures (1), (2), and (3)), as in the experiment, that in the zero-field-cooled (ZFC) case, the system follows a curve with lower values for the magnetization (squares in the figures). When cooling in a field (FC), once we have passed $T_c(H)$, the system follows only one curve while cooling or warming ($\times$ in the figures). The FC curve represents the equilibrium value for M for a given value of the field.

In Figs. (1), (2) and (3) we show the magnetization as a function of temperature: The corresponding values for the parameters are: $H = 0.01$, $\delta = 0.1$ (1); $H = 0.015$, $\delta = 0.1$ (2); and $H = 0.02$, $\delta = 0.1$ for (3). H is measured in units of Φ_0 and $M(T)$ is normalized by the number of lattice sites N and the area of the fundamental plaquette $S = a^2 = 1$. The plots correspond to a lattice of 16×16 and each point is an average of five configurations of δ's with an average of $60,000 MCS$/angle. Thus, a complete run of warming-cooling-warming took, for each set of parameters, about 7 million MCS/angle. We notice that the results are very similar to those obtained in the real experiments, in particular those of reference [6]. As the field increases, for fixed δ, we see that the size of the metastability decreases monotonically while the magnetization tends to zero.

Furthermore, to check on the dynamics of the trapped flux as a function of temperature, we take an equilibrium point in the FC branch, turn the field off, for a given δ configuration, and watch for the time decay of the magnetization. The results of this type of experiment are shown in figure (4). The parameters

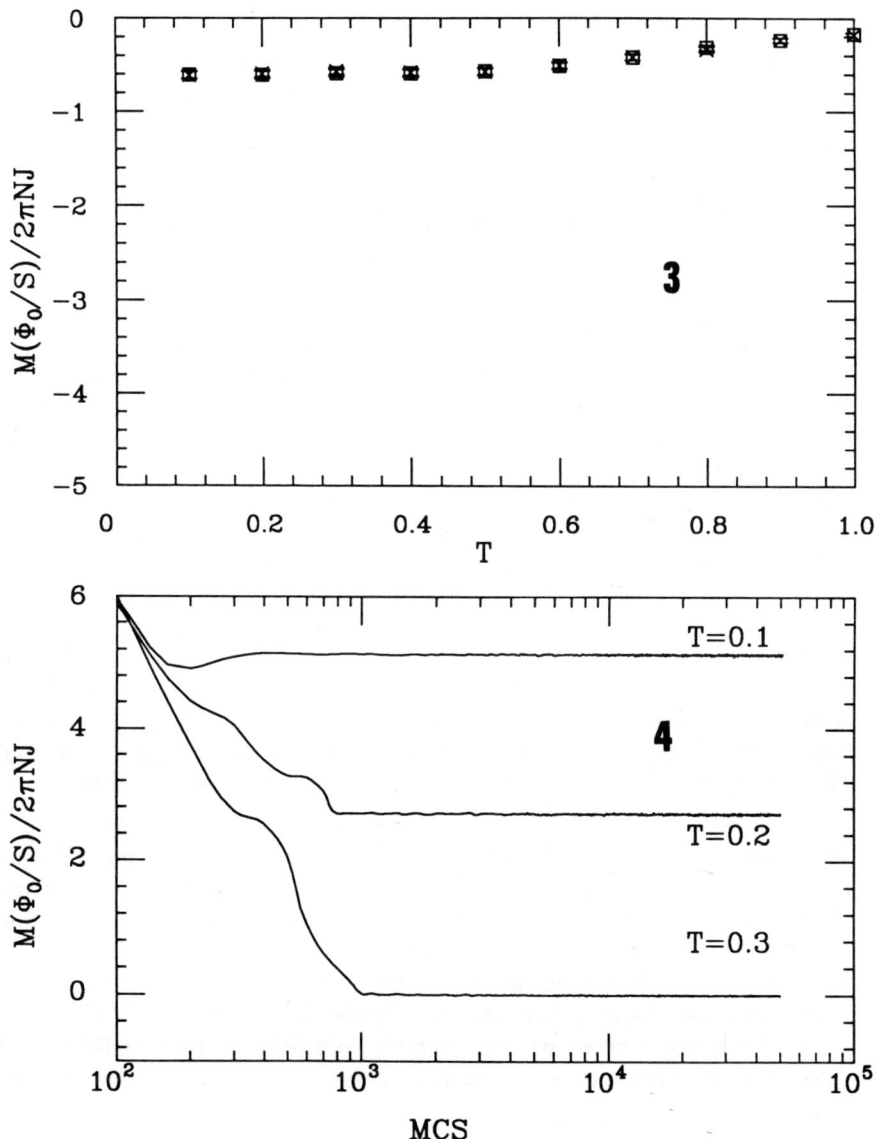

in this figure are; $H = 0.02$, $\delta = 0.2$ and $L_x \times L_y = 16\ times16$. We see, first, that as soon as the field is switched off the magnetization changes sign, as in the experiments. Initially, the magnetization seems to decay logarithmically, and then seems to stay constant up to 10^4 to 10^5 MCS/angle. We have done longer runs for selected temperatures, and the constant behavior seems to persist. The remanent value is largest the lowest the temperature, and the initial slope increases monotonically with temperature, as intuitively expected. For $T = 0.3$ the magnetization decreases logarithmically all the way down to zero. There is a critical value of the temperature for which the magnetization does show the constant behavior and it is around $T \sim 0.23$, but we need to sharpen this number further.

The results for the helicity modulus, specific heat, magnetic hysterisis loops, will be detailed in reference [7].

DISCUSSION

I have presented preliminary results obtained from Monte Carlo simulations of the model defined by Eq(2), and the particular type of disorder discussed above. This model tries to mimic some of the properties that seem to be essential in describing the metastable magnetic properties found in the oxide superconductors. Except for the specific type of disorder considered, this model is analogous to that considered recently by Morgensten et. al. [20]. It is not yet fully clear if the type of disorder is essential to the results, but some evidence was found in ref. [20] that this may be the case. We note that in our calculation we do see the ZFC branch, whereas in ref. [7] results for this branch were not discussed. With regard to the FC branch we find that it is thermodynamic stable against cooling or warming. An important difference between the results shown in ref. [7] and those discussed here relate to the specific time decay of the magnetization. Here we see a non exponential decay whereas for the results shown in ref. [7] the decay appears to be exponential. It is important to study in full detail different models of disorder to ascertain if the results do depend on the type of disorder.

With regard to the applicability of the model itself to the oxides we make the following comments. The qualitative properties of the magnetization as a function of cooling or warming conditions are strikingly similar. Of course, more definitive experimental confirmation of the existence and properties of the micro Josephson junctions is needed, since otherwise the model may not have an experimental basis. Although the history dependent and remanent properties are similar to those of a SG, there are important differences that make the specific study of this model necessary. For example, in a SG the magnetic field couples directly to the order parameter (the spin operator), while in the gauge-glass case the magnetic field is not the thermodynamic conjugate variable to the phase. The magnetization in the SG phase is zero while in the gauge-glass problem it is different from zero, at least for small fields. The magnetic susceptibility has a cusp in the SG case whereas here it is an smooth function of T. Thus the similarities are there, but are not sufficient to make a complete analogy. More work is needed to unravel all the specific properties of the gauge glass model discussed here.

All in all we see that the results obtained by Morgensten et al [20] and the preliminary ones shown here, do resemble those found in the experiments on the oxide superconductors. However, much more comparative work needs to be done to ascertain if this type of description is the correct one.

The calculations presented in this paper were carried out in a Cray1 XMP from the Pittsburgh Supercomputing Center supported by the National Science Foundation Office for Advanced Scientific Computing. The partial support for this work by NSF grant DMR-8640360 is gratefully acknowledged.

REFERENCES

[1] K. A. Muller, M. Takashige and J. G. Bednorz, Phys Rev. Lett. **58**, 1143 (1987).

[2] A. C. Mota et.al., Phys Rev. **B36**, 4011 (1987).

[3] G. Giovanella, et. al. Europhys. Lett. **4**, 109 (1987).

[4] M. Oussena, S. Senoussi and G. Collin, Europhys. Lett. **4**, 625 (1987).

[5] F. Carolan et. al. Solid State Commun. **64**, 717 (1987).

[6] M. Tuominen, A. M. Goldman and M. L. Mecartney. Phys. Rev. **B37**, 548 (1987).

[7] J. Choi and J. V. Jose (in preparation).

[8] P. W. Anderson, Phys. Rev. Lett. **9**, 309 (1962).

[9] M. Beasley et. al. Phys. Rev. **181**, 682 (1969)

[10] M. Tinkham, *Introduction to superconductivity* (R. E. Krieger Publ. Co. Inc. N. Y. (1980)).

[11] P. Esteve et. al. Europhys. Lett. **3**, 1237 (1987).

[12] G. Deutcher and K. A.M. Muller, Phys Rev. Lett. **59**, 1745 (1987).

[13] S. Teitel and C. Jayaprakash Phys. Rev. Lett. **51**, 1999 (1983) and Phys. Rev. **B27**, 598 (1983).

[14] W. Y. Shi and D. Stroud Phys. Rev. **28**, 6575 (1983) **30**, 6774 (1984), and **B32**, 158 (1985). M Y. Choi and S. Doniach ibid **B31**, 4516 (1985) T.H. Halsey ibid **B31** 5728 (1985). M. Yosefin and E. Domany ibid **B32**, 1778 (1985). B. Berge, H. J. Diep, A. Ghuzali, and P. Lallemand, ibid **B34**, 3177 (1986).

[15] C. Lobb, D.W. Abraham and M. Tinkham, Phys. Rev. **B27**, 150 (1983). B. Pannetier, J. Chaussy, R. Rammal, and C.V. Villegier, Phys. Rev. Lett. **53**, 1845 (1984).

[16] J. M. Kosterlitz and D. Thouless J. Phys. **C6**, 1181 (1983). V. L. Berezhinskii Zh. Theor. Fiz. **59**, 907 (1970) [Sov. Phys. JETP **34**, 610 (1971)]

[17] J. B. Sokoloff, Phys. Rep. **126**, 189 (1985).

[18] C. Ebner and D. Stroud, Phys. Rev. **B31**, 165 (1985).

[19] S. John and T. C. Lubensky, ibid **B34**, 4815 (1986).

[20] I. Morgensten, K. A. Muller and J. G. Bednorz, Z. Phys **B69**, 33 (1987).

[21] J. Villain J. Phys. **C10**, 4793 (1977); E. Fradkin, B. Huberman, and S. Shenker Phys. Rev. **B18**, 4789 (1978). J. V.José ibid **B20**, 2167 (1979), and *Proc. Kyoto Summer Institute (1979)* ,Ed. Y. Nagaoka and S. Hikami. Publ. Prog. of Theor. Phys. p.p. 127.

[22] L. Civale et.al., Solid State Commun. **65**, 129 (1988).

[23] P. L. Gammel et.al., Phys. Rev. Lett. **59**, 2592 (1987).

STRUCTURAL DEFECTS IN THE 1:2:3 PHASES OF HIGH Tc SUPERCONDUCTORS

DAVID RIOS-JARA
Instituto de Investigaciones en Materiales, UNAM, Apdo. Postal 70-360, 04510 México, D. F., MEXICO.

ABSTRACT

Several kinds of structural defects have been observed in the orthorhombic and tetragonal phases of 1:2:3 high-Tc superconducting compounds: Reflection twins on (110) planes, rotation twins on (001) planes, anti-phase domain boundaries, superstructures, dislocations, stacking faults, grain boundaries, amorphous strips and precipitated phases. Some of these defects are common to both, the tetragonal and orthorhombic phases of 1:2:3 compounds. However, defects such as twins, antiphase boundaries and superstructures are intrinsic to the oxygen-vacancy sublattices of the orthorhombic superconducting phase. They are formed while cooling the sample as grow-in defects during the tetragonal to orthorhombic transition observed at about 650°C, and result from an ordered addition of oxygen atoms into the structure. Among all the above mentioned defects, twins are the more studied ones and some relationships with the superconducting transition have been proposed.

In this work we present a review of the main characteristics of all these structural defects and we discuss the mechanisms associated to the formation of some of them.

INTRODUCTION

Critical superconducting temperatures (Tc) of about 90K (and slighly higher) are actually a well known characteristic of $R_1Ba_2Cu_3O_{7-y}$ systems with R=Y or rare-earth (or rare-earth combinations, except for Pr, Ce and Tb). These systems have another important common feature: All of them show a reversible phase transition from a tetragonal to an orthorhombic structure, at temperatures of about 650°C. The tetragonal phase (Fig. 1-a) being no superconducting, while the orthorhombic phase (Fig. 1-b) being the high-Tc superconducting one. On cooling, the tetragonal to orthorhombic phase transition is produced by the absorption and ordering of oxygen atoms, which align along one of the two equivalent <100> directions of the Cu(1) type basal planes of the tetragonal structure. The formation of these O-Cu-O chains increases the lattice parameter in this direction (b_o in Fig. 1-b) and reduces the lattice parameters in the two other perpendicular directions (a_o and c_o in Fig. 1-b), leading to a total effective reduction in the volume of the resulting orthorhombic unit cell.

The dependance of oxygen absorption and lattice parameters as a function of temperature have been reported by A. M. Manthiram et al. [1] and J. D. Jorgensen et al. [2] respectively. On cooling the oxygen content of the tetragonal phase increases continuously, from almost the stoichiometric value of 6.0 to reach about 6.4 close to the transformation temperature and finally attaining about 6.9 at room temperature. Above the transition temperature oxygen atoms are in a disordered state and tetragonality is mantained. Below this temperature the oxygen content increases rapidly, in the ordered manner mentioned above, and goes trough a maximum at 450°C. Measurements of the variation of lattice parameters with temperature suggest a second order transition [3]. It is interesting to note that, in well oxigenated samples, the oxygen content of the orthorhombic phase has been reported to be about 6.95 [1,4], which is below the expected stoichiometric value of 7.0 and seems to be the maximum equilibrium value for an external O_2 partial pressure of 1 atm. The reason for this difference still remains an open question.

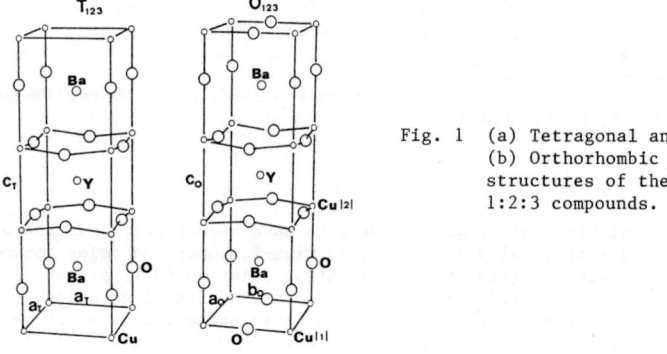

Fig. 1 (a) Tetragonal and (b) Orthorhombic structures of the 1:2:3 compounds.

DEFECTS COMMON TO THE TETRAGONAL AND ORTHORHOMBIC PHASES

We describe in this section those defects which are common to the two crystallographic structures of high Tc superconductors. Although their relationship with the superconducting transition seems to be only secondary, these defects are very important in the different technological developments with superconducting compounds. Unless otherwise specified, all the following results were obtained in $Y_1Ba_2Cu_3O_{7-y}$ compounds by Transmision and Scanning Electron Microscopies (TEM and SEM) techniques.

Grain boundaries

Grain boundaries are necessarily obtained in the production of bulk and thin film samples of high Tc superconductors. The role of these defects in technological applications of superconductivity is highly relevant, since they can be a limiting factor on the superconducting current density following through the material. in this context, there have been some reports of amorphous intergrain-boundaries layers with thickness from 1-5 nm [5] to 100 nm [6]. Furthermore, amorphous layers (10-100 nm thick) at the interior of grains in the a-b planes have been reported [6,7]. However, these informations must be taken with care, since there have been some reports indicating that amorphisation can be caused by water contamination and electron beam damage during the observation in the electron microscope [8]. It is worthwhile noting here that threshold energies of 131 $\pm$7 KV have been measured for the removal of oxygen atoms by electrons in the orthorhombic structure of YBaCuO compounds [9]. The same type of amorphisation of grain boundaries, with a measured increase in the room temperature resistivity, has been observed in thin films as a result of radiation damage with He, O and As ions [10]; at high does the whole grains of the Y-Ba-Cu-O compound became amorphous.

Another interesting observation is that grain boundaries are found often to be coplanar with the (001) basal plane in at least one of the bordering grains [11]. After S. Nakahara et al. [12], grains in well sintered samples are platelets with minimum thickness in the c-axis direction; grain boundaries are also found to be usually perpendicular to the c-axis and some segregation of Cu is detected. Three different types of boundaries were observed by Nakahara et al.: 1) Boundaries with a network of dislocations (tight boundaries), 2) Boundaries with dislocation loops (on the c-axis side only) and depth of penetration of the loops of 50-500Å, depending on the local strain and 3) Boundaries where small voids and substantial plastic deformation have been produced by severe local strains, probably due to the highly anisotropic thermal contraction experimented by the sample during cooling.

Also of interest is the drastic deterioration of the critical superconducting current density (Jc) observed by T. H. Tiefel et al. [13] during the cycling from -196 to 850°C of yttrium-1:2:3 samples. A decrease of the critical current density from 400 to 70 A/cm^2 was observed after 5 cycles; probably due to the formation of microcracks at the grain boundaries, produced by the large anisotropic thermal deformations introduced by the tetragonal to orthorhombic transformation. This is consisten with the small decrease in Jc (400 to 370 A/cm^2) observed by cycling between -196 and 600°C.

In the case of thin films, L. A. Tietz et al. [14] have observed that many grains have their c-axis perpendicular to the surface. A small rotation about the c-axis from grain to grain was measured, which produces small angle grain boundaries. Some grains were found to have a or b axis perpendicular to the surface and c-axis on the surface; adjacent grains of this type have their c-axis perpendicular to each other [15]. Similar relationships between axis orientations were observed by H. W. Zandbergen et al. [16] in bulk materials. They explain these observations as resulting from the three equivalent stacking of BaO and Y planes which are possible in the cubic framework formed by the copper and oxygen atoms.

Precipitated phases

There have been several reports of precipitation of minor phases in yttrium-1:2:3 compounds. S. Ikeda et al. [17] detected the presence of Y_2BaCuO_5 as spherical isolated grains and $BaCuO_2$ as small grains (100 nm) along grain boundaries between clusters of grains or at the three fold nodes of quasi-boundaries. $BaCuO_2$ grains included other minor Y-rich phases and small quantities of Cu were also detected. W. T. Lin et al. [18] reported the presence of Y_2BaCuO_5 as minor precipitated phases plus some amorphous aggregates. S. Ikeda et al. [19] reported minoritary untwinned grains with 50% plus of yttirum content, showing precipitation of Cu_2O from 290°C in insitu heating experiments at the electron microscope. Here, twinned grains were the last to transform into the tetragonal phase during the heating process, indicating some higher stability of the oxygen content in twinned regions. The presence of Ba-rich compounds at the surfaces of the grains have been detected in Auger spectroscopy experiments [20].

Some explanations for the segregation of these minor phases have already been proposed. W. H. Philipp et al. [21] explored the advantages of using barium peroxide (BaO_2) instead of barium carbonate ($BaCO_3$) in the preparation of the samples. In the latter case, they detected the precipitation of unreacted $BaCO_3$ and an incongruent melting from the apparent formation of a $BaCO_3/BaCuO_2/CuO$ eutectic at 890°C; further reaction of $BaCO_3$ was observed to produce porous and non-uniform samples.

The effect of CO_2 during the preparation of the samples has been studied by G. J. Fisanick et al. [22]. They found a strong tendency for the formation of $BaCO_3$ at the surface of the samples due to reaction with CO_2 in the atmosphere.

Hyde et al. [23] reported the aggregation of $BaCO_3$ at the surface in water-vapour saturated air. Also, Z. P. Zhang et al. [8] studied the effect of water and electron beam radiation. They found that water vapour leads to the decomposition into BaO, the green phase (Y_2BaCuO_5), a brown phase and a barium deficient bulk structure with BaO whiskers growing epitaxially on the surface. On the other hand, the electron beam first induces amorphisation, followed by barium oxide production at the surface. Both water-vapour and electron beam damages indicate that, with a little oxygen loss, the Ba atoms can readily diffuse out of the material, destroying superconductivity. Electron beam damage at 400 KV and 1 MV was also reported by Y. Mtsui et al. [24]. They observed the formation of regions with cubic structure and a = 0.41 nm lattice parameter, and proposed a disordered state of BaO and Y layers to explain their results.

Diffraction spots from a similar body centered cubic lattice with a = 0.385 nm were observed by H. W. Zandbergen et al. [16]. They explain this observation by noting that a random filling of the centers of the copper-oxygen framework by Y and Ba atoms would produce an body centered cuic unit cell of metal atoms.

Dislocations

Very few reports exist in the litterature about the observation of dislocations in these structures. However, the existance of these defects is a very important fact in technological applications of high Tc superconducting ceramics, since they indicate the possibility for the material to accept some degree of plastic deformation.

Long and straight dislocations (i.e. contained on the plane of the surface) are usually observed in TEM samples; specially in grains with a c-axis perpendicular to the surface [25,26]. It is well known that this type of dislocations are currently observed in TEM samples; they result from the introduction of some plastic deformation during the preparation of the thin regions needed for observation.

According to M. Suenaga et al. [27], some dislocations are formed in the material by large internal stresses generated by the introduction of oxygen at low temperatures (400°C), since no such defects were observed in samples heated at high temperatures (850°C).

S. Ikeda et al. (a) [17] determined the Burgers vector of some dislocations, by the usual contrast extintion technique, as b = [100] and [010]. They conclude that some dislocations are introduced at the grain boundaries during the tetragonal to orthorhombic transformation. Other are likely to be formed during the preparation of the TEM samples and other are introduced during the sintering process. Finally, as mentioned before, loops of dislocations have also been observed [12,16].

Stacking faults

These defects are currently observed in 1:2:3 compounds. Although no definitive characterization of them has been made, all authors concide however in placing these defects on (001) basal planes, but their exact nature is still controversial.

Extra layers of yttrium alternating with barium layers have been proposed [28], additional layers of Y or Cu have been mentioned [24] and double layers of CuO reported [7,16,17]. In addition, a heavily faulted structure on (001) planes, was observed in electron beam damaged TEM samples [9].

J. Tafto et al. [6] have determined the displacement vector of the faults. They report a displacement vector that switches from $(a/2,0,c/6)$ to $(b/2,0,c/6)$ since the stacking faults are observed to continue through the twin boundaries. They propose two possible models for the stacking faults, consistent with the determined displacement vector. The first model is equivalent to taking out one layer of BaO in the basal plane and displacing the whole crystal on one side of the fault by $[a/2,0,0]$ (or $[b/a,0,0]$ in the neighbouring twinned related region). The second model is constructed by removing an yttrium layer and displacing the crystal as before. The first model leads to the formation of Cu(1)-Cu(2) neighbouring layers with Cu atoms in a fourfold coordination, while the second model leads to Cu(2)-Cu(2) neighbouring layers with Cu atoms octahedrally coordinated as in $La_{2-x}Ba_xCuO_4$ [29].

Y. Hirotsu et al. [3] have also determined the displacement vector of the faults as being $(a/2,b/2,c/6)$. They propose a model which is equivalent to the introduction of an extra BaO layer close to an existing one (i.e. without any CuO layer in between) and displacing the whole crystal on one side of the fault

by a vector (a/2,b/2,0).
Stacking faults on (110) planes seems also to exist in these compounds, which could be associated to a mismatch of Y and BaO layers [7]. Streaking on [110] directions has been recently reported [20], which could be associated to these faults.
Finally, it is interesting to mention here the work by J. P. Zhang et al. [31] who have observed stacking faults in the $Gd_1Ba_2Cu_3O_{7-y}$ compound. They conclude that Cu and Gd/Ba sites can be interchanged on (001) planes. This result indicates a smaller ordering energy in the 1:2:3 compound with Gd, than in the 1:2:3 compound with Y, where no such substitution is observed. This conclusion is in good agreement with studies of the degradation of Gd-1:2:3 compounds, where a more rapid degradation than in Y-1:2:3 compounds was observed, possibly due to the higher movility of atoms in the structure even at room temperature [32].

DEFECTS ASSOCIATED TO THE OXYGEN-VACANCY SUBLATTICE

The second group of structural defects is that associated to the ordering characteristics of the oxygen-vacancy sublattice, on the Cu(1) type basal planes of the orthorhombic structure of 1:2:3 compounds. Within this group of defects, twins and superstructures are the more relevant and direct relationships with the superconducting transition have been proposed; however, the nature of these relationships is still no clear at the present time. These defects will be described below.

Twins

From the earlier reports on the existence of twins in the orthorhombic structure (ex. [28]), a large ammount of observations of these defects have been reported and are summarized in the following discussion.
S. Sueno et al. [33] have derived the possible modes of twinning of the orthorhombic structure, by considering the rotation axes and mirror planes of the tetragonal phase as qualified to become twin operations. They conclude that only 3 types of independent twin orientations are possible and are obtained by the following rotations with respect to the original crystal: 1) $\pm$ 90° in [110] direction with respect to the c-axis (rotation twins or 90° twins), 2) $\pm$ 180° in [110] direction and 3) $\pm$ 180° in [$\bar{1}$10] direction, the two latter are called reflection twins and the rotation is made with respect to the corresponding perpendicular direction (i.e. [$\bar{1}$10] and [110] respectively). Type 1 has its twin boundary on the (001) basal plane and type 2 and 3 have their twin boundaries on {110} planes. The three types of twin orientations involve a switching of the O-Cu-O and V-Cu-V chains (V = vacancy) across the corresponding twin boundary, producing an interchange of a and b axes between the crystal and the twin related region. Therefore, an antiphase boundary is observed on the oxygen-vacancy sublattice, coinciding with the twinning plane [34] and producing δ-type fringes when observed in the electron microscope [18,16].
Only the first two types of orientations were observed by S. Sueno et al. in X-ray studies of single crystals. However, the existance of the three types of twins has already been stablished by other authors, in single crystals and polycrystals, by X-ray and TEM techniques (ex. [25-28]). It is interesting to note that single crystals containing only one of the three different twin orientations and showing high Tc (>90K) in all cases, have been prepared by H. You et al. [38]. This result indicates that there is no clear need for a given type of twin in order to get high Tc superconductivity.
EELS measurements revealed a high stability of the oxygen content in regions were twins were present, as compared to the relative fast loss of oxygen measured in twin-free zones [39].

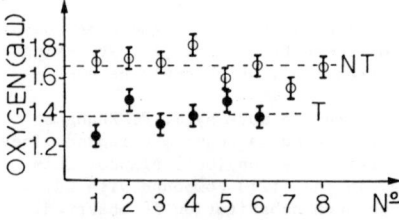 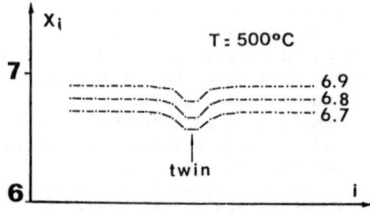

Fig. 2 Relative oxygen content as a function of the number of EELS measurements for twinned (T) and untwinned (NT) regions.

Fig. 3 Calculated oxygen concentration profiles across the twin boundary, for global oxygen contents of 6.7, 6.8 and 6.9 at 500°C.

This result is in good agreement with in-situ heating TEM studies of the orthorhombic to tetragonal transformation, were a higher stability of the oxygen content in twinned regions was deduced [19]. Therefore, it seems that the oxygen atoms removed by the electron beam are accomodated easily at the twin boundaries and remain there in stable positions until higher temperatures are reached. This conclusion is consistent with the lower room-temperature value of the relative oxygen content measured by EELS in twinned regions, than the same value measured in untwinned ones [39], as shown in Fig. 2. This result is in good agreement with the measured decrease of the orthorhombic deformation in heavily twinned regions as compared with that measured in large orthorhombic domains [16].

A lattice gas model for the oxygen positions in the Cu(1) type basal plane [40] was used to calculate the oxygen concentration profile across the twin boundary in mean field approximation [39]. The results for $O_{6.7}$, $O_{6.5}$ and $O_{6.9}$ with T = 500°C are shown in Fig. 3. In all cases a decrease in the oxygen content at the twin boundary is observed, which is consistent with the above mentionned EELS measurements and could explain, at least partially, the observed difference in oxygen content between the ideal stoichiometry with O_7 and well oxigenated samples with $O_{6.95}$.

It is interesting to note that, by considering an additional oxygen vacancy per unit cell at the twin boundary (i.e. O_6) and the stoichiometric oxygen content in the bulk (i.e. O_7), a simple calculation shows that a separation between twin boundaries of 103Å is required to obtain a global oxygen content of $O_{6.95}$. This is in good agreement with the measured spacement between microtwins in well oxygenated samples [30,41].

A model for the formation of twins on (110) planes has been proposed recently [39], and is summarized in Fig. 4. This considers the formation of a small orthorhombic region within a single crystalline powder particle of tetragonal phase with an oxygen content of about $O_{6.4}$, which corresponds to that measured at the transformation [42,44]. The growth of such an orthorhombic region introduces a certain ammount of compresional and tensional stresses in the tetragonal matrix, which can be released by the growing of a second orthorhombic region in a 90° configuration. Further growing and proliferation of this self-accommodating pair of twin related regions follows from an autocatalytical process that leads to a decrease of the elastic stresses.

The actual shape and critical size of the first formed orthorhombic domain (i.e. the maximum size before the second domain is formed), should depend on the distribution of stresses, oxygen content and ordering of the oxygen-vacancy sublattice [45,46], associated to the powder particle. Nevertheless, this critical size is expected to be smaller than the final size of the twin domains usually observed in TEM (100-1000Å). For instance, a simple calculation

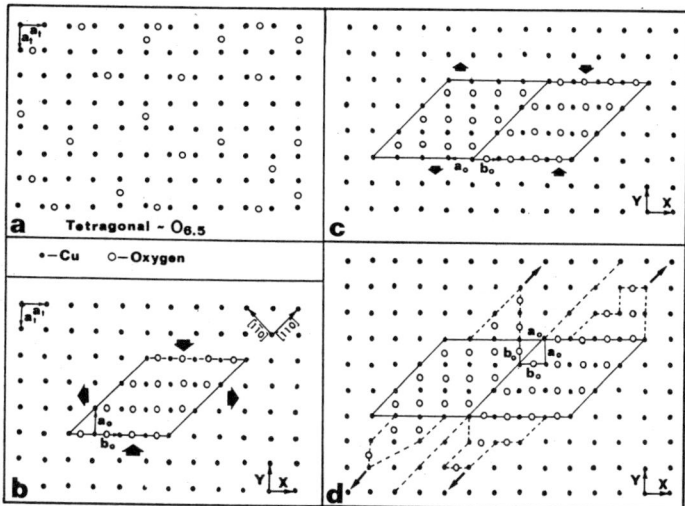

Fig. 4 Model for the formation of (110) twins. (a) Oxygen-rich tetragonal matrix, (b) Formation of a small orthorhombic region, compresional and tensional stresses appear (arrowed), (c) A neighbouring orthorhombic region is formed in a 90° configuration; this self-accomodating pair releases most of the stresses, (d) Auto-catalytical growing of the twin-related self-accommodating pair of orthorhombic regions.

shows that 2Å displacement of atoms in the matrix is obtained when a 200Å uncompensated orthorhombic region is formed on it. This displacement is about half of the lattice parameter of the tetragonal cell.

Interesting enough is the prediction of this model for the formation of untwinned orthorhombic single crystals. This situation might appear when the growing of the first formed orthorhombic domain occurs near the surface of a stress-free tetragonal single crystal (grain). In this case there is no need for stress compensation since the deformation introduced can be relaxed at the surface.

It must be noted that in Fig. 4 one could expect the Cu sublattice to be rhombohedrally distorted at the twin boundary, because of the non-equilibrium distribution of neighbouring oxygen atoms for these Cu sites. Distorted configurations of this type could help to explain, at least partially, the differences obtained between the calculated and observed values for the splitting of the characteristic twin spots in TEM diffraction patterns (i.e. the so-called twin obliquity), which has been associated with local strains [16,20,30,47] and oxygen deficiency [48]. Also, the missalignment of the C-axis ($\frac{1}{2}$°) at the twin boundary, measured by P. C. Gibbons et al. [49], could be related with this type of distorsions.

It should also be noted that the oxygen content at the twin boundaries shown in Fig. 4-C is O_7. However, as explained before, it seems that a lower oxygen content exists in these regions. The extra vacancies distribution at the twin boundary region is not expected to be a simple one; since for instance, twin boundaries 5 to 50Å wide have been reported [7,20,47,49]. A possible configuration for introducing extra vacancies at the twin boundary ($O_{6.5}$) has been mentioned by J. B. Goodenough et al. [46]. More complicated arrangements can be found, which could lead to a lower oxygen content at the twin boundaries.

The orthorhombic to tetragonal trnasformation has been observed in in-situ heating and cooling TEM experiments [16,19,37,50] that have centainly apported interesting clues to the understanding of the kinetics of formation of twinned structures and the different superstructures observed in these compounds. The twin contrast disappeared gradually as the temperature of the sample rised. At temperatures slighly higher than the orthorhombic to tetragonal transition, the sample was observed to show a "mottled" structure [37], which could be associated to diffusion processes in the material at these high temperatures. A further lowering of the temperature of the sample induced the tetragonal to orthorhombic transition with the nucleation of new twinned regions. Repeated cycling produces narrower twin domains, which sometimes appeared interpenetrated with a second twin system on the equivalent {110} type plane at 90°, then producing a "checker-board" like contrast [37]. No reversibility of the orthorhombic to tetragonal transformation was observed after some transformation cycles, because of the gradual loss of oxygen during cycling.

Superstructures

Several types of superstructures, associated with the oxygen-vacancy sublattice, have been identified in the 1:2:3 compounds.

Diffuse streaking in the [100] direction and extra weak spots at $[\frac{1}{2},0,0]$ positions in TEM diffraction patterns are currently observed in samples with an oxygen deficient orthorhombic structure [16,37,41,51,52]. Obviously, in diffraction patterns obtained from twinned regions diffuse streaks in the [010] direction and extra weak spots in the $[0,\frac{1}{2},0]$ positions are also observed. The extra weak spots have been associated with a short range order, produced by a unit cell having extra alternated rows of vacancies along the b_o direction and $2a_o$, b_o, c_o lattice parameters. The oxygen content of this ordered unit celd is $O_{6.5}$ although M. A. Alario-Franco et al [45], observed the existence of the extra weak spots in the range of $O_{6.3}<O_y<O_{6.6}$. In any case a highly oxygen deficient orthorhombic structure is involved. For oxygen contents O_y in the range of $O_{6.5}<O_y<O_7$, the extra weak spots become more and more delocalized as the oxygen content increases, resulting in diffuse streaks in the [1,0,0] direction [16]. These streaks are associated with some random distribution of the extra V-Cu-V and the normal O-Cu-O rows parallel to the b_o parameter.

It must be noted that diffuse streaking along the [001] direction has also been observed for oxygen contents as high as $O_{6.6}$ [52]. This fact has been used to propose more complicated arrangements of oxygen and vacancies that explains diffuse streaking. However, it should be clear that a random distribution of V-Cu-V and O-Cu-O rows along the b_o parameter, as proposed by H. W. Zandbergen et al. [16], is enough to explain streaking for a large range of values of oxygen content.

More complicated effects in TEM diffraction patterns, like diffuse streaking in the [001] direction, have been observed. M. A. Alario-Franco et al. [45] observed this [001]-streaking for oxygen contents in the range $6.85<O_y<7.0$. They associated this effect to an uncorrelated arrangement along the [001] direction of the long range order of oxygen and vacancies in the (001) basal plane. They found this result to suggest a unit cell with parameters $2\sqrt{2}\ a_c$, $2\sqrt{2}\ a_c$, c with a_c equal to some mean lattice parameter associated to the cubic Cu-O framework. H. W. Zandbergen et al. [16] also explain the streaking in the [001] direction by considering a random arrangement of O-Cu-O and V-Cu-V rows in different (001) planes. For oxygen contents in the range of $6.0<O_y<6.2$, M. A. Alario-Franco et al. [45] observed again diffuse streaking in the [001] direction, consistent with an ordered unit cell having the same parameters as for the oxygen-rich orthorhombic structure (i.e. $2\sqrt{2}\ a_c$, $2\sqrt{2}\ a_c$ and c).

J. B. Goodenough et al [46] have constructed a preliminary phase diagram that includes the stability ranges for the different superstructures mentionned above. However, more work is needed in order to fully understand the nature of oxygen-vacancies order in these structures.

Highly interesting is the work by R. J. Cava et al. [53], who measured the dependence of Tc as a function of the oxygen content. they found a stepwise behavior of Tc, with values of about 90K in the $O_{6.96}$-$O_{6.9}$ range to about 60K in the $O_{6.72}$-$O_{6.52}$ range and of about 25K below $O_{6.3}$. The occurrence of this stepwise behavior of Tc has been associated with the formation of superstructures in the orthorhombic 1:2:3 phase. In particular the superstructure proposed by H. W. Zandbergen et al. [16] is believed to be responsible for the 60K critical temperature [54]. The formation of this superstructure has been investigated by M. Inoue et al. [54]. By considering nearest and two nearest-neighbor interactions. They calculated the phase diagrams associated to the formation of different superstructures with oxygen contents close to O_7 and $O_{6.5}$. Their results allow to identify the superstructure of Zandbergen et al. as the more favourable energetically for an $O_{6.5}$ content.

REFERENCES

1. A. Manthiram et al., J. Am. Chem. Soc., 109, 6667 (1987).
2. J. D. Jorgensen et al., Phys. Rev. B36, 3608 (1987).
3. W. J. Weber et al., MRS-1987 Fall Meeting (AA7.6), Boston, Mass., USA.
4. J. D. Jorgensen. XVIII Int. Conf. on Low Temp. Phys. (1987). To be published in Jpn. J. Appl. Phys., and references therein.
5. R. A. Camps et al., Nature 329, Sept. 1987.
6. J. Tafto et al. Submitted to Appl. Phys. Lett.
7. B. Raveau et al., Proc. IX Winter Meet. Low Temp. Phys., (this volume).
8. J. P. Zhang et al., MRS-1987 Fall Meeting (AA7.101). Boston, Mass., USA.
9. M. A. Kirk et al., MRS-1987 Fall Meeting (AA6.6). Boston, Mass., USA.
10. G. J. Clark et al., MRS-1987 Fall Meeting (AA3.12). Boston, Mass., USA.
11. H. W. Zandbergen et al., MRS-1987 Fall Meeting (AA4.92). Boston, Mass.,USA
12. S. Nakahara et al., MRS-1987 Fall Meeting (AA4.99). Boston, Mass., USA.
13. T. H. Tiefel et al., MRS-1987 Fall Meeting (AA4.40). Boston, Mass., USA.
14. L. A. Tietz and C. B. Carter, MRS-1987 Fall Meet. (AA7.30). Boston, Mass.
15. L. A. Tietz et al., MRS-1987 Fall Meeting (AA7.31). Boston, Mass., USA.
16. H. W. Zandbergen et al., Phys. Stat. Sol. (a) 103, 45 (1987).
17. S. Ikeda et al., MRS-1987 Fall Meeting (AA6.7), Boston, Mass., USA.
18. W. T. Lin et al., MRS-1987 Fall Meeting (AA7.92), Boston, Mass., USA.
19. S. Ikeda et al., J. Electron Microsc., 36, 237 (1987).
20. M. J. Yacamán et al., IX Winter Meet. Low Temp. Phys., (this volume).
21. W. H. Philipp et al., MRS-1987 Fall Meeting (AA7.1), Boston, Mass., USA.
22. G. J. Fisanick et al., IX Winter Meet. Low Temp. Phys., (this volume).
23. B. G. Hide et al., High Temp. Superc. in Nature (Vol. 327, June 1987).
24. Y. Matsui et al., Jpn. J. of Appl. Phys., 26 L1183-1185 (1987).
25. S. Ikeda et al., Jpn. J. of Appl. Phys., 26, L729-731 (1987).
26. D. Ríos-Jara and A. Huanosta, unpublished work.
27. M. Suenaga et al., MRS-1987 Fall Meeting (AA4.82), Boston, Mass., USA.
28. A. Ourmazd et al., High Temp. Superc. in Nature (Vol. 327, May 1987).
29. J. D. Jorgensen et al., Phys. Rev. Lett. 58, 1024 (1987).
30 Y. Hirotsu et al., Jpn. J. of Appl. Phys. 26, L1168-1171 (1987).
31. J. P. Zhang et al., MRS-1987 Fall Meeting (AA4.94), Boston, Mass., USA.
32. T. Akachi et al., to be published in J. of Phys. C.
33. S. Sueno et al., Jpn. J. of appl. Phys., 26, L842-844 (1987).
34. C. H. Chen et al., Phys. Rev., B35, 8767 (1987).
35. S. X. Dou et al., Appl. Phys. Lett., 51, 535 (1987).
36. A. Ono and T. Tanaka, Jpn. J. of Appl. Phys., 26, L825 (1987).
37. G. Van Tendeloo and S. Amelinckx, Phys. Stat. Sol. (a) 103, K1(1987).
38. H. You et al., MRS-1987 Fall Meeting (AA3.3), Boston, Mas., USA.
39. D. Ríos-Jara et al., MRS-1987 Fall Meeting (AA6.10), Boston, Mass., USA.
40. C. Varea and A. Robledo, Submitted to Rapid Comm. in High Temp. Superc.

41. K. Hiraga et al., J. Electron Microsc., $\underline{36}$, 261-269 (1987).
42. D. G. Hinks et al., MRS-1987 Fall Meeting (AA1.3), Boston, Mass., USA.
43. P. Grant. IX Winter Meet. Low Temp. Phys. (this volume).
44. E. D. Specht et al., Submitted to Phys. Rev. B.
45. M. A. Alario-Franco et al., MRS-1987 Fall Meeting (AA2.2.), Boston, Mass.
46. J. B. Goodenough and A. Manthiram, IX Winter Meet. Low Temp. Phys. (this volume).
47. J. G. Pérez-Ramírez et al., MRS-1987 Fall Meet. (AA4.105), Boston, Mass.
48. S. Ijima et al., Jpn. J. of Appl. Phys. $\underline{26}$, L1478-1481 (1987).
49. P. C. Gibbons et al., MRS-1987 Fall Meet. (AA7.100), Boston, Mass., USA.
50. K. Sasaki et al., J. Electron Microsc., $\underline{36}$, 232-236 (1987).
51. M. Tanaka et al., Jpn. J. of Appl. Phys. $\underline{26}$, L1237-1239 (1987).
52. Y. Matsui, Jpn. J. of Appl. Phys., $\underline{26}$, L2021-2022 (1987).
53. R. J. Cava et al., Phys. Rev. B$\underline{36}$, 5719 (1987).
54. M. Inoue et al., Jpn. J. of appl. Phys. $\underline{26}$, L2015-2017 (1987).

ACKNOWLEDGMENTS

Financial support from the Programa Universitario de Investigación en Superconductores de Alta Tc is fully acknowledged.

CRYSTALLINE STRUCTURE IN THE Cu BASAL PLANE AND MICROSTRUCTURE IN SUPERCONDUCTING $YBa_2Cu_3O_{7-y}$

C. VAREA[1] AND A. ROBLEDO[2]
1 Facultad de Química, Universidad Nacional Autónoma de México, México 04510, D.F., México.
2 Instituto de Física, Universidad Nacional Autónoma de México, Apdo. Postal 20-364, México 01000, D.F., México.

ABSTRACT.

The crystal structure of $YBa_2Cu_3O_{7-y}$ is discussed by means of a layered oxygen lattice gas model in equilibrium with an external source of O_2. The model is capable of reproducing quantitatively the tetragonal to orthorhombic transition at high temperature. It also suggests the possible phase behavior at low temperature and indicates how twin formation, and proliferation, is driven by elastic strains.

The detailed structure of the high T_c superconductor $YBa_2Cu_3O_{7-y}$ has been recently obtained from neutron diffraction measurements[1] It has also become clear, via numerous determinations of oxygen content, that the framework of cations and six oxygens readily absorbs or desorbs oxygen ($0<y<1$) in response to external conditions. The mechanism behind the unusually high critical superconducting temperature is not yet fully understood. However, it is known that its value appears to be directly related to the number of (hole-like) charge carriers produced by oxygen dopping[2,3]. The ordering of oxygen atoms, into alternating rows of oxygen and vacant sites on the (two-dimensional) Cu basal planes, introduces anisotropic strains responsible for the observed tetragonal (T) to orthorhombic (O) structural transition. The small size of the measured coherence lengths[4] suggests that the superconductivity order parameter is likely to be coupled to the crystalline order[5]. Nevertheless, important aspects of microstructural behavior and its relationship to superconductivity in this material remain unclear.

The high-temperature structural transformation is now well understood[5,6,7] in terms of simple two-dimensional order-disorder models. To obtain a bona fide description of the concomitant transitions it is necessary that the order-disorder model takes into account explicitly the variation of lattice spacings with temperature and with oxygen content and distribution. This is accomplished through the consideration of a free energy term describing the elastic deformations of the crystal[6]. Absorption isobars, similar to those measured when samples are cooled under fixed oxygen partial pressures, can be obtained, too, if the model is set in equilibrium with a compressible gas containing molecular oxygen[6]. Samples with high oxygen content ($y \leq 0.1$) are always densely twinned, typically along the (011) (or ($0\bar{1}1$)) directions. These crystal imperfections are formed during the T-O transformation process which generates two out-of-register O nucleation domains and disrupts the order along the oxygen chains (See Fig. 1a). Another way of breaking these chains has been suggested in Ref. 8, where the signs of three different oxygen stoichiometric regions, with different unit cells, are identified. Here the chains appear disrupted for low oxygen content, $1 > y > 0.8$. (See Fig. 1b).

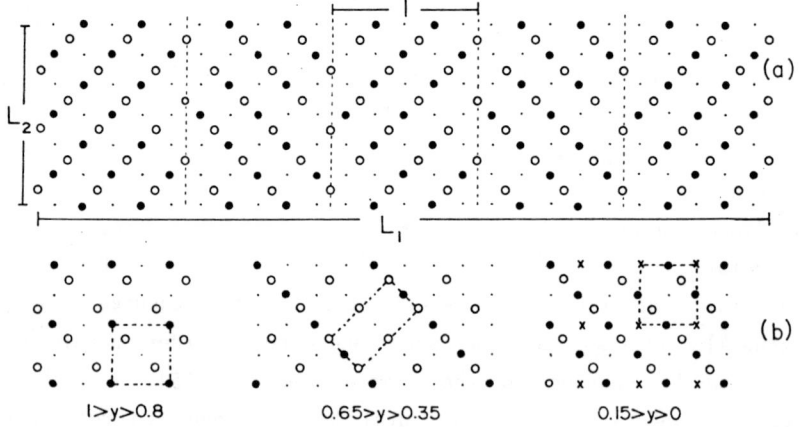

Fig. 1. Broken Cu-O chain structures in $YBa_2Cu_3O_{7-y}$ basal planes. a) Twinned structure of Cu (o) and O (•) chains with vacancies (.). The domains have length ℓ and L_1 and L_2 are dimensions of the crystal. b) Unit cells proposed in Ref. 8 for the ordering of oxygen in the Cu basal planes. The crosses X represent nearly empty sites.

The role of the elastic energy of the crystal in the formation of twin domains, when the structural transformation takes place at high temperatures, has been discussed elsewhere[9] Here we briefly recall the formulation of our model and describe its associated phase diagrams. We then indicate how these defects might proliferate through thermal manipulation at considerably lower temperatures leading to an even denser twinned microstructure.

We consider a rectangular (oxygen) lattice gas consisting of two interwoven sublattices with oxygen average occupancies n and m, respectively. Oxygen atoms are attracted to each site by a potential V and feel a nearest-neighbor repulsion U. The mean-field grand potential (per cell) due to these interactions is

$$\Omega_o^{MF} = kT[(n)\ln(n)+(m)(\ln n)(m)+(1-n)\ln(1-n) + (1-m)\ln(1-m)]$$
$$+ U(n)(m) - (\mu+v)(n+m), \qquad (1)$$

where μ is the oxygen chemical potential. Elastic deformation and thermal expansion are written in the small strain limit as

$$F_{el} = 1/2 \; \Sigma_{ik\ell m}\lambda_{ik\ell m}u_{ik}u_{\ell m} - 1/3 \; \Sigma_{ik}\alpha_{ik}u_{ik}(T-T_o), \qquad (2)$$

where u_{ik} is a component of the strain tensor, and the undeformed reference state is at a temperature T_o where the equilibrium state has tetragonal symmetry. With this choice the elastic modulus tensor λ has six nonzero components and the thermal expansion tensor α is diagonal with two different components. Crystal expansion due to oxygen absorption is taken into account in the term

$$F_{exp} = -\varepsilon_a [u_{aa}(n-p_o) + u_{bb}(m-p_o)] - \varepsilon_c u_{cc}(p-p_o), \qquad (3)$$

where the subindexes a, b and c refer, respectively, to the directions of oxygen vacant rows, Cu-O chains, and the perpendicular to the basal Cu-O plane. ε_a and ε_c are compositional expansion coefficients, $p = (n+m)/2$ and p_o is the oxygen average content at the reference state, $7-y = 6 + 2p$. Because experiments are often performed at constant O_2 pressure, the labile oxygen atoms change their concentration with temperature. We therefore set the oxygen lattice gas in equilibrium with a reservoir of O_2 (a compressible two-component gas of O_2 concentration x). The O_2 chemical potential μ' at constant pressure P is given by $\mu' = kT\,[\ln(xP/kT)] + \omega P$, where ω is an average molecular volume that we set equal to one. $\mu' = 4\mu$ since there are two sites per cell for labile oxygen atoms and two atoms per molecule. The Euler-Lagrange equations that result from minimizing $\Omega_{tot} = \Omega_o + F_{e\ell} + F_{exp}$ with respect to n and m are identical[6] to those corresponding to a two-dimensional lattice gas (without elastic $F_{e\ell}$ and F_{exp} terms) with first U' and second U" nearest neighbor interactions and with an effective temperature-dependent external potential $V' = V + v(T/T_o)$.

When first neighbor repulsions U'>0 dominate over second neighbor interactions U" a second order order-disorder transition produces the above-mentioned oxygen chains. To proceed with our comparison with experiment, we fit the data of lattice parameters and oxygen occupancies in Ref. 1 to the linear relationships that result from minimizing Ω_{tot} with respect to the strain tensor components[6] This yields

$$u_{aa} = 0.0062\,\Delta n - 0.0155\,\Delta m + 0.0128\,\Delta t,$$

$$u_{bb} = 0.0062\,\Delta m - 0.0155\,\Delta n + 0.0128\,\Delta t,$$

$$u_{cc} = -0.0335\,\Delta p + 0.1189\,\Delta t, \qquad (4)$$

where $\Delta n = n-p_o$, $\Delta m = m-p_o$ and $\Delta t = (T-T_o)/T_o$. T_o and p_o are the temperature and concentration at the T-O transition at 100% O_2 atmosphere. This leaves us with three independent parameters that we chose to be the ratio $R = U"/U'$ and the two effective site potentials V and v. These constants were again determined from neutron scattering data[1] at 100% O_2. A good fit to them in the range 0°C to 900°C is shown in Fig. 2 where we show also our calculated

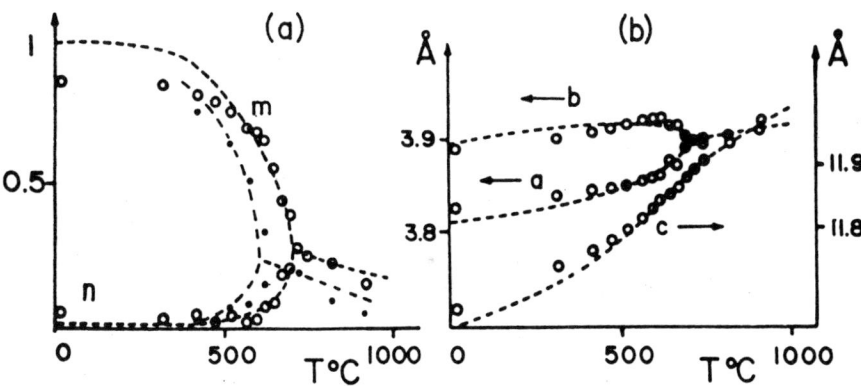

Fig. 2. Temperature dependence of a) Sublattice oxygen site occupancies in the basal planes of $YBa_2Cu_3O_{7-y}$, open circles for 100% oxygen and points for 2% oxygen. b) a-, b- and c- lattice parameters for samples heated in 100% oxygen. The dashed lines are from the theory, and the data is taken from Ref. 1.

isobar occupancies at 2% O_2. There R = -0.01 with first neighbor repulsions and second neighbor effective attractions. We have found that we can produce good fittings to the oxygen occupancies with different sets R, V, v, with R small but of either sign. This is unfortunate because the phase behavior of the model for R>0, is quite different from that for R<0 at low temperatures [6,10]

When R<0 second neighbor attractions reinforce the T-O transition which becomes first order below a tricritical temperature T_t (with $T_t \to 0$, $p_t \to 0.31$ when R→0). On cooling below T_t an orthorhombic sample with $p > p_t$, regions of the tetragonal phase will grow coherently from the orthorhombic phase, and this process induces additional strains[11] that can be reinterpreted as renormalized second neighbor repulsions with effective strength $|U'''| < |U''|$. Since T_t is a decreasing function of R, the tricritical temperature for coherent growth is depressed and there is an additional region in (T,p)-space for the metastable orthorhombic phase (see Fig.3). This phase accumulates elastic energy some of which may be released

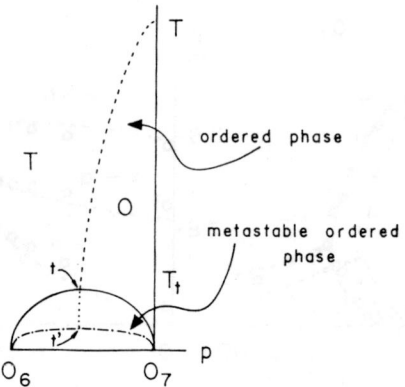

Fig. 3. Phase diagram for the T-O transition with R<0. The solid lines represent incoherent equilibrium, dashed-dotted lines represent coherent metastable equilibrium, and dashed and dotted lines second order transitions. t and t' are tricritical points.

by twin-pair formation. To promote the proliferation of these defects, orthorhombic samples can be repeatedly taken in and out of the metastable region shown in Fig. 3 by means of temperature cycling. The attainable twin average density, formed with domains of width ℓ, can be estimated by minimization of the total excess (strain plus interfacial) energy that is left in the system after the release of strains along the direction that corresponds to the length L_1 shown in Fig. 1a[9]. The elastic energy under consideration is proportional to the number of oxygen chains crossing the twin boundary, i.e. ℓ/L_2 (see Fig. 1a), to an elastic constant λ, to the square of the strain u, and to the volume of the sample. This is

$$F_{e\ell} = c\lambda u^2 (\ell/L_2)(L_1 L_2 L_3). \qquad (5)$$

The interfacial energy is proportional to the number of twins L_1/ℓ, to the twin boundary energy σ, and to the twin boundary area $L_2 L_3$,

$$E_{int} = \sigma(L_1/\ell) L_2 L_3 \qquad (6)$$

We obtain[12].

$$\ell = \left(\frac{\sigma L_2}{c\lambda u^2} \right)^{1/2}. \tag{7}$$

Close to the tricritical point the above expression for ℓ may not give meaningful answers, because, there, the thickness ξ of the twin boundary becomes larger than the size of the domain ℓ. We have

$$\ell/\xi \sim t^{[(d+1)\nu-2\beta]/2}. \tag{8}$$

To obtain the expression above we have used the scaling forms, $\sigma \sim t^{(d-1)\nu}$, $u \sim t^\beta$ and $\xi \sim t^{-\nu}$, with $t = (T-T_t)/T_t$, and d dimensionality. For $d = 2$, $\nu \simeq 1.05$ and $\beta \simeq 0.58$,[13] and we have $\ell/\xi \sim t^{0.68}$.

When R>0 second neighbor repulsions frustrate the ordering induced by first neighbor interactions. This case has been studied in detail through Monte Carlo simulations[10]. A highly degenerate state of broken chains appears always at T = 0K, and for R<1/2 it becomes the lowest energy state for the composition interval 0.38<y<0.706. It consists mainly, in our model, of Cu^{++} clusters ($O_{6.5}$) in which two Cu atoms share one basal-plane oxygen. Because this state is formed by rows of alternating oxygen and oxygen-vacancy occupations separated by unoccupied rows, its associated order-disorder temperature is 0K (that of a one-dimensional system). Further neighbor interactions can stabilize this phase at nonzero temperature, and it may correspond to that reported in Ref. 8 to occur within the range 1>y>0.8. Monte Carlo simulations[10] that represent slow cooling from the disordered phase, find the above-described arrangement (see Fig. 4) a metastable phase with frozen-in domain walls at T>0K. For this composition interval, 0.38<y<0.706, the equilibrium picture is that of a second order transition from the disordered phase into a particular type of ordered phase, shown in Fig. 4, that can be obtained from the usual orthorhombic-CuO chains phase by taking proliferation of twins to its close packing limit. At higher oxygen concentrations, in the interval 0<y<0.38, the usual order-disorder transition induced by 1st. neighbor repulsions and which forms the usual chains, takes place provided

$R<1/2$. This is of the second order always and its concentration threshold at $T = 0$ is $y = 0.38$ irrespective of R. See fig. 4.

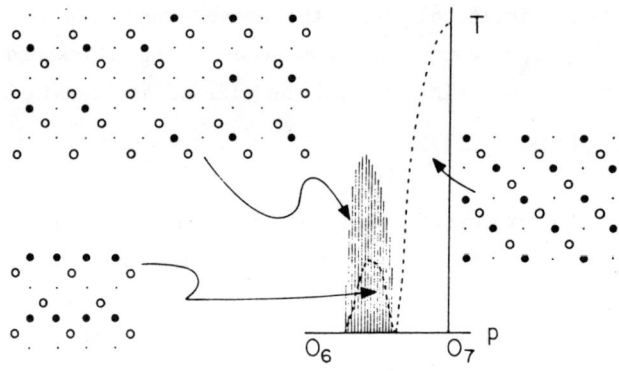

Fig. 4. Phase diagram for oxygen ordering with $R>0$. The insets represent the specific orderings occurring in different regions (oxygen (●), vacancy (.)). In the shaded region appears the metastable phase mentioned in the text.

We have reviewed a model for the ordering of oxygen and the structural transition in $YBa_2Cu_3O_{7-y}$. The consideration of the elastic deformations of the crystal have allowed us to obtain quantitative agreement with experiment. In the two cases discussed, $R<0$ and $R>0$, we found some of the characteristic features observed in this material, amongst them metastability and twin formation.

References.
1. J.D. Jorgensen, M.A. Beno, D.G. Hinks, L. Soderholm, K.J. Volin, R.L. Hitterman, J.D. Grace, and I.K. Schuller, Phys. Rev. B36, 3608 (1987).
2. R.J. Cava, B. Batlogg, C.H. Chen, E.A. Rietman, S.M. Zahurak, and D. Werder, Phys. Rev. B36, 5719 (1987).
3. E. Takayama-Muromachi, Y. Uchida, M. Ishii, T. Tanaka, and K. Kato, Jpn. J. Appl. Phys. 26, L1156 (1987).

4. R.J. Batlogg, R.B. van Dover, D.W. Murphy, S. Sunshine, T. Siegrist, J.P. Remeika, E.A. Rietman, S. Zahurak, and G.P. Espinosa, Phys. Rev. Lett. $\underline{58}$, 1676 (1987).
5. C. Varea and A. Robledo, in Novel Superconductivity, S.A. Wolf and V.Z. Kresin, eds. (Plenum Press, N.Y. 1987) pp 1033-1039; A. Robledo and C. Varea, Phys. Rev. $\underline{B37}$, 631 (1988).
6. C. Varea and A. Robledo, Proceedings of the MRS Symposium on High-Temperature Superconductors, Boston, Mass., Nov. 30-Dec. 1, 1987; C. Varea and A. Robledo, Mod. Phys. Lett. $\underline{B1}$ Nos. 11-12 (to be published; C. Varea and A. Robledo, Rev. Mex. Fis. $\underline{33}$, 311 (1987).
7. D. de Fontaine, L.T. Wille, and S.C. Moss, Phys. Rev. $\underline{B36}$, 5709 (1987); H. Bakker, D.O. Welch, and O.W. Lazareth, Sol. St. Comm. $\underline{64}$, 237 (1987); A.G. Khachaturyan and J.W. Morris, Phys. Rev. Lett. $\underline{57}$, 2776 (1987); K. Nakamura and K. Ogawa, Jpn. J. Appl. Phys. (to be published).
8. C. Chaillout, M.A. Alario-Franco, J.J. Capponi, J. Chenavas, J.L. Hodeau and M. Marezio, Phys. Rev. $\underline{B36}$, 7118 (1987).
9. D. Rios-Jara, C. Varea, A. Robledo, A. Huanosta, J.M. Dominguez, J. Omaña, T. Akachi, and R. Escudero, Proceedings of the MRS Symposium on High-Temperature Superconductors, Boston, Mass. Nov. 30-Dec. 1, 1987.
10. K. Binder and D.P. Landau, Phys. Rev. $\underline{B21}$, 1941 (1980).
11. S.M. Allen and J.W. Cahn, Acta Met. $\underline{23}$, 1017 (1975).
12. G.R. Barsch, B. Horovitz, and J.A. Krumhansl, Phys. Rev. Lett. $\underline{59}$, 1251 (1987).
13. D.P. Landau, Phys. Rev. Lett. $\underline{28}$, 449 (1972).

ON THE STRUCTURAL CHARACTERISTICS OF THE HIGH-Tc SUPERCONDUCTORS $YBa_2Cu_3O_{7-x}$ AND $HoBa_2Cu_3O_{7-x}$

R. PEREZ, J.G. PEREZ-RAMIREZ, J. REYES-GASGA, U. OSEGUERA, J. CRUZ AND M. JOSE-YACAMAN

INSTITUTO DE FISICA-UNAM
Apdo. Postal 20-364
01000 MEXICO, D.F.

ABSTRACT

A study of the superconducting phases of $YBa_2Cu_3O_{7-x}$ and $HoBa_2Cu_3O_{7-x}$ is carried out. The temperature dependence of the structure in these compounds has been accounted by following in-situ heating experiments in a TEM. The structural phase transition in both compounds is always characterized by large number of twins suggesting that the most important deformation mechanism is twinning. A magnetic phase transition from a paramagnetic to a non-magnetic state has also been detected. Image processing of HREM images suggest the presence of a superstructure in these twinned crystals.

INTRODUCTION

Structural characterizations of high-Tc superconductor compounds have been widely reported in the literature [1-3]. Measurements of resistivity, low field ac and dc magnetization [4-5] have given superconducting transition temperatures commonly of the order of 90°K. X-ray powder diffraction studies [6] have revealed that room temperature structure of these ceramics is a modified Perovskite structure. The work presented in this article explores some of the structural properties of the superconductors $YBa_2Cu_3O_{7-x}$ and $HoBa_2Cu_3O_{7-x}$. The structure of these compounds as a function of temperature have been studied with in-situ heating experiments in a TEM. Large number of twinned bands are commonly observed in the orthorhombic grains. Image processing techniques applied to HREM images of these type of twinned crystals show indications of the presence of superstructures.

EXPERIMENTAL PROCEDURES

Samples have been prepared by mixing the prescribed amounts of Y_2O_3, $BaCO_3$ and CuO_2 for the $YBa_2Cu_3O_{7-x}$ case and H_2O_3 was used for the preparation of the $HoBa_2Cu_3O_{7-x}$ compound. The reacted powders were first calcined at 950°C for 12 hours and subsequently pressed into tablets; the tablets were finally annealed at the same temperature for 12 hours and cooled down gradually to about 200°C in approximately 6 hours. The structural characterization was carried out in a scanning diffractometer with steps of 0.05° and CuKα radiation. The resistivity measurements were obtained with the usual 4-terminals technique, the observations were carried out in a JEOL-100CX and a high-resolution JEOL 100C. Some of the HREM images have been digitized using a standard TV camara, the numerical data were processed in a VAX-1178 and displayed in a Innovion graphic system.

EXPERIMETAL RESULTS.

The standard four probe technique has been used for resistivity measurements. The temperature dependence of the resistivity for the samples used in this work are illustrated in Fig. 1. Both compounds ($YBa_2Cu_3O_{7-x}$ and $HoBa_2Cu_3O_{7-x}$) show a transition temperature of the order of 90°K. The identification of the superconducting phase structure was carried out by powder x-ray diffraction patterns obtained using CuKα radiation. A typical example is illustrated in Fig. 2 which shows a diffraction pattern from $YBa_2Cu_3O_{7-x}$ with strong similarities to those recently reported in the literature [6]. The indexation of

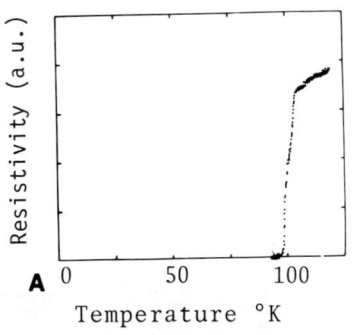
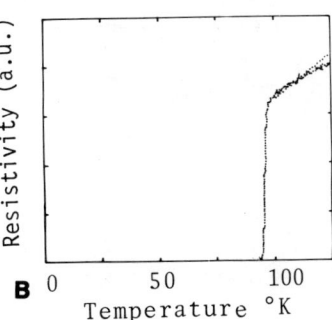
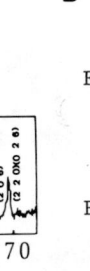

Fig.1. Temperature dependence of the resistivity for samples of : A) $YBa_2Cu_3O_{7-x}$, B) $HoBa_2Cu_3O_{7-x}$.

Fig.2. Powder x-ray diffraction pattern for a sample of $YBa_2Cu_3O_{7-x}$.

this pattern indicates an orthorhombic unit cell with parameters as
a=3.89, b=3.82 and c=11.7Å. Optical micrographs of polished specimens
show the presence of bands in the crystalline grains which strongly
suggest twinning mechanisms (Fig.3). The TEM observations also indicate the presence of these kind of bands. Fig.4A shows the BF image
of these bands under WB conditions in $YBa_2Cu_3O_{7-x}$. The same bands under SB conditions improve their image contrast as can clearly be seen
in Figs. 4B,C. These type of bands can also be commonly found in
$HoBa_2Cu_3O_{7-x}$. This can be seen from Figs. 5A and 5B which shows BF
and DF images from these type of boundaries. The displayed image contrast is characteristic of adjacent crystals in twin boundaries [8].
The diffraction conditions used to obtain the images of the boundaries
correspond to orientations closed to [001] zone axis. In this case,
the twin planes are not parallel to the [001] axis. Therefore, not
only the [110] plane can act as twin plane but others in the {110}
family.

This is illustrated in Fig.6 by HREM images of edge-on twin boundaries
with a twin plane parallel to a [110] zone axis.

The dependence of the structure of these compounds with temperature
has been obtained with in-situ heating experiments in a TEM. Fig. 7
show four different diffraction patterns from a crystal of $HoBa_2Cu_3$
O_{7-x}. They were obtained at room temperature and approximately at
400°C, 600°C and 800°C respectively.

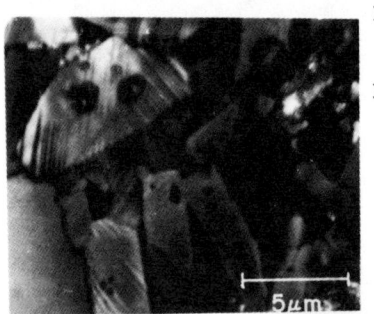

Fig.3. Optical micrograph of the superconducting oxide (Y-Ba-Cu-O).

Fig.4. TEM images of twins in YBa_2Cu_3
O_{7-x}. A) under WB conditions,
B) SB conditions (BF) and C) SB
conditions (DF).

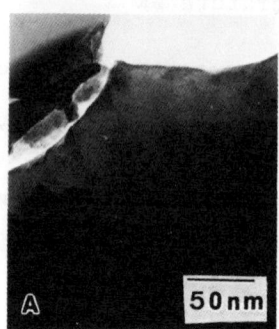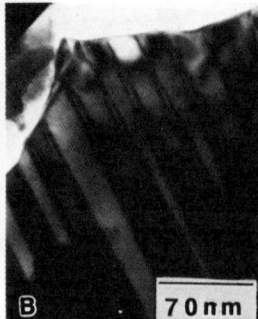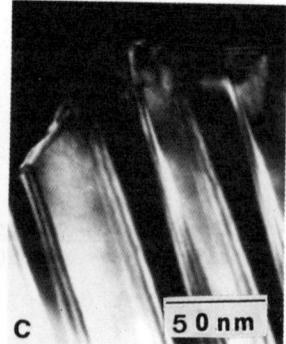

The splitting of the spots which is a common characteristic of twinning with [110] twin planes disappear at about 600°C indicating the transition from the orthorhombic to the tetragonal phase [2]. The same behavior can be seen from the twin boundary images: Fig. 8 shows the image of these boundaries as a function of temperature in $YBa_2Cu_3O_{7-x}$ compound. They disappear at temperatures above 600°C also indicating the possible structural transition.

Magnetic susceptibility measurements in $YBa_2Cu_3O_{7-x}$ as a function of temperature also shows a transition from a paramagnetic state dependent on temperature to a constant value which is practically independent on this parameter. This is clearly illustrated in Fig. 9. On the other hand, the magnetization as a function of an external field shows also a paramagnetic variation to a linear section.

Digitized images of HREM edge-on twin boundaries in $YBa_2Cu_3O_{7-x}$ obtained under orientations closed to [001] diffraction conditions [1], [8] have also been computationally processed. This type of boundary is illustrated in Fig. 10. The computed diffraction patterns obtained at both sides of the boundary are illustrated in Figs. 11A and B. The most interesting feature displayed in these patterns are the appreciable superstructure reflections obtained around ½ (010). These reflections are rotated 90° one with respect to the other. A reconstructed image of one of the crystals bounded by the boundary has been obtained using a filter which allows the main reflections and the superstructure reflections to form the image. The resulting image is shown

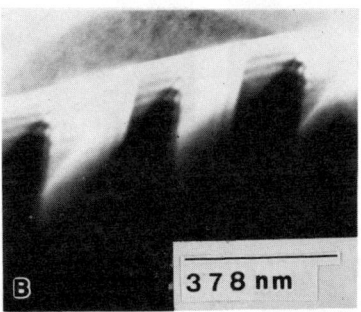
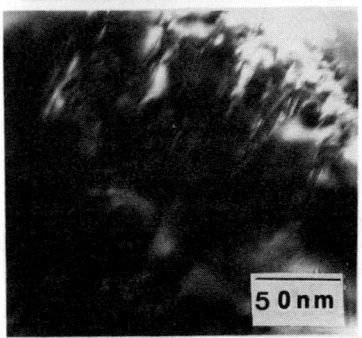

Fig. 5. TEM images of Twins in $HoBa_2Cu_3O_{7-x}$. A) SB conditions (BF), B) SB conditions (DF).

Fig. 6. HREM image of edge-on twin boundaries.

in Fig. 12 where the superstructure can clearly be seen.

CONCLUSIONS

The A15 compounds (until recently the superconductors with the highest measured Tc) are known to undergo a "martensitic" structural phase transition a few degrees above the superconducting transition temperature [9].

The $YBa_2Cu_3O_{7-x}$ and $HoBa_2Cu_3O_{7-x}$ experienced a structural transformation in the range of 600°C. This transformations shows characteristics similar to martensitic transformations. Both compounds have the same kind of structure and similar superconducting critical temperatures. The family of {110} planes can act as twin planes. Also the crystals adjacent to the twin boundaries display image contrast features similar to the presence of super structures.

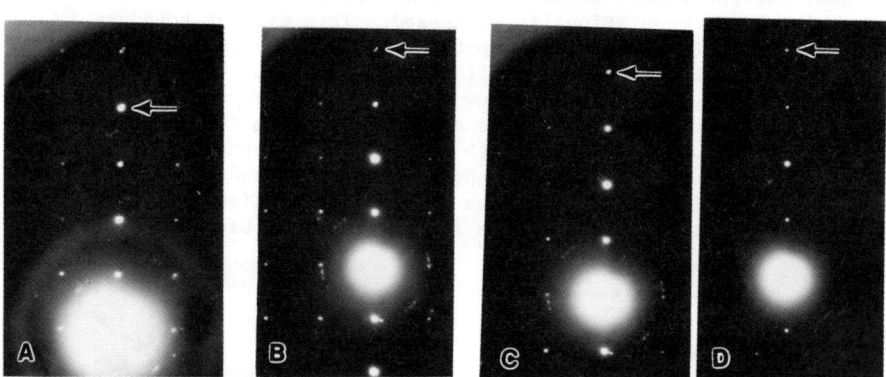

Fig.7. Temperature dependence of the structure in $HoBa_2Cu_3O_{7-x}$. A) ∿ Room Temp., B) - ∿400°C. C) - ∿600°C. D) ∿800°C.

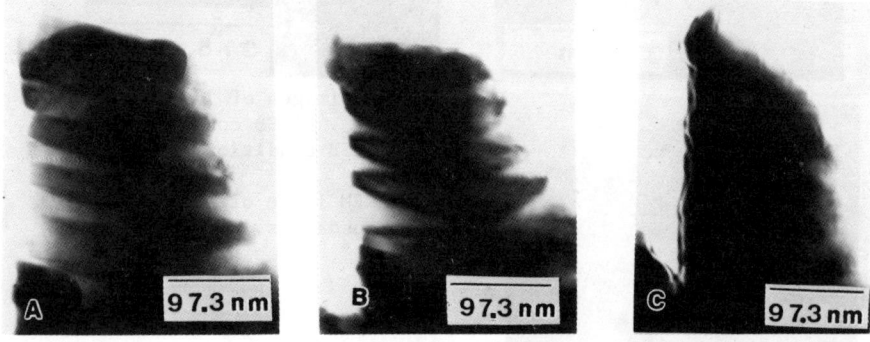

Fig.8. Temperature dependence of the structure in $YBa_2Cu_3O_{7-x}$. A) ∿ Room Temp. B) ∿400°C and C) ∿600°C.

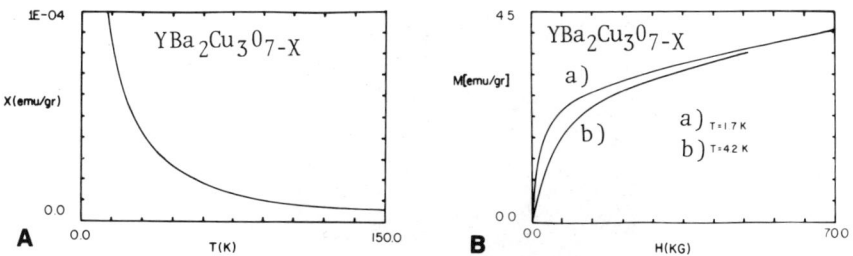

Fig.9. Magnetic measurements in $YBa_2Cu_3O_{7-x}$. A) X vs. T, B) M vs. H at T = 1.7 and 4.2°K

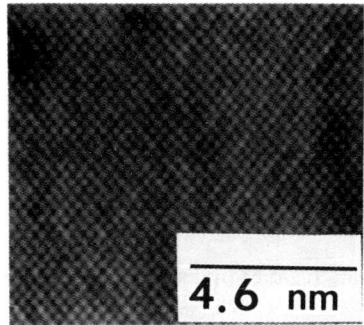

Fig. 10. HREM image of a edge-on twin boundary in $YBa_2Cu_3O_{7-x}$.

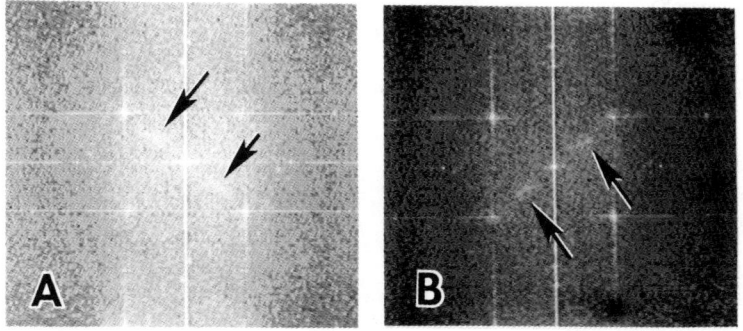

Fig. 11.
A) Computed diffraction pattern from one of the crystals adjacents to the boundary.
B) Diffraction pattern from the other crystal.

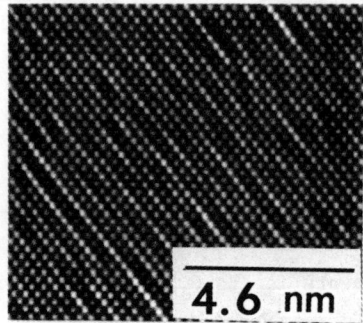

Fig.12. Reconstructed image allowing the superstructure and main reflections to form the image.

ACKNOWLEDGEMENTS

The authors would like to thank the Programa Universitario de Superconductividad of the UNAM for financial support.

REFERENCES

[1] G. Van Tendeloo and S. Amelinckx, Phys.Stat.Sol. in the press.
[2] M. Tanaka, et.al., Jpn.J. Appl.Phys. 26 N°7, L1237 (1987)
[3] J.G. Pérez-Ramírez, et.al., submitted to Phys.Stat.Sol.
[4] J.G. Bednorz and K.A. Müller, Z.Phys. B64, 189 (1986)
[5] S. Ushida, et.al., Jap.J. App.Phys. 26 N°4, L443 (1987)
[6] E. Takayama-Muromachi, et.al., Jpn. J.Appl.Phys. 26 N°4, L476 (1987)
[7] J.M. Corbett and S.S. Sheinin, Phys.Stat.Sol.(a), 38,151,(1976)
[8] Y. Syono, et.al., Jpn.J.Appl.Phys. 26 N°4, L498 (1987)
[9] G.A. Held, et.al., submitted to Europhysics Letters.

SUPERCONDUCTIVITY AND STABILITY IN Y-Ba-Sr-Cu-O COMPOUNDS

EDUARDO CARRILLO [1], A. MENDOZA [1], G. PACHECO [2], J.L. ALBARRAN [1], J. FUENTES [1], L. MARTINEZ [1], AND E. OROZCO [1]

1 Instituto de Física,UNAM. Apdo Postal 20-364,01000 México,D.F.,MEXICO
2 Facultad de Química,UNAM. División de Estudios de Posgrado,Edificio B Ciudad Universitaria, 04510 México,D.F., MEXICO

ABSTRACT

We report the effect of ageing time at room temperature on the resistive transition of mixed metal oxides $YBa_mSr_nCu_3O_y$ with Y:Ba:Sr in 1:1:1, 1:1:2, 1:2:1, and 1:2:2 nominal ratios. In all the samples a resistive transition width of 10 K or greater indicates the presence of several phases. After 3500 hours the lower effect in the degradation is observed in the $YBa_2SrCu_3O_y$ samples, where higher onset temperatures have been measured as a function of the ageing time. Optical microscopy observations and electron microprobe analysis are also reported.

Introduction

After the discovery of Wu, Chu and coworkers [1,2] of superconductivity above 90 K in single phase Y-Ba-Cu-O system, significant work has been done involving substitutions of elements on the Y sites giving rise to new superconducting perovskites [3-7] with similar T_c values.
On the other hand, substitutions on the Ba site have not been investigated as extensively in spite of the depression of T_c for its partial substitution. A wide variety of compounds and heat treatments have been used for the preparation of compounds in the Y-Ba-Sr-Cu-O system [6-12]. It has been reported that a full replacement of Ba with Sr produces a sharp superconducting transition in $Y_{1-x}Sr_xCu_3O_{3-\delta}$ [8], with 0.375< x< 0.5, whereas an insulator response has been observed in compounds with composition $YSr_2Cu_3O_y$ [9,10].
In mixed Ba-Sr samples several values have been measured for the onset and offset transition temperatures, from a highest onset of 96 K [12] to a lowest offset value of 65 K [10] according to the preparation process. Some authors have determined a region of solid solubility for the compounds $YBaSrCu_3O_{7-\delta}$ and lower Sr concentrations [9]. They have also found a monotonic depression of T_c with Sr content and they

considered it as a consequence of the local distortion of the lattice in the neighborhood of the Sr state and the introduction of additional oxygen vacancies. However very scanty information is available in relation to the stability and metallographic structure in this kind of polycrystalline compounds.

In the present work we report the behavior of the resistive transition as a measure of the stability of aged samples with four different compositions. The metallographic structure was characterized by optical microscopy and an estimation of the homogeneousness was done by electron microprobe analysis in different metallographic regions.

Experimental procedure

$YBa_mSr_nCu_3O_y$ compounds were prepared by mixing Y_2O_3, $BaCO_3$, $SrCO_3$, and CuO according to the nominal values for m:n ratio: 1:1, 2:1, 1:2, and 2:2. Each mixture was ground in an agate mortar and heated in an alumina boat at 900 °C in air for 24 hours. The resulting black powder was reground and heated again under similar conditions and furnace cooled. For measurement purposes, the powder was cold-pressed into pellets and sintered at 910 °C for 3 hours under oxygen flow. Later the pellets received a last treatment at 500 °C for one hour under oxygen flow and slow cooling. The samples were kept at room temperature in a dry container.

Four-probe technique was employed in the dc electrical resistance measurement where the electrical contact was made through copper wires attached to the sample with conducting silver paint. A closed cycle refrigerator was used to cool the samples down to 15 K and a chromel - alumel resistance thermometer was used for temperature determinations with a maximum uncertainty of 3 K.

For metallographic studies the pellets were ground with coarse grain sand and polished with alumina up to 0.3 μm before their first resistive measurement. The grains were revealed without any chemical etching and the improvement of contrast at grain boundaries was done with polarized light in a Reichter MeF optical microscope.

A JEOL JSM-35CF microscope and a 5000-EG&G Ortec microprobe equipment were employed for local determination of Y,Ba,Sr, and Cu content in order to estimate the homogeneity of the four different compounds, using the $K\alpha$-Cu, $L\alpha$-Y, $L\alpha$-Ba, and $L\alpha$-Sr lines and an excitation voltage of 20 kV. Intrinsical restrictions for oxygen determination give a semiquantitative character to the homogenousness study.

Results

Table I shows the main characteristics of the resistive transition for the initial conditions and ageing times of 1450 and 3560 hours. For aged samples the highest transition temperatures are exhibited by those compounds where Y:Ba are in 1:2 ratio. In particular, the compounds $YBa_2SrCu_3O_y$ show an increasing in their transition temperatures as well as a reducing width as the ageing goes by.

TABLE I
Characteristics of the resistive transition

specimen	ageing time at room temp. (hours)	T_c^{onset} (K)	T_c^{end} (K)
$YBaSrCu_3O_y$	0	97 ± 3	81 ± 3
	1450	73	32
	3560	70	32
$YBa_2SrCu_3O_y$	0	85 ± 3	47 ± 3
	1450	99	87
	3560	102	94
$YBaSr_2Cu_3O_y$	0	100 ± 3	83 ± 3
	1450	87	34
	3560	85	< 15
$YBa_2Sr_2Cu_3O_y$	0	95 ± 3	85 ± 3
	1450	90	78
	3560	90	70

On the other hand those samples with composition $YBaSr_2Cu_3O_y$ (3560hr) exhibit a measurable resistance at 15 K, even with a dc value as low as 0.1 mA (other compositions showed a transition with currents between 10 - 90 mA).

The optical microscopy study revealed two compositions with a continuous surface layer: $YBaSrCu_3O_y$ and $YBa_2SrCu_3O_y$. However, the first one exhibits a smaller grain size in comparison with the samples of $YBa_2SrCu_3O_y$ (see figures 1a and 1b). For the other two kinds of compounds, we observed a "broken" layer aspect and it is possible to find some continuous regions in very small zones, greater for $YBa_2Sr_2Cu_3O_y$ than for $YBaSr_2Cu_3O_y$ (see figures 1c and 1d).

The grain morphology of the samples with same nominal ratio of Y and Ba presented smaller grain size. Prismatic grains were found in almost all continuous layer. Some equiaxed grains were observed in the $YBaSrCu_3O_y$ pellets but in a reduced rate.

Concerning to the electron microprobe analysis a rough estimation was done about the relative concentration of Y,Ba,Sr, and Cu. For the present report the "nearest" stoichiometries were obtained:

a) $YBaSrCu_3O_y$.- Equiaxed grains correspond to a 4:2:1:3 (Y:Ba:Sr:Cu) phase, whereas the prismatic grains have the nominal composition 1:1:1:3.

b) $YBa_2SrCu_3O_y$.- No strontium was detected in the larger grains, which presented a "phase" 1:3:4:5 (Y:Ba:Sr:Cu). The strontium was precipitated mainly in the grain boundaries.

c) $YBaSr_2Cu_3O_y$.- It was the most homogeneous surface including the small continuous "islands" with a composition 1:2:1.5:5.

d) $YBa_2Sr_2Cu_3O_y$.- The continuous "islands" showed two different

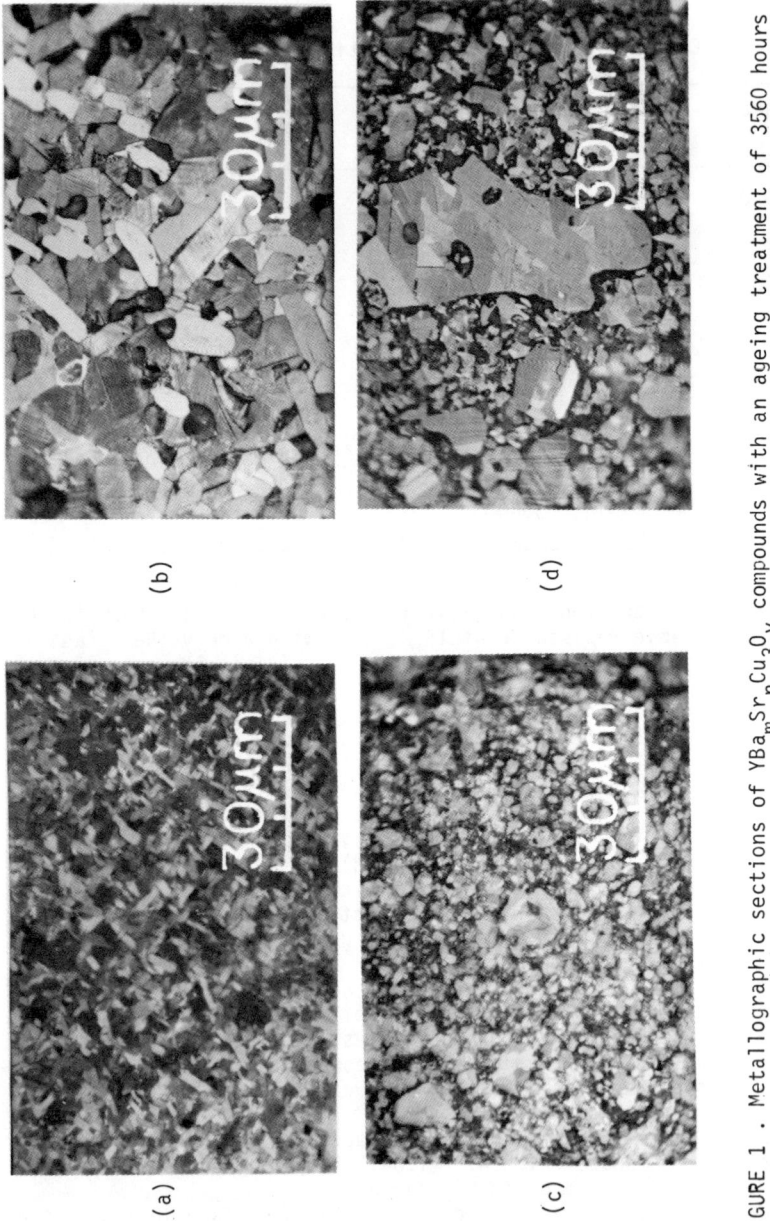

FIGURE 1. Metallographic sections of YBa$_x$Sr$_z$Cu$_3$O$_y$ compounds with an ageing treatment of 3560 hours at room temperature. (a) YBaSrCu$_2$O$_y$. A continudus and fine grain surface. In the few equiaxed grains Y is in greater concentration than Ba and Sr. (b) YBa$_2$SrCu$_3$O$_y$. The strontium is concentrated at the boundaries of the twined grains. (c) YBaSr$_2$Cu$_3$O$_y$. There are very small continuous "islands". The overall composition of the surface is quite homogeneous. (d) YBa$_2$Sr$_2$Cu$_3$O$_y$. Greater islands than (c).

compositions, the nominal 1:1:1:3 and 3:1.5 :1:4. The rest of the
surface exhibits a relative concentration 1:1.5 :1:6 .

Discussion

The highest density observed in $YBaSrCu_3O_y$ samples can be considered as an additional evidence of the solid solution conditions for Sr in $YBa_2Cu_3O_y$ as substitute for Ba ions [9].
The stability showed by $YBaSr_2Cu_3O_y$ and the precipitation of Sr to the grain boundaries could be an indication of the solid solution limit in the system $YBa_{2-x}Sr_xCu_3O_y$ for $x > 1$. In that case the ageing effect is actually reflecting the stability behavior of the $YBa_2Cu_3O_{7-\delta}$. It would be then interesting to study the role played by Sr as a soldering element in the grain boundary. At this point it has to be mentioned that in the present study it was determined an insulator behavior of the system $YSr_2Cu_3O_y$, in accordance to [9,10], so any precipitation of Sr in $YBaSr_2Cu_3O_y$ could be responsible for the formation of a non superconducting phase and a consequent reduction or even an annihilation of the high superconducting temperature.
The fast degradation process in $YBaSr_2Cu_3Oy$ samples and the great homogeneity in their surfaces indicate a high kinetic reactivity in this system and a considerable inestability of the superconductive phase in the presence of other Sr-insulating phases.
In relation with the $YBa_2Sr_2Cu_3O_y$ compounds it is important to point out the determination of a phase with the nominal stoichiometry for Y:Ba:Sr:Cu (1:1:1:3), just as in the samples of composition $YBaSrCu_3O_y$. However the higher stability found in the first system generates interesting questions about the role played by the other phases.

Conclusions

The resistive transition behavior was followed as a function of the ageing time in four different compositions of the Y-Ba-Sr-Cu-O system. After 3560 hours at room temperature, the most stable compound is found in the $YBa_2SrCu_3O_y$ composition.
The addition of Sr to the $YBa_2Cu_3O_y$ systems implied a reduction of the grain size which could be important for obtaining higher density oxides.

Acknowledgments

This work was sponsored by the Programa Universitario de Superconductividad de Alta Temperatura de Transición (PUSCATT) of the Universidad Nacional Autónoma de México.
The authors would like to thank Ing. Q.M. Federico Manrique for his technical assistance during the electron microprobe analysis.

References

[1] .- M.K. Wu, J.R. Ashburn, C.J. Torng, P.H. Hor, P.L. Meng, L. Gao, Z.J. Huang, Y.Q. Wang, and C.W. Chu. Phys. Rev. Lett. $\underline{58}$, 908 (1987)

[2] .- P.H. Hor, L. Gao, R.L. Meng, Z.J. Huang, Y.Q. Wang, K. Forster, J. Vassilious, C.W. Chu, M.K. Wu, J.R. Ashburn, and C.J. Torng. Phys. Rev. Lett. $\underline{58}$, 911 (1987)

[3] .- P.H. Hor, R.L. Meng, Y.Q. Wang, L. Gao, Z.J. Huang, J. Bechtold, K. Forster, and C.W. Chu. Phys. Pev. Lett. $\underline{58}$, 1891 (1987)

[4] .- S. Oshima and T. Wakiyama. Jpn. J. Appl. Phys. $\underline{26}$, L 815 (1987)

[5] .- S. Hosoya, S. Shamoto, M. Onoda, and M. Sato. Jpn. J. Appl. Phys. $\underline{26}$, L 456 (1987)

[6] .- D.W. Murphy, S. Sunshine, R.B. van Dover, R.J. Cava, B. Batlogg, S.M. Zahurak, and L.F. Schneemeyer. Phys. Rev. Lett. $\underline{58}$, 1888 (1987)

[7] .- E.M. Engler, V.Y. Lee, A.I. Nazzal, R.B. Beyers, G. Lim, P.M. Grant, S.S.P. Parkin, M.L. Ramírez, J.E. Vázquez, and R.J. Savoy. J. Amer. Chem. Soc., accepted (1987)

[8] .- Zhang Qi-rui, Cao Lie-Zhao, Qian Yi-tai, Chen Zu-yao, Guan Wei-yan, Zhao Yong, Pang Guo-qang, Zhang Han, Xia Jian-sheng, Zhang Ming-Jian, Yu Dao-qi, He-Zheng-hui, Sun Shi-fang, Fang Ming-hu, and Zhang Tao. Solid State Comm. $\underline{63}$, 535 (1987)

[9] .- B.W. Veal, W.K. Kwok, A. Umezawa, G.W. Crabtree, J.D. Jorgensen, J.W. Downey, L.J. Nowicki, A.W. Mitchell, A.P. Paulikas, and C.H. Sowers. Appl. Phys. Lett. $\underline{51}$, 279 (1987)

[10] .- B. Jayaram, S.K. Agarwal, A. Gupta, and A.V. Narlikar. Solid State Comm. $\underline{63}$, 713 (1987)

[11] .- T. Wada, S. Adachi, O. Inoue, S. Kawashima, and T. Mihara. Jpn. J. Appl. Phys. $\underline{26}$, L 1475 (1987)

[12] .- E.V. Sampathkumaran, P.L. Paulose, V. Nagarajan, N. Nambudripad, A.K. Grover, S.K. Dhar, R. Nagarajan, and R. Vijayaraghavan. Pramaña-J. Phys. $\underline{29}$, L 327 (1987)

NORMAL MODES OF VIBRATION IN HIGH Tc SUPERCONDUCTORS
A PICTORIAL VIEW

A. CALLES[1], E. YEPEZ[1,2], A. SALCIDO[1], J.J. CASTRO[3] AND A. CABRERA[1]

1 Facultad de Ciencias, UNAM, Apdo. Post. 70-646,
04510 México D.F., MEXICO.
2 On sabatical leave from ESFM, IPN. (supported in part by COFAA).
3 Depto. de Física, CINVESTAV, IPN, Apdo. Post. 14-740,
México D.F., MEXICO

Introduction.

The recent discovery of high critical temperature superconductivity in oxide perovskites [1,2] has made neccessary to review many of the physical mechanisms that could be responsible for superconductivity [3,4]. One of the main ingredients in any theoretical formulation is the proper knowledge of the lattice vibration characteristics of these complex structures.

The aim of the present work is to give some insight into the properties of the phonon spectrum on high Tc superconductors, that could serve as a starting point for the construction of theoretical models that could explain this new phenomenon. We present the results of a standard lattice dynamics calculation for the structure $YBa_2Cu_3O_7$.

Our model is a standard Born-Huang [5] model with a central potential for each pair of atoms in the harmonic approximation considering interactions up to second nearest neighbours. The force constant parameters are adjusted for the best possible fitting to the experimental phonon density of states reported by Rhyne et al [6].

The novelty of the present work is the versatility of the computer program that allows to determine the phonon dispersion relation for any structure with different potential for each species, whose only limitation is the memory and time facilities of the available computer and also the simulation on a personal computer (PC) of the lattice vibration giving a pictorial view of the phonon characteristics of the systems.

Phonon spectra of the $YBa_2Cu_3O_7$ compound.

The energies corresponding to the normal modes of vibration were calculated by solving the eigenvalue problem

$$\sum_{\nu} (B_{\mu\nu,\kappa} - E_{b\kappa} \delta_{\mu\nu}) C_{\mu b,\kappa} = 0, \qquad (1)$$

where

$$B_{\mu\nu,\kappa} = \sum_{j} V_{\mu\nu}(|R_i - R_j|) e^{i\kappa \cdot (R_i - R_j)}$$

and $V_{\mu\nu}(|R_j - R_i|)$ is the interaction potential between the particles μ and ν located at points R_i and R_j; $E_{b\kappa}$ is the energy corresponding to the b branch and the wave vector which is defined in the first Brillouin zone.

The equilibrium position for the structure of YBa_2Cu_3O, is the one determined by Izumi et al.[7].

The phonon spectra is determined by solving equation (1) for several κ values randomly selected in the first Brillouin zone until we reach approximatly 20,000 states.

The theoretical calculation of the phonon density of states (DOS) is straighforward through the definition

$$g(\omega) = (2\pi)^{-3} \sum_b \int \delta(\omega - \omega_b(\kappa)) \, d\kappa \qquad (2)$$

The force constant parameters are determined by comparing the theoretical DOS with the experimental one as reported by Rhyne et al. [6].

The theoretical result for the phonon DOS as a function of the energy is compared with the experimental result in Fig. 1 [6]. It is worth mentioning that the experimental curve was obtained from analysis of inelastic neutron scattering data using the incoherent approximation. As can be seen from this figure the theoretical curve shows at low energies the typical ω^2 dependence for a 3D system, it also predicts the small shoulder at 12 meV and also the peak at 20 meV, however from 40 to 45 meV the spectrum has a gap which does not appear experimentally. The peak at 70 meV is also reproduced and in general the structure above 70 meV has a good resemblance to the experimental spectrum

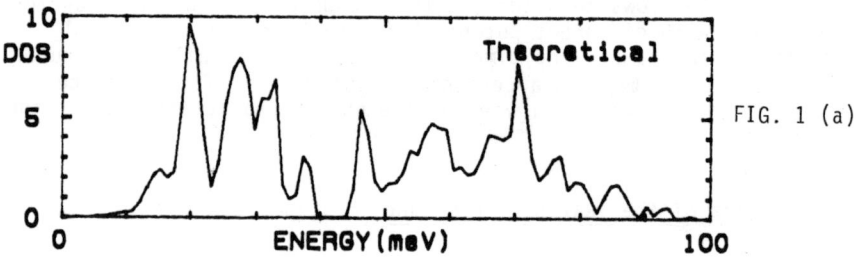

FIG. 1 (a)

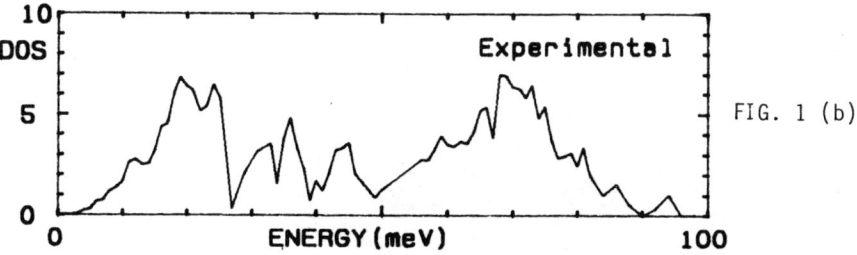

FIG. 1 (b)

A pictorial view of the lattice vibration.

With the purpose of getting some physical insight about how the different ions are moving in each mode we simulated the normal modes of vibration in a microcomputer. This was done for both the point group and space group symetries, "seeing" the molecular vibration and the phonon in branch b and wave vector k.

As an example we show in Fig. 2 a sequence of pictures as a function of time corresponding to the mode $\omega = 388$ cm^{-1}. This mode has been associated with a Raman mode [8].

This simulation has allowed us to identify the motion of the different O and Cu ions in several modes. In particular we can mention the phonons corresponding at 20 meV and 70 meV, the pictorial view shows how these modes are dominated by oxigen vibration which is in agreement with some authors [6,8].

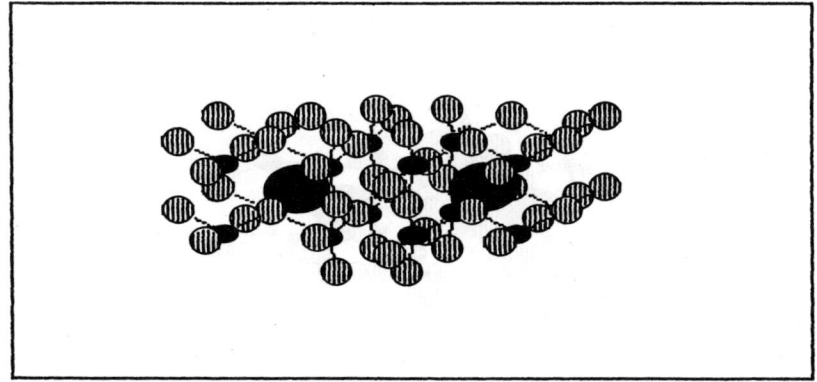

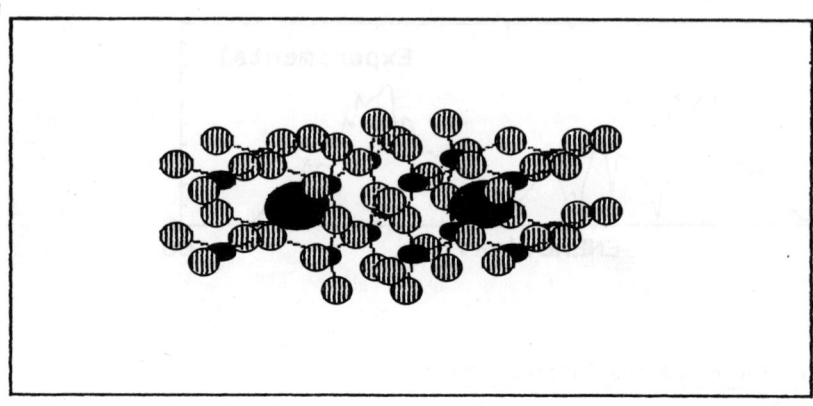

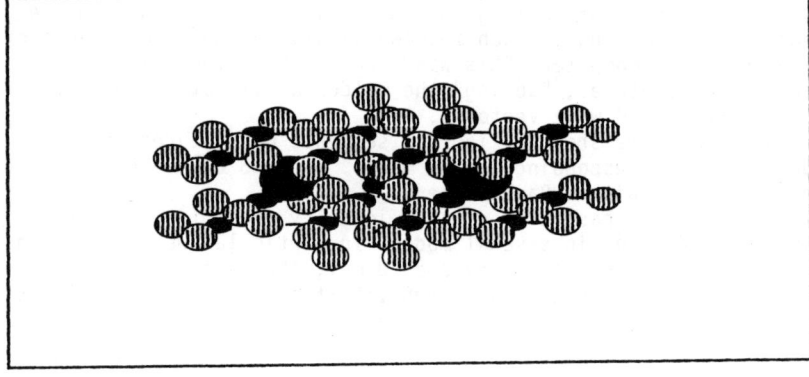

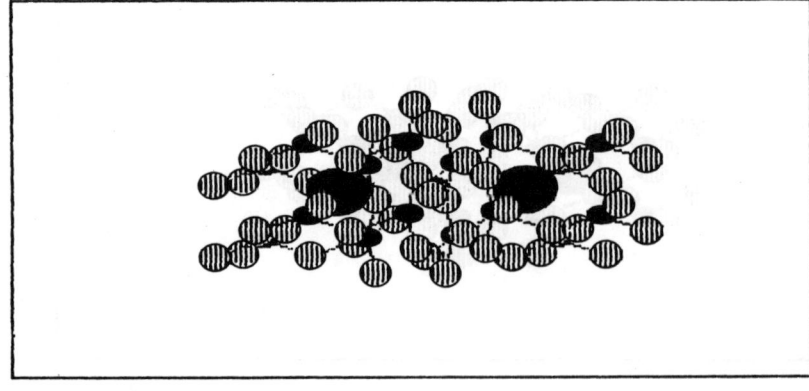

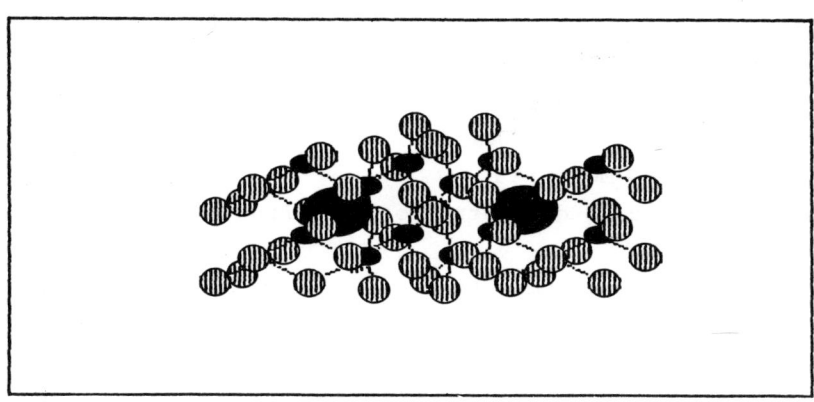

We would like to acknowledge R. Escudero, R. Barrio and T. Akachi (IIM - UNAM) for useful discussions concerning this work.

References

[1]. J.G. Bednorz and K.A. Muller, Z. Phys. B 64, 189 (1986)
[2]. C.W. Chu, D.H. Hor, P.L. Meng, L. Gao, Z.J. Huang and Y.Q. Wang Phys. Rev. Lett. 58, 405 (1987)
[3]. H. Kamimura, Jpn. J. Appl. Phys. 26, L627 (1987)
[4]. E.M. Engler, IBM Research Report RJ 5723 (57702), July 1987.
[5]. M. Born and K. Huang, Dynamical Theory of Cristal Lattices, Oxford Univ. Press N.Y. (1954).
[6]. J.J. Rhyne et al. Phys. Rev. B 36 (1987) 2294.
[7]. F. Izumi, H. Asano, T. Ishigaki, E. Takayama-Muromachi, Y. Uchida, N. Watanabe and T. Nishikawa, Jpn. J. Appl. Phys. 26 L649 (1987).
[8]. M. Stavola, D.M. Krol, W. Weber, S.A. Sunshine, A. Jayaraman, G.A. Kouroklis, R.J. Cava and E.A. Rietman, Phys. Rev. B 36, 850 (1987).

MICROSTRUCTURE CHARACTERIZATION OF SUPERCONDUCTIVE MIXED PHASES

J.M. DOMINGUEZ, C. FALCONY (#), O. GUZMAN, P. DEL ANGEL,
J. OMAÑA and A. MONTOYA
Instituto Mexicano del Petroleo, IBP, Apdo. Postal 14-805,
México 14, D.F.
(#) CINVESTAV-IPN, Apdo. Postal 14-740, México D.F.

ABSTRACT

Some structural aspects of the multiphase compounds $Y_{3-x}(BaSr)_xCu_3O_{7-y}$ and the Fe-substituted-YBCO-perovskites were characterized by means of microanalytical methods (Electron Energy-Loss Spectroscopy, Energy Dispersive Spectrometry, μμD). The valence state of iron in the orthorhombic lattice of $Y_1Ba_2(Cu_{3-x}Fe_x)O_{7-y}$ was found different from the iron state in the Fe_2O_3 type standards. Also, a network of crystalline mosaics free of defects was found characteristic of the series of $Y_{3-x}(BaSr)_xCu_3O_{7-y}$ type compounds which exhibited important resistivity losses [6].

INTRODUCTION

In the search of higher critical temperatures (T_c), in the series of Y-Ba-Cu-O compounds, both stoichiometric variations and partial ionic substitutions [1,2] have been made. In order to modify the transport properties of the solids, either the oxygen replacement or the foreign cation insertion into the orthorhombic lattice of $Y_1Ba_2Cu_3O_{7-y}$ has proven important. For example, a systematic decrease of T_c occurs concurrently with a lattice symmetry transition from orthorhombic to tetragonal, when copper is replaced by iron [3]. In comparison, a decrease of T_c without a major symmetry transition is observed [4] with the partial insertion of Sr in the Ba positions. Also, the possible presence of fluor [5] in the lattice might cause an enhancement of T_c towards the higher temperatures.

In view of this, other non-stoichiometric compounds were synthesized, as for example the $Y_{3-x}(BaSr)_x Cu_3 O_{7-y}$ series, which exhibited partial resistivity losses with a final value of R distinct from zero [6]. Thus, the purpose of this note is to report some microstructural aspects of these solids that might increase our knowledge on the partial phases involved in the transport propertied mentioned.

Also, as long as the chemical valence state of the constituents might play an important role in the transport properties, a series of experiments with the Fe-substituted- $Y_1 Ba_2 Cu_3 O_{7-y}$ - perovskites was made in order to determine the most probable valence state of iron in the lattice. For this, a direct comparison of the EELS spectra with respect to a reference spectrum of hematite (Fe_2O_3) allowed to verify whether or not the valence state of Fe corresponded to the most expected value, which was equal to +3. Other aspects like the aggregation state of the phases, compositional fluctuations and lattice parameters variation were studied by means of the microanalytical techniques incorporated in the electron microscope (EELS, EDS, μμD).

EXPERIMENTAL RESULTS

In the series of $Y_1 Ba_2 (Cu_{3-x} Fe_x) O_{7-y}$ compounds (ref.11) several experiments were conducted by EELS at 200 KV, in a JEOL-2000 electron microscope. The typical spectra were obtained in the transmission mode from areas thinner than 0.1 um; in this way, the collected spectra correspond to the bulk of the sample and, by extending the analysis to a number of distinct areas, a significant statistics is obtained. The EELS spectra revealed the characteristic absorption edges of both oxygen (ΔE = 531 eV) and iron (ΔE = 721 eV of L_{II} and ΔE = 708 eV for L_{III}). A rapid acquisition time was used in order to minimize the sample degradation during the analysis. As shown in figure (1), a sequence of EELS spectra of about ten seconds each demonstrate that there is a relative stability of the ceramic oxide with res-

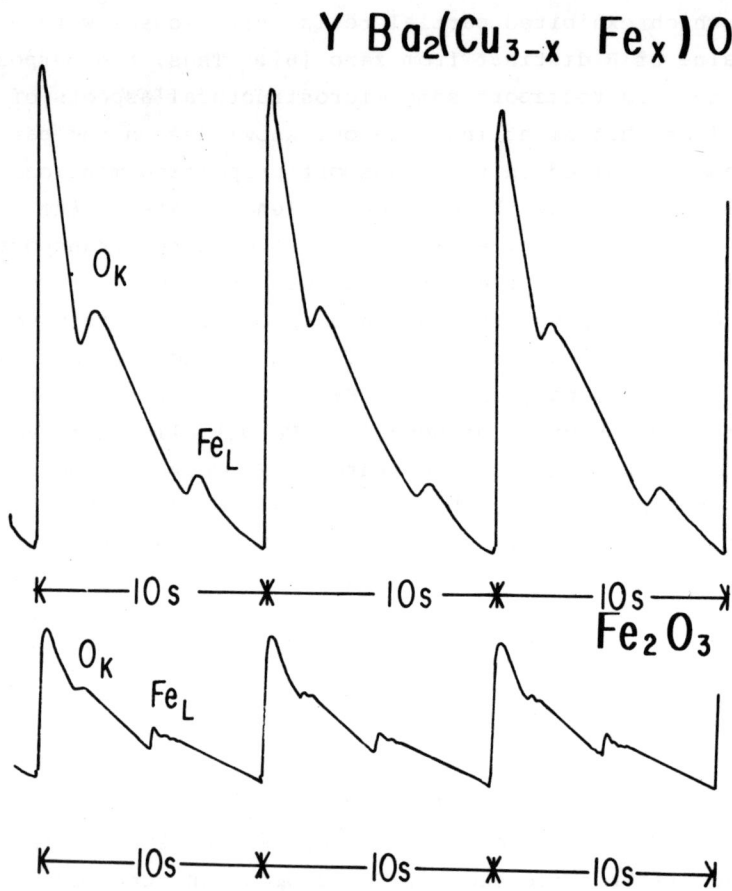

FIG. (1).- Sequence of three EELS spectra corresponding to the $Y_1 Ba_2 (Cu_{3-x}Fe_x) O_{7-y}$ ceramic oxide series, for x=0.125, and the pure $Fe_2 O_3$ standard. Note the distinct edge profile of Fe by comparison.

pect to the electron radiation exposure. These series of spectra correspond to the same sample area with dimensions of about 0.2 μm in diameter; as observed, the absorption edges of both oxygen and iron remain unchanged even for periods of several tens of seconds.

In contrast with the sharpness of the Fe_2O_3 (hematite) edges, the EELS spectra of the ceramic oxides show a modification of the profile of the L_{II} and L_{III} edges; furthermore, there is a convolution of peaks in the region between 708 and 721 eV. Although a deconvolution procedure for assigning the correct valence state should be necessary, at least it is evident from the present comparison, that the state of iron in the lattice of $Y_1Ba_2Cu_{3-x}Fe_xO_{7-y}$, with x=0.125, is different from the valence state of iron in the standards (+3).

On the other hand, a previous study of the multiphase compounds with general formula $Y_{3-x}(BaSr)_xCu_3O_{7-y}$ (x=1, Ba/Sr=0, 0.1, 0.3, 0.5 and 1), by means of x-ray diffraction analysis indicated that several phases were present, including the 1-2-3YBCO type phase and others similar to the ones reported elsewhere [8].

However, the aggregation state of phases as well as the presence of defects could only by delucidated by electron microscopy. In consequence, the series of solids with atomic ratios Ba/Sr from 0 to 1 were studied by means of the conventional bright field techniques and by high resolution methods.

It was found that the samples with Ba/Sr ratios equal to 0, 0.1 and 0.3 were constituted by similar grains of material, with major dimensions ranging from 2 to 10 μm; those aggregates were composed by a network of interconnected mosaics of 0.02 to 0.04 μm in average dimensions as shown in figure (2). Some of these sub-units are faceted, hexagonally shaped and have a random orientation. In comparison, the solids with Ba/Sr ratios of 0.5 and 1 show a distinct morphology corresponding to the spherulitic type, as illustrated in figure (3). These aggregates have been observed also in the 1-2-3-YBCO type compounds when synthesized under a temperature gradient [9].

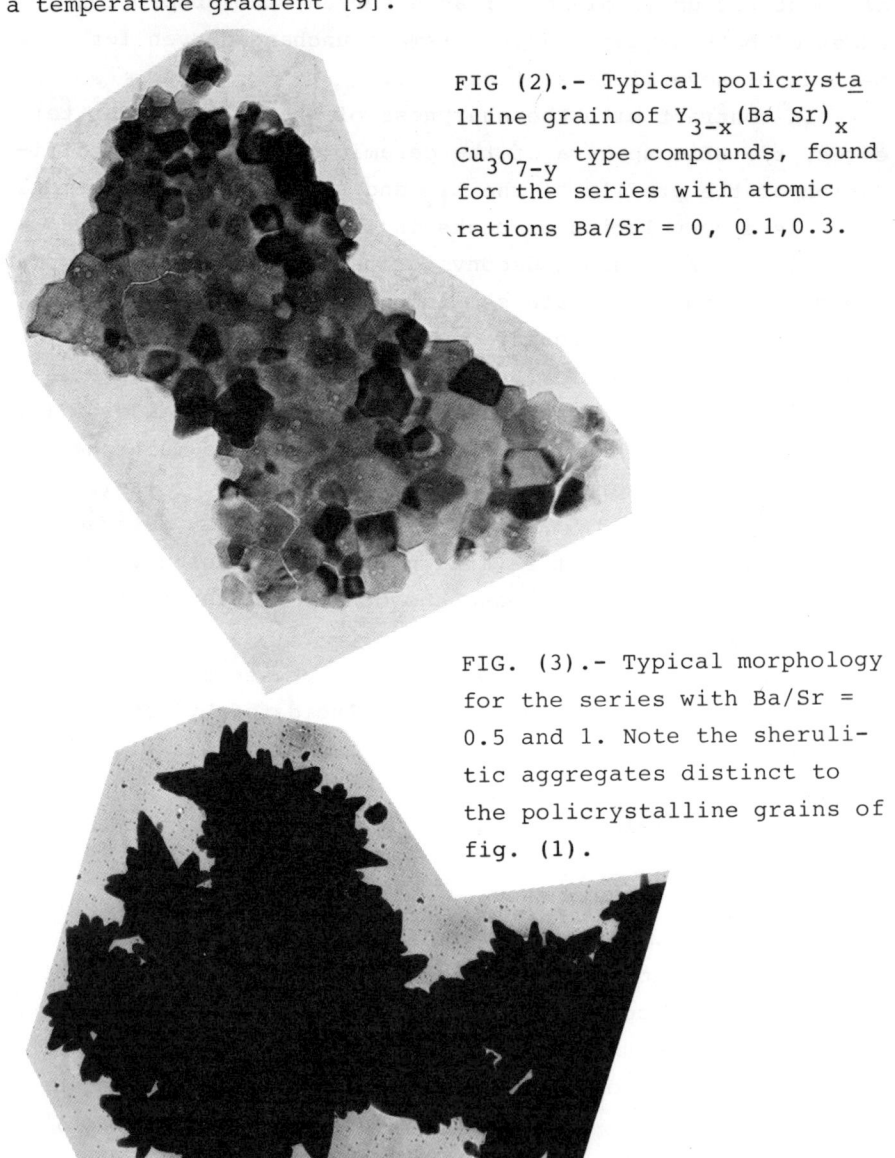

FIG (2).- Typical policrystalline grain of $Y_{3-x}(Ba\ Sr)_x Cu_3O_{7-y}$ type compounds, found for the series with atomic rations Ba/Sr = 0, 0.1, 0.3.

FIG. (3).- Typical morphology for the series with Ba/Sr = 0.5 and 1. Note the sherulitic aggregates distinct to the policrystalline grains of fig. (1).

A close view of the individual mosaics of the first series and their boundaries is shown in figure (4); the aggregation state is apparent. The set of parallel fringes across some of the mosaics are Moiré type fringe patterns, indicating the strong crystalline nature of the sub-units and the superposition of each other.

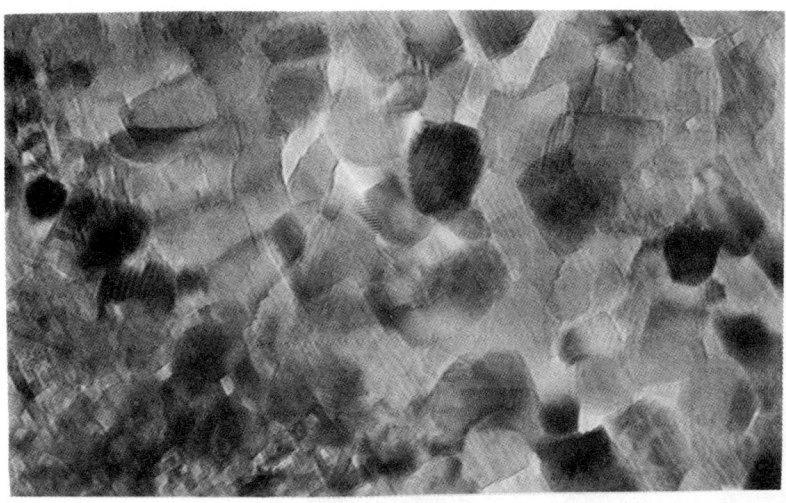

FIG. (4).- Intermediate magnification picture of the grain substructure. Note the aggregation state, as deduced from the Moiré type patterns.

The inner structure was investigated by means of the high resolution methods. The observation of the policrystalline grains along the axis perpendicular to the main face produces the type of images as shown in figure (5); that is the lattice resolution image of a single mosaic at almost atomic scale. As observed, there is a continuous sequence of planes in all directions with no defects. This lattice is defined by two axis with an interaxial angle of 100 degrees and lenght of 3.8 and 4.0 angstroms. A further study of the individual sub-units is underway to verify the symmetry of the lattice and the composition.

From the macroscopical point of view, the cross-section compositional parameter is an important one because the resistance measurements are usually made by means of the surface contacts on the top of the pellet. In consequence, a compositional gradient could conduct to misleading information on the bulk transport properties; to overcome this difficulty, it necessitates the control of the compositional profile by means of Energy Dispersive Spectrometry (EDS).

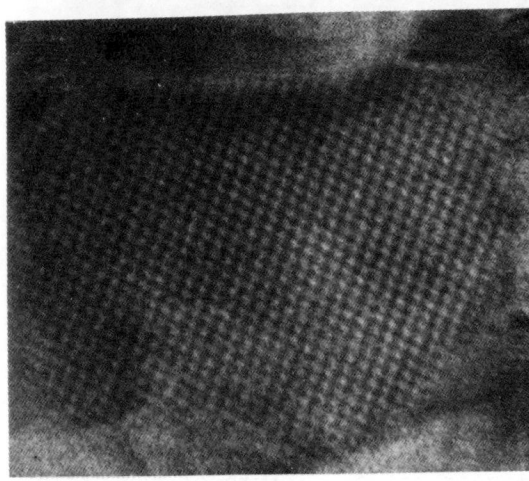

FIG. (5).- High Resolution image of a single mosaic unit. The projected structure is free of defects.

Although not included, the graphs of elemental composition obtained at low magnifications in the microscope, show the relative fluctuations along the pellet height, from 0 to about 1 mm at steps of 100 µm each. For this case, the standard deviation is the smallest one but, the samples with a low Ba/Sr ratio appeared to have higher fluctuations.

In summary, the multiphase compounds $Y_{3-x}(BaSr)_x Cu_3 O_{7-y}$ were constituted by policrystalline grains composed of single crystal units free of defects. Some of those mosaics are identical to each other but distinct entities were also found. The policrystalline aggregates are thus characterized by heterogeneous portions of material. This result is consistent with the initial explanation relative to the macroscopical transport properties of these solids, which speculated on the presence of superconductive clusters, i.e. mosaics in the present case, which could be responsable of the partial resistance drops and those should be embedded into a non-superconductive matrix [6]. A further characterization of the structure and composition of the individual sub-units and their relatioship with the matrix will give more information soon.

Also, the influence of compositional gradients in the pellets was evaluated as a possible source of error in the resistance measurements; a general correct mixing in the series of multiphase solids was found, but there is a trend towards important fluctuations as the ratio Ba/Sr decreases. Finally, other features which could be of relevance for the understanding of the superconductivity mechanism, as for example the local valence coordination of iron in the substituted perovskites, were found susceptible of being studied by microanalytical methods, like EELS, in the electron microscope; in particular, the Fe-substituted samples, for x=0.125, presented a valence state different from the pure standards (+3) but more similar to a mixed state of Fe(+2)-Fe(+3) (10)

REFERENCES

(1).- Y. Maeno et al., Jpn. J. Appl. Phys. 26, (1987),L774.
(2).- S. Kagoshima et al., J. J. A. Phys. 26, 1355 (1987).
(3).- Y. Maeno et al., Jpn. J. Appl. Phys. (1987).
(4).- B. W. Veal et al., Appl. Phys. Lett. 51(4), 279 (1987).
(5).- S.R. Ovshinsky et al., Phys. Rev. Lett. 58,2579 (1987).
(6).- J.M. Domínguez et al., MRS Fall Meet., Boston,Nov.1987.
(7).- K. Kitazawa et al., MRS Fall Meet.,Boston Nov. (1987).
(8).- K. Matsuzaki et al., J.J. Appl. Phys. 26,5, L624 1987.
(9).- D. López et al., J. Mat. Res. Bull., Submitted, 1988.
(10).- R. Gomez et al.., Chem. Rev. vol. 36, 13, 7226 (nov 1987).
(11).- G. Gonzalez et al., MRS Meet., Boston, Mass.,Nov 29-dec 4, 1987.

We acknowledge to Dr. David Ríos, from IIM-UNAM, for useful discussions, to L. Ramírez Sch.(IMP), for the technical assistance, and to D.Thompson(IMP) for text corrections.

MECHANISMS
OF
SUPERCONDUCTIVITY

ORTHORHOMBIC CELL

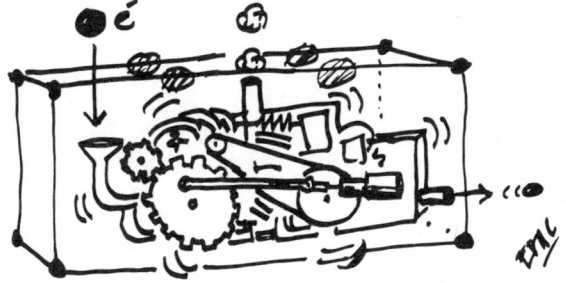

THE ELECTRONIC STRUCTURE AND SUPERCONDUCTIVITY OF THE HIGH T_c COPPER OXIDES

M. SCHLUTER
AT&T Bell Laboratories, Murray Hill, New Jersey 07974, USA

ABSTRACT

The electronic structure and possible superconductivity mechanisms of the new high T_c copper oxides are discussed. The conventional band picture is contrasted with the localized orbital Mott-Hubbard picture for the electronic groundstate. The nature of quasiparticles introduced by doping of the insulating parent compounds depends on assumptions made for the underlying groundstate. Possible pairing mechanisms of these particles that have been proposed to lead to high temperature superconductivity are analyzed.

INTRODUCTION

The discovery [1] of superconductivity with transition temperatures well above "traditional" high values of ≈20K has spawned an enormous number of theoretical attempts to identify mechanisms which would allow for high T_c's of order 100K [2], [3]. Now, about one year after the original findings the theoretical situation is still rather unsettled. This review attempts to address what is known so far by comparing the variety of independently proposed models and by putting them into perspective with our knowledge of the basic electronic structure of the copper oxide materials. The paper starts by emphasizing the essential structural features of the new materials; it then proceeds to discuss pictures for the electronic groundstates of the (undoped) semiconducting parent materials, continues with the description of quasiparticles (e.g. introduced by doping or by non-stoichiometry) and finally addresses different possible pairing mechanisms leading to superconductivity. At this stage most theoretical pictures are still quite raw and their fate depends ultimately on experimental findings. The amount of experimental data has been overwhelming, but due to the involved nature of these oxide compounds (anisotropy, non-stoichiometry, structural instability), clear and universally accepted data are only gradually appearing. Single crystal measurements are ultimately needed in most cases to draw information useful for theoretical interpretation.

THE CRYSTAL LATTICE STRUCTURE

We briefly review the lattice structures of the two classes of compounds of interest here. There is the original La_2CuO_4 or "(214)" compound crystallizing in the K_2NiF_4-structure (Fig. 1) which if appropriately doped on the La-sites leads to ~35K superconductivity and there is the next generation $YBa_2Cu_3O_{6+x}$ or "(123)" compound

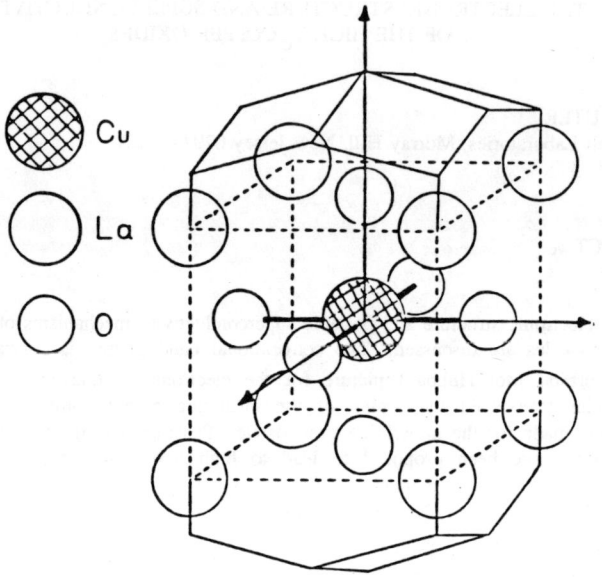

Fig. 1   Structure of the La_2CuO_4 "(214)" compound.

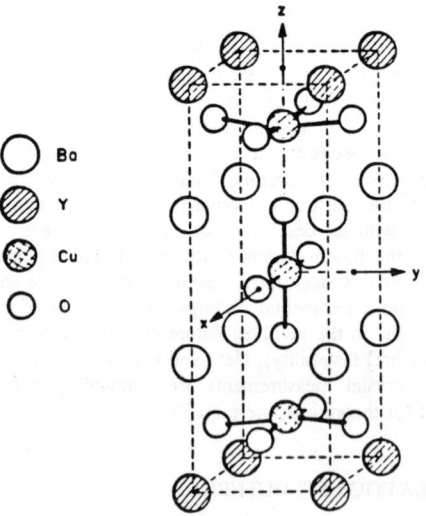

Fig. 2   Structure of the $YBa_2Cu_3O_7$ "(123)" compound.

(Fig. 2) which for $0.5 \leq x \leq 1$ leads to superconductivity up to ~95K. Many aspects of the structure and its stability have been investigated, see e.g. refs. [2], [3]. Here we merely review some of the features thought to be essential for the understanding of the basic electronic structure of these compounds. The (214) material can be thought-of as "networks" of square Cu-O planes with additional oxygens attached to each Cu above and below the planes. The planes are stacked inside a stabilizing cation (La) lattice in a staggered fashion. The conduction occurs via holes in the Cu-O planes upon doping on the cation lattice. The Cu-O interatomic distances are remarkably short in the plane (1.90Å) and considerably longer for the out-of-plane oxygens (2.40Å). Each Cu may thus also be viewed as surrounded by strongly Jahn-Teller distorted (elongated) oxygen octahedra which in turn are connected to each other sharing corners in two dimensions. In the (123) structure Cu-O planes are connected pairwise in the third dimension by the insertion one-dimensional Cu-O chains. Running parallel between the planes (along x in Fig. 2). In addition now the Cu-atoms in the planes are missing one of their octahedral oxygen neighbors while those in the chains are missing two. In each case, however, the Cu atoms maintain their strong four-fold coordination. Incomplete oxygenation (x<1) of the (123) structure creates vacancies in the chains which may undergo various ordering transitions [3]. T_c usually drops. Two points of view can be taken when asking what constitutes the difference between the (214) and (123) materials leading to a 60K difference in T_c. One may argue that the chains themselves are important elements (e.g. due to their restricted dimensionality) for the superconducting state. Studies of vacancy ordering [4] and Cu NQR measurements [5] seem to support this point of view. On the other hand one may argue the essential elements are the Cu-O planes and that the chains are just a particular way to chemically stabilize an elevated electron (or hole) count in the planes which may be impossible by merely doping on the cation lattice. This point of view is unifying in that it considers the occurrence of superconductivity in both the (214) and (123) phases to be connected solely to the conducting Cu-O planes. Regardless of the answer to this particular question, however, most theoretical studies limit themselves to the physics of the conducting Cu-O planes. We will restrict our discussion here to the (214) compound as well.

THE BAND STRUCTURE PICTURE OF THE ELECTRONIC STRUCTURE OF La_2CuO_4

Soon after the experimental confirmation of high T_c superconductivity, calculations of the electronic band structure of $La_{2-y}X_y CuO_4$ appeared [6],[7] (and references in [2],[3]). Generally, the doping with element X was simulated by a rigid motion of the Fermi level. The bandstructure calculated by Mattheiss using the LAPW method within the local density functional scheme is reproduced in Fig. 3. The complex of 17 bands between -8 eV and $+2$ eV corresponds to mixtures of Cu(d) and O(p) orbitals. The unoccupied bands above 2 eV involve La orbitals including the flat 4f bands at ~4 eV. Only two (labeled A and B in Fig. 3) of the 17 bands of the Cu(d)-O(p) manifold are of primary interest. These arise from strong nearest neighbor (pdσ) interactions between Cu 3d orbitals of (x^2-y^2) symmetry (with some z^2 admixture) and neighboring O 2p orbitals that are directed along the short Cu-O bonds in the planes. The antibonding band (A) is half-filled in this zeroth-order picture of La_2CuO_4. The 15 intermediate bands correspond to more weakly bonding states including (pdπ) bonds and (pdσ) bonds along z, the longer Cu-O distance. The dispersion of bands A and B can well be reproduced in a simple nn tight binding model with two parameters ($E_d = E_p = -3.2$ eV and t(pdσ) = -1.85 eV)

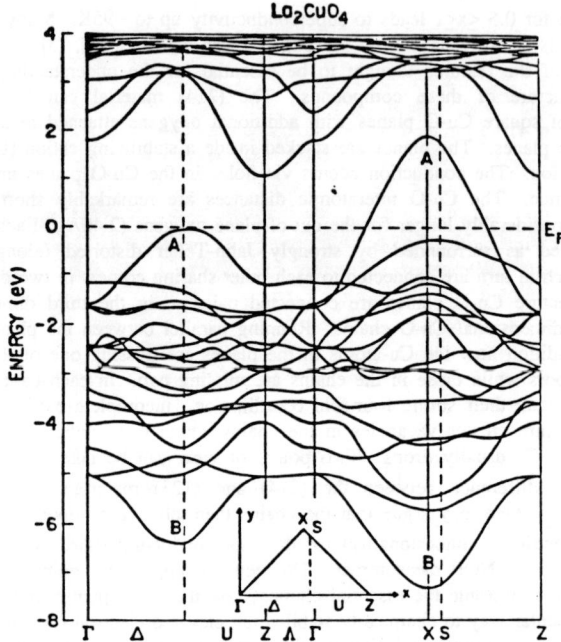

Fig. 3   LAPW energy bandstructure of La_2CuO_4 as calculated by Mattheiss [6].

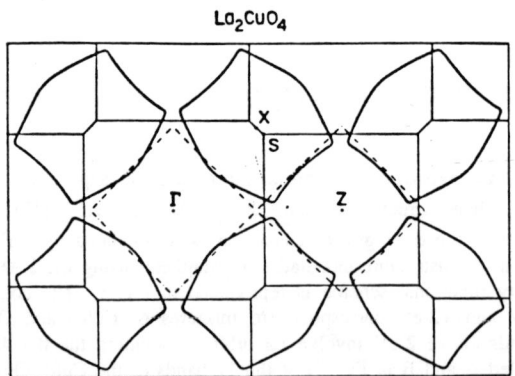

Fig. 4   Calculated [6] La_2CuO_4 Fermi surface displayed in an extended zone scheme. The dashed lines derive from a simple tight binding model.

collapsing the 15 bands into a single degenerate level at -3.2 eV.

When band A is half-filled the two-dimensional Fermi surface consists of squares centered at X for holes that nest perfectly those centered at Γ for electrons (dotted lines in Fig. 4). The actual three-dimensional calculations give results close to that simple picture (full lines in Fig. 4). Thus the bandstructure would predict metallic behavior for La_2CuO_4 in the tetragonal structure. Experimentally, a particular orthorhombic distortion is reported due to a small tilting of the oxygen octahedra. This, however, leaves all copper atoms equivalent and does *not* open a gap at the Fermi surface. However, since pure La_2CuO_4 is presumably an insulator the possibilities are either that the crystal has a lower symmetry due to a charge (CDW) - or spin density wave (SDW) or that there exists a correlation gap à la Mott-Hubbard without change of symmetry which would signal a breakdown of the band picture.

In fact, the model square type Fermi surface is highly unstable due to both the perfect nesting properties giving a $(\ln)^2$ type divergence in the susceptibility a $2k_F = (1/2\ 1/2\ 0)$ and a $(\ln)$-type van Hove singularity in the density of states. The small 3D interactions remove the exact divergences but strong susceptibility enhancements remain. It has been first argued by Mattheiss [6] and later shown by Weber [8] that a zone-boundary O-Cu breathing mode couples strongly and drives La_2CuO_4 into a CDW distorted state. Whether this state is truly insulating or a semi-metal remains unclear. However, in contrast to this bandstructure prediction no breathing mode distortions have been observed experimentally. On the contrary, under certain conditions (specific oxygen annealing) an antiferromagnetic (AF), insulating phase has been observed for La_2CuO_4 [9]. Spin-resolved band calculations using the local density LSD formalism fail to stabilize a SDW state [10]-[12] in spite of the large susceptibility value at $2k_F$. The reason is probably two-fold: three-dimensionality yields a band dispersion near E_F larger than the small spin-split gap and this gap is rather small itself because of the strong Cu-O hybridization suppressing AF moment formation on the Cu atoms. It may be argued that this is an artifact of the local density approximation which overemphasizes hybridization and therefore favors CDW formation over SDW formation.

Based on the band-picture it can furthermore be argued that doping (i.e. a shift of the Fermi level) will spoil the perfect nesting and either the CDW or the SDW would be suppressed allowing the formation of a superconducting state near the instability. Several theories starting from this weak coupling point of view have been proposed and shall be discussed below.

The nature of quasiparticles (QP) cannot directly be derived from density functional calculations. To use the E(k) bandstructure as QP dispersion fails even for simple metals like Na [13] or for semiconductors like Si [14]. This is also the case for La_2CuO_4 as demonstrated by the (unsuccessful) comparison of UPS photoemission structures [15] with peaks in the calculated density of states spectra. More troublesome than peak shifts of order 1-2 eV is, however, the fact that the UPS data indicate a significantly lower density of states near E_F than LDA calculations predict. This points at possible problems in the description of the ground state.

Summarizing the results of the bandstructure picture, calculations predict a metallic state for La_2CuO_4 which is strongly unstable towards CDW formation. A CDW is, however, not observed experimentally. On the other hand, the observed AF ordering cannot be stabilized as SDW within the band picture. This seems to indicate that the LSD band theory is inadequate to describe states near the Fermi level and that a Mott-Hubbard type picture may be more appropriate. Another result of the band picture that may be significant is the near degeneracy of the "renormalized" levels of Cu and O i.e. $E_p \approx E_d$.

This leads to a stronger hybridization than normally encountered in transition metal oxides where $E_p < E_d$ and where the band intersected by the Fermi level is narrow and mostly d-like. The hybridization in La_2CuO_4 as calculated within the LSD leads to a bandwidth of W ≈9 eV which is comparable to typical Coulomb repulsion energies U (see below), thus possibly rendering the simple Mott-Hubbard picture also marginal.

THE MOTT-HUBBARD PICTURE OF THE ELECTRONIC STRUCTURE OF La_2CuO_4

The individual dynamics of electron correlation is emphasized by considering Hubbard type Hamiltonian of the form:

$$H = \sum_{ij\sigma} \varepsilon_{ij} C^+_{i\sigma} C_{j\sigma} + \frac{1}{2} \sum_{ij\sigma\sigma'} U_{ij} C^+_{i\sigma} C_{i\sigma} C^+_{j\sigma'} C_{j\sigma'} \tag{1}$$

where i,j label oxygen or copper sites and where $C^+_{i\sigma}$ creates a hole with spin σ either in the Cu (x^2-y^2) or O(x,y) states which are assumed to be Wannier-type states of unspecified but appropriate range. The parameters of the model are the one-electron terms $\varepsilon_{ii} = \varepsilon_d$, $\varepsilon_{jj} = \varepsilon_p$, $\varepsilon_{ij} = t$ and the on-site $U_{ii} = U_d$, $U_{jj} = U_p$ and intersite $U_{ij} = U_{pd}$ Coulomb repulsion energies.

We may e.g. choose as the vacuum reference state Cu^+ (all d-states occupied) and O^{--} (all p-states occupied). The one-electron on-site energy difference $\varepsilon = \varepsilon_p - \varepsilon_d$ and the three Coulomb repulsion terms then refer to this ionic ground state. The final physics does depend on the choice of reference state, and so do the individual parameters. The energy of the system may be viewed as depending to second order on the occupation numbers. The dynamics of all electrons not explicitly treated by H are implicitly incorporated into the model parameters. This can be easily illustrated by comparing the free atom/ion Coulomb energies obtained from term-values; e.g. U (Cu^{++}) = 16.5 eV exceeds U (Cu^+) = 12.6 eV because of the absence of an extra screening electron. In addition to the appropriate choice of reference state, the orbital range associated with the $C^+_{i\sigma}$ is crucial. It is not clear whether ionic or atomic wave functions or longer range Wannier orbitals describe the dynamics of the valence electrons more appropriately. Neither is it clear whether parameters derived from XPS and Auger electron spectroscopic data are a good choice. In a simple one-band Hubbard model, where U >> t and $U_p = U_{pd} = 0$ these questions are not relevant but in a multiple band situation the relative size of parameters can be crucial.

With these caveats in mind, there are several ways to estimate the values of the parameters: XPS and Auger data analysis yields [16] e.g. U_d ≈9 eV, U_p ≈6 eV, U_{pd} ≈2.5 eV. These values can be used in a mean-field solution of Eq. 1 which when compared to the LDA bandstructure results give t ≈ 1.5 eV and ε ≈ 1.5 eV. Similar, but somewhat different values have been assumed in Ref. [17]. Parameter values (with the exception of t) may also be derived from total energy calculations within the LDA. The idea is to create localized, impurity-like charge (or spin) fluctuations of particular symmetry (e.g. Cu d) and calculate the energy of the system including self-consistent screening by imposing a constraining (localized) external potential [18]. The calculations are "groundstate"-like in the spirit of Density Functional Theory but for a restricted Hilbert space. The detailed form of the constraining external potential can become a rather involved projection operator but simplifying approximations are possible [19]. The result is a set of energies $E_i\{n_i\}$ belonging to a set of different charge distributions $n_i(r)$.

These can now be mapped onto a Hubbard Hamiltonian of the form (1) and the parameters can be extracted. Preliminary results [19], [20] indicate U-values similar, but somewhat larger than those derived spectroscopically in [16]. The size of the oxygen U_p is worth further studies. It can vary from near zero for an assumed ionic O^{--} configuration to values comparable to the copper U_d for neutral O^0. The LDA calculations probably tend to overestimate U_p because of the LDA's tendency to prefer more neutral Cu-O configurations. The intersite Coulomb repulsion U_{pd} is generally found to be smaller than the corresponding on-site values by a factor of about 3-5.

Although studied for many years, solutions to the Hamiltonian (Eq. 1) are known only in a few very limiting cases (single band, half-filling, reduced dimensionality, etc.). Reviews of certain aspects of the problem can e.g. be found in [21], [22]. For the present case of La_2CuO_4, a simple one-band picture involving only Cu^{++} is believed to result in antiferromagnetism as observed (at least in 3D). The energy scale is set by the super exchange coupling $J = \tilde{t}^2/U_d$ where $\tilde{t}$ is an effective Cu-Cu hopping matrix element. Similar arguments can be made for the two-band model with an effective J involving both copper and oxygen parameters [23].

The question is what are the low lying excitations in the system which in turn control the physics of J? To discuss the chemical nature of quasiparticles it is convenient to consider the (t=0) impurity model. For the insulating groundstate (La_2CuO_4), two configurations d^9p^6 and $d^{10}p^5$ are competing with a charge transfer energy difference $\Delta = \varepsilon + U_{pd}$. Doping will create holes on oxygen ($O^{--} \rightarrow O^-$) if $\varepsilon + 2U_{pd} < U_d$ or on copper ($Cu^{++} \rightarrow Cu^{+++}$) otherwise. The charge transfer excitation energy in the presence of oxygen holes is $\Delta = U_d - \varepsilon - 2U_{pd}$ while in the presence of copper holes it would be $\Delta = \varepsilon + 2U_{pd} - U_d$. These excitation energies of the two-band model have to be compared to the excitation energy of U_d for the one-band model. Zaanen et al. [24] have studied the competition between U_d and Δ for a large class of transition metal (TM) compounds and have derived a schematic metal-insulator phase diagram shown in Fig. 5. They concluded that the on-site Coulomb repulsion U_d controls the physical behavior of the early TM compounds while the charge transfer energy Δ becomes important for the late TM compounds. This is also in line with the parameter estimates given above. One may therefore conclude that Δ is important in determining the low-lying excitations and the nature of quasiparticles in La_2CuO_4 [25]. This point of view has also been expressed by Varma, et al., [26] who assume the doping holes to be created on the copper, while Jennison [16], Emery [17], and Hirsch [27] favor the existence of oxygen holes, citing experimental evidence arising from photoemission UPS data [28] and energy loss EELS data [29].

The picture for La_2CuO_4 that seems to emerge from Hubbard-type studies is that of an AF insulator with $J \sim t^4/\Delta^2$ $(1/\Delta + 1/U_d)$ [23], [25]. The value of J deduced from neutron [30] or light scattering [31] on 2D magnetic fluctuations is extremely large (~1000 cm^{-1}), conceivably due to a small Δ. The 3D ordering temperature is, however, relatively low because of frustration and weak inter-plane coupling and it is influenced by oxygen vacancies which may act as pinning centers. The size of J is rather unusual for a localized moment material and more reminiscent of itinerant bandlike SDW fluctuations. This borderline point of view is reinforced by the observation of nuclear quadrupole (NQR) signals [5] which are believed to survive only if either local electron moments fluctuate very fast or if the moments are of itinerant character.

When La_2CuO_4 is doped the AF phase is destroyed and a superconducting state is

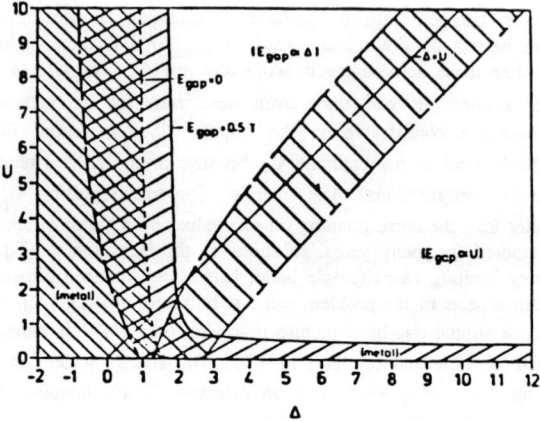

Fig. 5 Schematic phase diagram for the groundstate of transition metal (TM) compounds [24]. U is the TM on-site Coulomb repulsion and Δ is the charge transfer energy difference. The shaded areas are metallic.

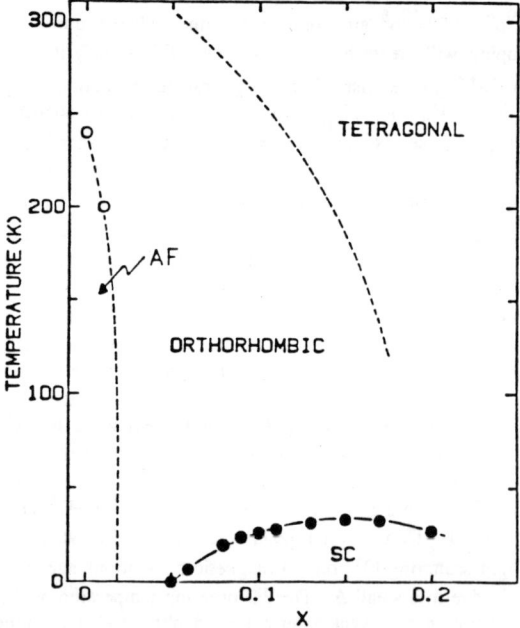

Fig. 6 Schematic phase diagram for the copper oxides as a function of carrier concentration x. AF denotes the antiferromagnetic phase, SC the superconducting phase.

facilitated. This is schematically shown in a phase diagram (Fig. 6) where x describes the concentration of holes. The exact nature of the state for finite x and the transition from the AF to the SC phase are not known. As mentioned above, chemically, the quasiparticles are likely to possess oxygen character. However, their exact spin structure involving the Cu atoms is much debated.

P. W. Anderson's resonant valence band (RVB) derived quasiparticles have soliton-like nature and allow for a separation of spin and charge [22], [32]. Electron-like quasiparticles ($c=e$, $s=1/2$) can be decomposed into "holons" ($c=e$, $s=0$) and "spinons" ($c=0$, $s=1/2$) which follow their own Bose or Fermi statistics. They are derived from a simple one-band Hubbard model and are therefore independent of the detailed parameter space discussed above. However, the legitimacy of a simple effective one-band model is not universally accepted for the copper oxides. Proposals by Jennison [16], Emery [17], Hirsch [27], and others all start from two-band models making specific assumptions about the relative size of the electron parameters. This leads to a variety of quasiparticles with exotic spin and charge structures which compliment the original ferromagnetic spin-polarons [33].

Theoretically the situation is thus rather confuse, since not much is known about the solution space of Eq. (1) away from half-filling. Experimentally, quasiparticle and collective excitation spectra should be accessible by optical reflectivity, light scattering, neutron scattering, etc. However, here too, no clear picture has emerged yet, in particular near the crossover (in Fig. 6) from the AF to the SC state.

SUPERCONDUCTIVITY IN THE COPPER OXIDES

Most theoretical discussions of possible mechanisms of superconductivity in the copper oxides are based on the assumption of two-particle pairing. Indeed, recent flux decoration experiments unambiguously show the quantization into pairs of charge 2e [34]. The general question that remains has then two aspects: i) Is the pairing BCS-like in the sense that the coherent length $\xi >> r_s$ the mean distance between carriers or are the particles so tightly paired that they pre-exist above T_c and undergo a Bose condensation as the temperature is lowered. If the pairing is BCS-like questions about strong vs. weak coupling and about the symmetry (s,d, etc.) of the pair-wave function are of importance. The second general question is: ii) What mechanism does the pairing? Is it through electron-phonon-interaction, through the exchange of electronic excitations (charge fluctuations) or through magnetic correlations (spin fluctuations)? The interpretation of thermodynamic measurements that usually answer question i) is severely complicated by the strong anisotropy of the material (factors of 100 in some cases), by the possible existence of multiple bands near E_F and by lack of material reproducibility. Few facts are accepted like the short coherence length $\xi \approx 10\text{-}30\text{Å}$ which coupled with a magnetic penetration depth of $\lambda > 1000\text{Å}$ makes the materials extreme type II superconductors.

Traditional electron-phonon coupling has been thoroughly investigated for $La_{2-y}X_yCuO_4$ for varying doping concentration [8]. The calculations were done based on the bandstructure picture. It was found that a large contribution to T_c (~30K) can under extreme conditions originate from coupling of electrons to Cu-O breathing mode vibrations. However, the coupling was found extremely strong ($\lambda \approx 3$) such that the underlying theory (Eliashberg) was used beyond its limits of validity: the phonon lifetime due to coupling to the electrons had become shorter than its vibrational period. Also, the same theory predicts pure La_2CuO_4 to become CDW distorted which, as discussed above is not observed.

The very strong coupling to phonons has also been described as the formation of polarons, their binding into bi-polarons with subsequent Bose condensation [35], [36], [37]. This scenario encounters the same difficulty of having to explain the absence of a CDW distortion in pure La_2CuO_4. Moreover, general arguments have been given as to why such coupling should always give a lower T_c than BCS-type coupling [38]. At any rate, a T_c of 95K for the (123) materials is most certainly outside the range of conventional electron-phonon coupling. Arguments for non-conventional coupling to structure inherent soft-modes have, however, been proposed [39]. Experiments on the isotope effect of O^{18} on T_c show a moderate effect in $La_{2-y}X_yCuO_4$ but virtually none in $YBa_2Cu_3O_7$ [2], [3].

The use of purely electronic excitations to couple electron pairs has been proposed in the past [40]-[43]. The idea leads naturally to high temperature superconductivity if the frequency ω_0 of the exchange boson in $T_c \sim \hbar \omega_0 \exp(-1/\lambda)$ is increased from a phonon Debye frequency to an electronic frequency of order (0.1-1.0) eV [26]. The concept is best discussed in terms of the electronic dielectric function $\varepsilon(q,\omega)$ which enters the screened interaction $V(q,\omega) = 4\pi e^2/q^2\varepsilon(q,\omega)$ between electrons. For pairs to form via electronic polarizations, $1/\varepsilon(q,\omega)$ has been attractive (overscreening) in a large enough part of momentum-frequency phase space without violating stability rules such as $\varepsilon(q,\omega) > 0$ for $\omega \rightarrow 0$ [43]. The attractive strength also has to overcome the long-range Coulomb repulsion, vertex corrections to the interaction have to be considered and a good coupling strength has to be achieved. These are difficulties which need quantitative discussion. Various scenarios of two spatially separated electron systems [40], the existence of strong local field modulations [42], [43] and the emphasis of transverse resonances [26] have been proposed to model a dielectric function $\varepsilon(q,\omega)$ with the appropriate characteristics. A strong coupling point of view involving charge and spin fluctuations within a two-band Hubbard model has been recently proposed by Hirsch [45]. In this model electronic bi-polarons where found stable for a limited range of U_d, U_p, U_{pd} and ε parameters $(U_d >> t$, $U_{eff} = U_p + 2\varepsilon - 2U_{pd} < 0)$. Whether these attractive excitations exist in the copper oxide systems is not clear at this point. If they do, more experimental evidence should become available from optical studies on single crystals.

The idea of effective electron pairing via magnetic fluctuations occurs for $U_{eff} > 0$ and follows from a Hubbard-type description with $U/t >> 1$ [22]. At large values of U/t the Hubbard Hamiltonian can be transformed into a Heisenberg Hamiltonian with a superexchange energy scale J. It was argued in the context of heavy fermion superconductivity that a superconducting state may exist near half-filling using the superexchange J as the BCS interaction [46]. This state has anisotropic d-wave or extended s-wave type gap functions which leads to particular experimental fingerprints [46] (the gap is zero in parts of k-space) which have been verified for heavy fermion compounds. Moreover, $T_c \approx 1K$ in these compounds.

The question is, can J be simply scaled up to give $T_c \approx 100K$ or can a different condensation occur for quasiparticles with strong magnetic interactions? The RVB theory of Anderson with its three types of quasiparticles, i.e. electrons, holons and spinons develops formally into a three-component condensate with charges 0, e and 2e. In this state gap-less spinon excitations continue to exist (they carry no charge) while it is argued that the condensation of holons (which together with spinons form electrons) is the "major phenomenon accounting for T_c." [22]. Because of the interconnection of three particles, some of which are Bosons, some of which are Fermions, the situation is very complicated and not yet transparent. More direct contact with experiment is needed. In particular the existence in the superconducting state of gap-less spin excitations is unique to this model. Some low temperature experimental work is in progress to investigate this question [3].

In other strong coupling model studies [16], [17], [45] the main emphasis is put on finding individual pairs to bind. Condensation is then usually discussed within the BCS-limit [17]. The essential feature in these models is the introduction of extra degrees of freedom in form of two-bands to allow for J values larger than the usual t^2/U and to permit coupling similar to the exchange of collective modes. The large U in one band creates the local moments and the second band acts as a low energy scattering channel. Bad news for all these models, however, comes from direct quantum Monte-Carlo simulations [27], [45] of few particles described by the Hubbard Hamiltonian (Eq. 1). No enhanced pair susceptibilities, which are thought to be indicative of the tendency to become superconducting, could be observed for a variety of one- and two-band Hubbard models. The reason for this is still unclear, maybe the simulations didn't go to low enough temperatures, maybe the sizes of the samples were too small or maybe there is an underlying physical reason that inhibits pairing at high T_c in simple models with large repulsive U.

CONCLUSIONS

The discovery of high T_c superconductivity in the copper oxides has caught most theorists by surprise, although it was never really clear why superconducting transition temperatures had to be so much lower than those of spin-or charge density waves. Since the discovery by Bednorz and Muller there are now many proposals that claim to offer reasonable explanations for the high T_c's. These are based on rather different point of views of the basic electronic structure of the oxides. It seems that simple pictures like the bandstructure picture or the extreme (large U, single band) Hubbard picture are insufficient to give an adequate description. More complex pictures, accounting more faithfully for the chemistry of these materials may be necessary. The materials are, aside from having the high T_c, unusual in many respects. They probably are structuraly metastable in their oxygen-intercalated superconducting phases; the corresponding insulating parent compounds show unusual magnetism; the materials are bad metals before they become superconducting and their conductivity does not follow simple Drude-like behavior. The observation of any of these general features is then used as argument to back up specific models for superconductivity, such as unusual electron-phonon coupling, coupling to magnetic fluctuations or coupling to charge fluctuations. The correct explanation is, however, far from evident and the challenge remains.

References

[1] J. G. Bednorz and K. A. Muller, Z. Phys. B *64*, 189 (1986); C. W. Chu, P. H. Hor, R. L. Meng, L. Gao, Z. J. Huang and Y. Q. Wang, Phys. Rev. Lett. *58*, 405 (1987).
[2] Proceedings of the 1987 Spring MRS meeting, Anaheim, Symposium S, ed. D. Gubser and M. Schluter.
[3] Proceedings of the Workshop on "Novel Superconductivity", Berkeley 1987, ed. S. A. Wolf and V. Z. Kresin, Plenum, New York, 1987.
[4] B. Batlogg, R. J. Cava, C. H. Chen, G. Kourouklis, W. Weber, A. Jayaraman, A. E. White, K. T. Short, E. A. Rietman, L. W. Rupp, S. M. Zahurak and D. J. Werder, ref. 3, p. 653.

[5] W. W. Warren, Jr., R. E. Walstedt, G. F. Brennert, G. P. Espinosa and J. P. Remeika, Phys. Rev. Lett. *59*, 1860 (1987).
[6] L. F. Mattheiss, Phys. Rev. Lett. *58*, 1028 (1987).
[7] J. Yu, A. J. Freeman and J. H. Xu, Phys. Rev. Lett. *58*, 1035 (1987).
[8] W. Weber, Phys. Rev. Lett. *58*, 1371 (1987).
[9] D. Vaknin, S. K. Sinha, D. E. Moncton, D. C. Johnston, J. Newsam, C. R. Safinya and H. E. King, Jr., Phys. Rev. Lett. *58*, 2802 (1987).
[10] T. C. Leung, X. W. Wang and B. N. Harmon, Phys. Rev. B., in print.
[11] P. A. Sterne, and C. S. Wang, to be published.
[12] O. Jepsen, J. Zaanen and O. K. Andersen, to be published.
[13] E. Jensen and E. W. Plummer, Phys. Rev. Lett. *55*, 1918 (1985); J. E. Northrup, M. S. Hybertsen and S. G. Louie, Phys. Rev. Lett. *59*, 819 (1987).
[14] L. J. Sham and M. Schluter, Phys. Rev. Lett. *51*, 1888 (1983).
[15] P. D. Johnson, S. L. Qiu, L. Jiang, M. W. Ruckman, M. Strongin, S. L. Hubbert, R. F. Garrett, B. Sinkovic, N. V. Smith, R. J. Cava, C. S. Jee, D. Nichols, E. Kaczanowicz, R. E. Salomon and J. E. Crow, Phys. Rev. B *35*, 8811 (1987).
[16] E. B. Stechel and D. R. Jennison, to be published.
[17] V. J. Emery, Phys. Rev. Lett. *58*, 2794 (1987).
[18] P. H. Dederichs, S. Blugel, R. Zeller and H. Akai, Phys. Rev. Lett. *53*, 2512 (1984).
[19] M. S. Hybertsen, N. S. Christensen and M. Schluter, to be published.
[20] C. F. Chen, X. W. Wang, T. C. Leung and B. N. Harmon, to be published.
[21] D.Vollhard, Rev. Mod. Phys. *56*, 99 (1984).
[22] P. W. Anderson in "Frontiers and Borderlines in Many Particle Physics", Varenna, July 1987, preprint.
[23] F. C. Zhang and T. M. Rice, to be published.
[24] J. Zaanen, G. A. Sawtzky and J. W. Allen, Phys. Rev. Lett. *55*, 418 (1985).
[25] Z. Shen, J. W. Allen, J. J. Yeh, J. S. Kang, W. Ellis, W. Spicer, I. Lindau, M. B. Maple, Y. D. Dalichaouch, M. S. Torikachvili and J. Z. Sun, to be published.
[26] C. M. Varma, S. Schmitt-Rink and E. Abrahams, Sol. Stat. Comm. *62*, 681 (1987) and ref. [3].
[27] J. E. Hirsch, Phys. Rev. Lett. *59*, 228 (1987).
[28] P. Thiry, G. Rossi, Y. Petroff, A. Revcholevschi and J. Jegaudez, to be published.
[29] M. Nucker, J. Fink, B. Renker, D. Ewart, C. Politis, J. W. P. Weijs and J. C. Fuggle, Z. Phys. B, in print.
[30] G. J. Shirane, Y. Endoh, R. J. Birgeneau, M. A. Kastner, Y. Kokada, M. Oda, M. Suzuki and T. Murakami, Phys. Rev. Lett. *59*, 1329 (1987).
[31] K. B. Lyons, P. A. Fleury, J. P. Remeika and T. J. Negran, to be published. K. B. Lyons, P. A. Fleury, L. F. Schneemeyer and J. V. Waszczak, to be published.
[32] P. W. Anderson, Science *235*, 1196 (1987).
[33] W. F. Brinkman and T. M. Rice, Phys. Rev. B *2*, 4302 (1970).
[34] P. L. Gammel, D. J. Bishop, G. J. Dolan, J. R. Kwo, C. A. Murray, L. F. Schneemeyer and J. Waszczak, Phys. Rev. Lett. *59*, 2592 (1987).
[35] B. K. Chakraverty, J. de Physique Lett. *40*, 99 (1979).
[36] P. Prelovsek, T. M. Rice and F. C. Zhang, J. Phys. C *20*, L229 (1987).
[37] H. Kamimura, Jpn. J. Appl. Phys. *26*, L627 (1987).
[38] P. Nozieres and S. Schmitt-Rink, J. Low Temp. Phys. *59*, 195 (1985).
[39] J. C. Phillips, to be published.
[40] W. A. Little, Phys. Rev. *134*, A1416 (1964).
[41] V. L. Ginzburg, Sov. Phys. Usp. *13*, 335 (1970).

[42] D. Allender, J. Bray and J. Bardeen, Phys. Rev. B 7, 1020 (1973).
[43] M. L. Cohen and P. W. Anderson in "Superconductivity in d- and f-Band Metals", ed. D. H. Douglas (AIP, New York, 1972).
[44] J. Ruvalds, Adv. Phys. 30, 677 (1981) and ref. [3], p. 455.
[45] J. E. Hirsch, to be published.
[46] J. E. Hirsch, Phys. Rev. Lett. 54, 1317 (1985).
[47] S. Schmitt-Rink, K. Miyake and C. M. Varma, Phys. Rev. Lett. 57, 2575 (1986).

PROPERTIES OF HIGH T_c PHONON-EXCITON SUPERCONDUCTORS

J.P. CARBOTTE
Department of Physics, McMaster University,
Hamilton, Ontario, Canada L8S 4M1

ABSTRACT

We have studied the properties of superconductors with phonon, exciton or some other boson exchange mechanism. A combined mechanism is also considered. Comparison with experimental results is attempted with a view to getting some information on the mechanism. Present uncertainties in data make such comparisons inconclusive.

I - INTRODUCTION

The mechanism responsible for the superconductivity of the high T_c oxides is not yet known, although many suggestions have been put forward.[1] The discovery of a significant isotope effect in La Sr Cu O[2,3] suggests that the phonons are playing an important role. This is an attractive possibility since it is the electron-phonon interaction that is responsible for the superconductivity of all conventional superconductors. On the other hand, the near zero isotope effect[4,5] in Y Ba Cu O favours a new mechanism which has its origin in electronic processes since these do not depend directly on the phonons. A suggestion which is worth studying is that of a combined phonon and exciton mechanism described within the framework of conventional Eliashberg theory. Such a suggestion has the attractive feature that a single theory would apply to the conventional as well as oxide superconductors. The differences would arise solely from the relative weight of phonon and exciton (or some other boson) spectral contribution.

In this lecture, we study the properties of superconductors that can be described within Eliashberg theory, at least as a first approximation. We do not specify the origin of the kernels in these equations and leave them general. It is found that a single parameter can usefully be introduced which simply measures the average boson exchange frequency ($\omega_{\ell n}$) involved and describes reasonably well superconducting properties. Results for thermodynamic and other properties are given as a function of $T_c/\omega_{\ell n}$, for the weak coupling limit $T_c/\omega_{\ell n} \sim 0$, the conventional strong coupling regime $T_c/\omega_{\ell n} \lesssim 0.25$ and the very strong coupling limit $T_c/\omega_{\ell n} \simeq 1.0$. When possible, comparisons with experiment are presented. Section II is devoted to the conventional phonon case while Section III deals specifically with La Sr Cu O and examines the possibility that it might be a conventional strong coupling system. In Section IV, we give results for the very strong coupling limit and consider Y Ba Cu O. In Section V, we treat a combined phonon-exciton mechanism while conclusions can be found in Section VI.

II - CONVENTIONAL PHONON CASE

The fundamental equations that underlie all the work to be described here are the Eliashberg equations for the pairing energy $\tilde{\Delta}(n)$ and renormalized frequency $\tilde{\omega}(n)$ in Matsubara representation.[6] The kernels in these equations involve the microscopic parameters responsible for superconductivity which are the Coulomb repulsion pseudopotential μ^* opposing superconductivity and the electron-boson exchange spectral density $\alpha^2 F(\omega)$ that gives the required effective electron-electron attraction.

For conventional superconductors the bosons exchanged are the phonons and $\alpha^2 F(\omega)$ describes a phonon frequency (ω) distribution in which each phonon mode ω is properly weighted by the strength of its coupling to the electrons through the electron-phonon vertex. For the high T_c oxides this may no longer be appropriate and other excitations may be involved. It has been suggested that plasmons[7,8] or excitons[9,10] play a role. For our purposes here it is not important to favour one over the other since we plan to model the necessary kernel $\alpha^2 F(\omega)$. The spectral density would then have two parts, one describing phonon exchange and the other the exchange of some second boson centered around its appropriate boson characteristic frequency.

For many conventional superconductors the electron-phonon spectral density is well known from tunneling data. The current voltage characteristic I(V) contains an image of both $\alpha^2 F(\omega)$ and μ^* which can be recovered by inversion of the Eliashberg equations. The technique developed by McMillan and Rowell[11] is widely used and now standard.

For the many superconductors for which tunneling data has been inverted, we have calculated thermodynamic and other properties which follow directly from the Eliashberg equations at finite temperature. For thermodynamic properties all that is needed in addition to $\tilde{\Delta}(n)$ and $\tilde{\omega}(n)$ is a free energy formula[6] which is well known.

Results for the normalized specific heat jump at T_c are presented in Fig. 1. The dark solid points indicated are for the various conventional systems as labeled.[12] For further identification of the materials involved see the original work.[12] While they do not all fall on a smooth curve the general trend is clear when $\Delta C(T_c)/\gamma T_c$ is plotted against the strong coupling parameter $T_c/\omega_{\ell n}$. Here γ is the Sommerfeld constant, and ΔC the difference between superconducting and normal state electronic specific heat. The parameter $\omega_{\ell n}$ is related to the boson exchange spectral density $\alpha^2 F(\omega)$ through a logarithm average[13] and is interpreted as some appropriate average boson energy in which the low frequency part of $\alpha^2 F(\omega)$ is given more weight than is the high frequency part. It is extremely important to realize that Fig. 1 relates to good approximation, the specific heat jump to the characteristic boson energy involved in the attractive effective electron-electron interaction. Thus, in principle, a measurement of this quantity can reveal information on $\omega_{\ell n}$. Note that the BCS limit (1.43) corresponds to $T_c/\omega_{\ell n} \to 0$ which for T_c of the order of 100K would imply a mechanism for which the characteristic $\omega_{\ell n}$ becomes quite large (excitons).

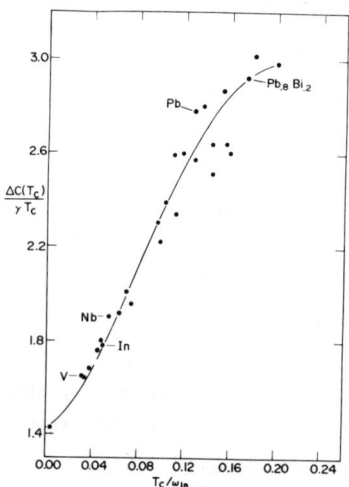

Fig. 1 The normalized specific heat jump $\Delta C(T_c)/\gamma T_c$ as a function of $T_c/\omega_{\ell n}$ for many conventional superconductors.

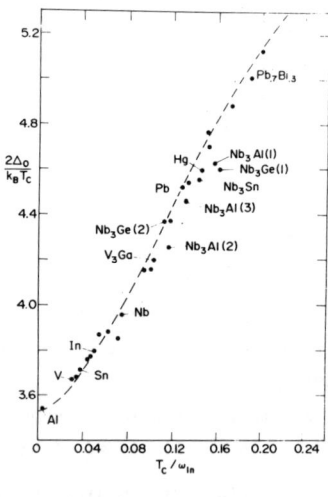

Fig. 2 Same as figure 1 but for the dimensionless ratio $2\Delta_0/K_B T_c$.

Other similar results have been found for the gap in the excitation spectrum Δ_0 divided by T_c[14] and the dimensionless ratio $\gamma T_c^2 | H_c^2(0)$ [12] where $H_c(0)$ is the zero temperature thermodynamic critical magnetic field The normalized gap $(2\Delta_0 | K_B T_c)$ is shown in Fig. 2. In principle the measurement of any of the above quantities could give $T_c/\omega_{\ell n}$ and so $\omega_{\ell n}$ for that particular system. Measurements of several such quantities would give consistency conditions imposed by the Eliashberg equations.

III - THE CASE OF $La_{1.85} Sr_{0.15} Cu O_4$

Isotope effect experiments for $La_{1.85} Sr_{0.15} Cu O_4$ in which oxygen O^{16} is replaced by O^{18} gives a partial isotope effect of 0.16 ± 0.02 in one experiment and 0.2 ± 0.1 in another.[3,4] This is very significant and implies a large phonon contribution. While tunneling data does not exist on the electron-phonon spectral density $\alpha^2 F(\omega)$ for $La_{1.85} Sr_{0.15} Cu O_4$ with $T_c = 36K$, W. Weber[15] has calculated it from first principles. His results are reproduced in Fig. 3. If only the oxygen mass is changed in this spectral density, the partial isotope effect that results is about 0.3 which is not so far from the experiment. This leads us to raise the question: could $La_{1.85} Sr_{0.15} Cu O_4$ largely be a conventional electron-phonon superconductor? This question can be partially answered by consideration of the predicted thermodynamic properties resulting from the spectral density of Fig. 3 and comparison with the experimental data.

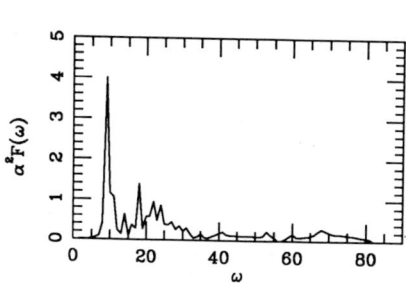

Fig. 3 The electron-phonon spectral density for $La_{1.85} Sr_{0.15} Cu O_4$ calculated by W. Weber.

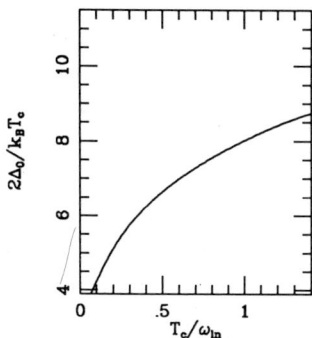

Fig. 4 The dimensionless ratio $2\Delta_0 | K_B T_c$ as a function of $T_c/\omega_{\ell n}$

The electron phonon mass enhancement λ (twice the inverse moment of $\alpha^2 F(\omega)$) corresponding to Weber's[15] spectrum is 2.6 and $\omega_{\ell n}$ = 14.0 mev which gives $T_c/\omega_{\ell n}$ = 0.23. This falls just at the edge of the conventional strong coupling range of Figs. 1 and 2. If we calculate the normalized gap ratio and the specific heat jump, we obtain respectively (for T_c = 36.0K)[16] $2\Delta_0/K_B T_c$ = 5.3 and $\Delta C(T_c)/\gamma(0)T_c$ = 2.8.

Before comparing with experiment some explanation is required about the value of $\gamma(0)$ used. The Sommerfeld $\gamma(0)$ cannot be obtained directly from low temperature work on normal La Sr Cu O because its superconductivity cannot be quenched by a magnetic field. Indirect methods exist. One such method is to measure the slope of the second upper critical magnetic field $H_{c2}(T)$ at T_c. Some experimental results for the slope are summarized in Table 1.[17] We see considerable variation from experiment to experiment and great differences between single crystal results and polycrystals. This implies serious differences in $\gamma(0)$ which is proportional to this slope and to the inverse of the resistivity in the dirty limit. Use of pure limit formulas can lead to further differences. The conclusion is that $\gamma(0)$ is not well known, so that the normalized jump is not well defined. On the other hand, $\Delta C(T_c)$ itself is fairly well determined experimentally as can be seen from Table I.

We can use our prediction for $\Delta C(T_c)/\gamma(0)T_c$ and the measured $\Delta C(T_c)$ of approximately 17 mJ/mole K^2 to deduce $\gamma(0)$. We get $\gamma(0)$ = 6mJ/mole K^2 a value that is not very different from some of the measured values. This is reassuring. The Sommerfeld $\gamma(0)$ is related to the single spin electronic density of states at the Fermi surface (N(0)) through the formula $\gamma(0)$ = (2/3) $\pi^2 K_B^2$ N(0) (1+λ) with K_B the Boltzmann constant. This formula implies a value for N(0) = 0.36 states/ev Cu atom which is in serious disagreement with the theoretical value of Freeman et al.[18] of 0.9 states/ev Cu atom.

TABLE I

$2\Delta_0/K_B T_c$	FIR	2.4	1.6-2.7	2.5			
	TUNNEL	2.9-4.5	5.2-9.1	<4.5	5-8.7	4.1-4.8	4.5-5.8
$\Delta C(T_c)/T_c$		7.6 ± 1.8	20.0 ± 5.0	16.8	8.8	22.0-26.0	
$H_{c2}(T_c)$ (T/K)		2.2-5.0	1.51	1.7	2.7-6.0	1.8	$(.3_{\parallel}, 4_{\perp})$

Some properties of $La_{1.85} Sr_{0.15} Cu O_4$ determined in various experiments.

We conclude that a pure phonon mechanism is not very consistent with the band structure results. This could mean that a pure phonon mechanism needs to be abandoned or perhaps that strong correlation effects invalidate a band structure picture. An argument against band theory that is often given is that the parent stochiometric compound $La_2 Sr O_4$ is an insulator and shows magnetic properties.

The dimensionless gap ratio given in Table 1 shows great variation from one experiment to another. Note that, in general, the far infrared absorption experiments give a small gap while tunneling gives relatively larger values. A BCS value for the gap would rule out a pure phonon mechanism. On the other hand, some of the tunneling results are not inconsistent with our theoretical results. In this comparison, it should be kept in mind that our calculations deal with an isotropic superconductor while $La\ Sr\ Cu\ O_4$ is very anisotropic as can be seen from the few available single crystal results. It is clear that no definitive conclusion can be made at this point largely because of uncertainties in experimental data. If some consensus could be reached as to the value of the energy gap, we could then deduce a value for $\omega_{\ell n}$ which would give valuable information on the mechanism. Better experiments are clearly needed. It should be stressed, however, that there may not be a sharp gap in these superconductors. This represents a further complication in comparing theory and experiment.

IV - THE VERY STRONG COUPLING REGIME

While La Sr Cu O would be at the borderline of the conventional strong coupling regime if it turned out to be an electron phonon system, Y Ba Cu O would be very different. First, the isotope effect is virtually nonexistent,[4-5] a result which would need to be explained and secondly, its $T_c/\omega_{\ell n}$ ratio would be unusually large. No spectral density for Y Ba Cu O is known to us but it is not unreasonable to assume that its characteristic phonon energy would not be so different from that of La Sr Cu O. This implies that $T_c/\omega_{\ell n} \simeq 0.6$, a value well beyond the conventional strong coupling regime. We will call this the very strong coupling limit and study the thermodynamic and other properties of superconductors in this limit.

As a model spectral density, we take Weber's $\alpha^2 F(\omega)$ for La Sr Cu O and multiply it by a constant B.[19] At the same time, we introduce a scaling δ on the frequency axis so as to soften or stiffen the phonon spectrum. To be definite in what follows, we adjust B so as to fix T_c at 96K. At the same time, we vary δ to change the ratio $T_c/\omega_{\ell n}$ from weak coupling $T_c/\omega_{\ell n} \simeq 0$ to conventional strong coupling $T_c/\omega_{\ell n} \leq 0.25$ and finally to very strong coupling with $T_c/\omega_{\ell n} \simeq 1.0$.

Results for the gap ratio $2\Delta_0/K_B T_c$ as a function of $T_c/\omega_{\ell n}$ are shown in Fig. 4. We see that this quantity keeps rising in the range considered and is of the order of 7 for $T_c/\omega_{\ell n} = 0.6$ which would be representative of Y Ba Cu O if the phonons are involved. In Table II, we show experimental results from infrared measurements, tunneling and NMR. All results except one are for polycrystaline samples. It is clear that there is little agreement at present between infrared measurement which give small BCS like gap and tunneling results which give much larger values. While the variation from experiment to experiment prevents us from making a definitive conclusion, it is clear that the theoretical estimate of 7 is not favoured. A smaller value for the average gap would be more consistent. This makes it unlikely that we are dealing with a pure phonon mechanism.

In Fig. 5, we show results for the specific heat jump (solid curve) at T_c as a function of $T_c/\omega_{\ell n}$. Initially, $\Delta C(T_c)/\gamma(0)T_c$ increases above the

TABLE II

$2\Delta_0/K_B T_c$	FIR	3.5	2.5-4.2	1.6-3.4	2.3-3.5	3.2	3.5	3.0	3.3
	TUNNEL	13	10	3.7-5.6	3.81-4.52	3.9	4.8	3.2	11
		4.8	(4.5-6.0)$_\perp$	and	(3.9-4.8)$_\parallel$				
	NMR	1.3							

The gap ratio for Y Ba$_2$ Cu$_3$ O$_{7-\delta}$

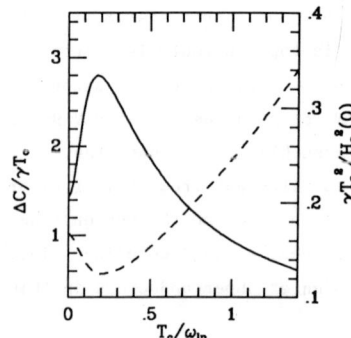

Fig. 5 The normalized specific heat jump $\Delta C(T_c) \mid \gamma T_c$ and $\gamma T_c^2 / H_c^2(0)$ as a function of $T_c/\omega_{\ell n}$.

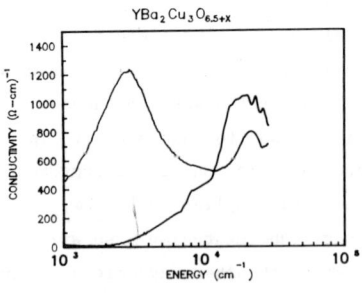

Fig. 6 The AC optical conductivity of Y Ba Cu O as a function of energy in the exciton region.

BCS value of 1.43 in accordance with our experience for conventional strong coupling systems. A maximum is, however, reached around $T_c/\omega_{\ell n} = 0.3$ and then the reduced specific heat drops back down and attains values less than 1.43 in the very strong coupling limit. This behaviour was unexpected. For $T_c/\omega_{\ell n} \sim 0.6$, we note that we are already below 1.43. This is confirmed in experiments.[20] While a BCS value can be obtained in two very distinct limits, namely for $T_c/\omega_{\ell n} \sim 0$ and ~ 0.6, the corresponding $2\Delta_0/K_B T_c$ are very different in these two cases and the weak coupling limit is favoured because of the lack of isotope effect and the small value of $2\Delta_0/K_B T_c$ observed in most experiments.

V - POSSIBLE COMBINED PHONON-EXCITON MECHANISM

A real possibility is that we have a combined exciton and phonon mechanism with the exciton getting progressively more important as we go to higher values of T_c. Such a possibility was already considered by Allender et al.[9] on theoretical grounds. There is experimental evidence for this. In Fig. 6, we show results of K. Kamaras and co-workers[21] for the optical conductivity of Y Ba Cu O as a function of energy. Note that a peak is observed indicating a strong absorption around 3.0×10^3 cm^{-1} (left hand side of Figure). This could only arise from some electronic mechanism.

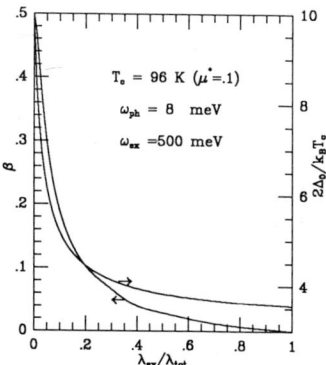

Fig. 7 The dimensionless ratio $2\Delta_0/K_B T_c$ and the isotope coefficient as a function of $\lambda_{ex}/\lambda_{tot}$.

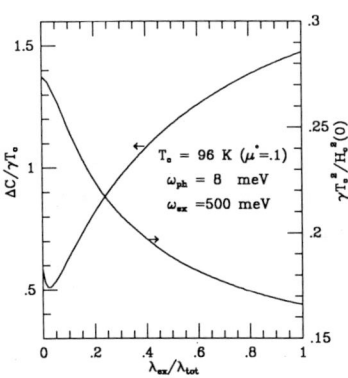

Fig. 8 The normalized specific heat jump $\Delta C(T_c)/\gamma T_c$ and $\gamma T_c^2/H_c^2(0)$ as a function of $\lambda_{ex}/\lambda_{tot}$.

Also, when enough oxygen is removed from Y Ba Cu O to destroy the superconductivity, the peak in the absorption is destroyed (curve with no absorption around 3.0×10^3 cm^{-1}) leading one to speculate about a possible connection between exciton peak and superconductivity.

The situation can be modeled by a spectral density having two parts, one at lower energy (8 mev) simulating phonon exchange and another at energies of the order of 500 mev to simulate exciton exchange.[22] The size of these peaks which we can take to be delta functions, can be characterized by the value of mass enhancement parameter λ_{ph} and λ_{ex} from phonon and exciton respectively with the total given by $\lambda_{tot} = \lambda_{ph} + \lambda_{ext}$

In Fig. 7, we show some results for the isotope coefficient β which is defined by $\beta = d \ln T_c / d \ln M$ with M the average ion mass in the system. The horizontal axis gives $\lambda_{ex}/\lambda_{tot}$ so that 0.0 corresponds to a pure phonon mechanism and 1.0 to pure exciton. In this latter case $\beta = 0$ and $2\Delta_0/K_B T_c$ is BCS like. As λ_{ph} increases towards the pure phonon region, β and $2\Delta_0/K_B T_c$ increase together, but not very rapidly at first. It is not until $\lambda_{ex}/\lambda_{tot} \cong 0.2$ that a sharp rise in these two quantities is observed. Small values of $2\Delta_0/K_B T_c$ favour the exciton mechanism as does, of course, β near zero. It is clear, however, that a considerable contribution from the phonons could be present in Y Ba Cu O without violation of the isotope effect results.

The situation is similar for the specific heat jump which we show in Fig. 8, along with the other dimensionless ratio $\gamma T_c^2 / H_c^2(0)$. It is clear that, for the pure exciton case, we have BCS results, but that, as the phonon contribution is increased, $\Delta C / \gamma(0) T_c$ can fall to values considerably below BCS and $\gamma T_c^2 / H_c^2(0)$ correspondingly increase way above the pure exciton limit. When $\lambda_{ex}/\lambda_{tot}$ is around 0.2 for which case β and $2\Delta_0/K_B T_c$ are still close to the pure exciton limit, the other two indices are not. In principle, such curves could be used to extract the value of λ_{ex} involved. In practice, the data is not sufficiently good to accomplish this. For the specific case considered, a value of λ_{ex} of order 0.5 is sufficient to explain the superconductivity of Y Ba Cu O. Such values are not unrealistic and within the estimates provided by Allender et al.[9] λ_{ex} decreases only slowly with increasing admixture of λ_{ph}. It is only when the pure phonon regime is approached that λ_{ex} decreases percipitously.

VI - CONCLUSIONS

We have calculated the thermodynamic and other properties of superconductors as a function of the strong coupling parameter $T_c/\omega_{\ell n}$ where $\omega_{\ell n}$ is a characteristic boson energy associated with the electron boson exchange spectral density. The weak coupling, conventional strong coupling, and very strong coupling regime were examined and comparison with experimental data attempted. In principle, it should be possible to determine from such comparisons a value for $\omega_{\ell n}$. In practice, we found that the data is, at the moment, not sufficiently consistent from experiment to experiment to allow a sharp conclusion. Still, some of the data indicates a possible combined phonon and exciton mechanism. In this instance, we find deviations from BCS values for various superconducting properties which are quite distinctive: a feature which could be exploited to determine the proper admixture of phonon and exciton contribution to T_c should the data become sufficiently well defined. I thank my colleagues, R. Akis, J. Blezius, F. Marsiglio, E. Schachinger, and M. Schossmann.

1) Many papers in *Novel Superconductivity* (edited by S.A. Wolf and V.Z. Kresin, Plenum Press New York) 1987.
2) B. Batlogg et al., Phys. Rev. Lett. 59, 912 (1987).
3) T.A. Faltens et al., Phys. Rev. Lett 59, 915 (1987).
4) B. Batlogg et al., Phys. Rev. Lett. 58, 2333 (1987).
5) L.C. Bourne et al., Phys. Rev. Lett 58, 2337 (1987).
6) J.M. Daams and J.P. Carbotte, Jour. Low Temp. Phys. 43, 263 (1981).
7) J. Ruvalds, Phys. Rev. B35, 8869 (1987).
8) V. Kresin, Phys. Rev. B35, 8716 (1987).
9) D. Allender, J. Bray and J. Bardeen, Phys. Rev. B7, 1020 (1973).
10) C.M. Varma, S. Schmitt-Rink, and E. Abrahams, Solid State Comm. 62, 681 (1987).
11) W.L. McMillan and J.M. Rowell in, *Superconductivity* (edited by R.D. Parks, Dekker, New York, 1969) Vol. 1, p. 562.
12) F. Marsiglio and J.P. Carbotte, Phys. Rev. B33, 6141 (1986).
13) P.B. Allen and R.c. Dynes, Phys. Rev. B12, 905 (1975).
14) B. Mitrovic, H.G. Zarate, and J.P. Carbotte, Phys. Rev. B29, 184 (1984).
15) W. Weber, Phys. Rev. Lett. 58, 1371 (1987).
16) M. Schossmann, F. Marsiglio and J.P. Carbotte, Phys. Rev. B36, 3627 (1987).
17) Tables 1 and 2 were compiled at McMaster. The experimental references used can be found in a review to appear in the proceedings of the NSERC-CAP 1987 Summer Institute in Theoretical Physics edited by F. Khanna and Z. Umezawa (to be published).
18) A.J. Freeman, J. Yu and C.L. Fu, Phys. Rev. B36, 7111 (1987).
19) F. Marsiglio, R. Akis and J.P. Carbotte, Phys. Rev. B36, 5245 (1987).
20) S.E. Inderhees, M.B. Salamon, T.A. Friedmann, and D.M. Ginsberg, Phys. Rev. B36, 2401 (1987).
21) K. Kamaras et al., Phys. Rev. Lett. 59, 919 (1987).
22) F. Marsiglio and J.P. Carbotte, Phys. Rev. B36, 3937 (1987).

Antiferromagnetic Interactions in High-T_c Superconductors

R.A. Barrio

Instituto de Investigaciones en Materiales
Universidad Nacional Autónoma de México
Apdo. Postal 70-360, 04510, México, D.F.

ABSTRACT

A brief review of the theoretical ideas floating around high-T_c superconductors is given. Stress is put onto the magnetic ordering and its possible relationship to superconductivity.

I. CONVENTIONAL IDEAS

Up to date, there is not a satisfactory theoretical explanation of 90 K superconductivity in ceramic materials. So far, it remains as an "Act of God", in the sense that nobody is to be blamed for it. Theorists are in an uncomfortable position, trying to predict results from experiments made in rather dirty materials, full with all sorts of defects as grain boundaries [1], twins [2], stacking faults [3], oxygen disorder [4] and deviation from stoichiometry [5]. However, it was established experimentally, quite promptly, that charge is carried in pairs [6] and therefore it was tempting to start looking at the Bardeen, Cooper and Schrieffer (BCS) theory [7], since all the theoretical elements have been already worked out there and only an investigation of the parameter space is needed to adjust the critical temperature and other measurable superconducting properties.

The naïve procedure of applying BCS is quite straightforward: First, one needs to identify a boson-like excitation, preferably with a high energy $\hbar\omega_{exc}$, then one assumes that the scattering amplitude for exchanging such a boson between a pair of fermions $(k\uparrow,-k\downarrow)$ is negative for energies below $\hbar\omega_{exc}$, (attractive interaction in the Frölich sense).

Then one assumes that the effective coupling constant λ is greater than the effective Coulomb repulsion μ^*, and one concludes directly that a BCS type of relation $T \sim \hbar\omega_{exc} \exp[1/(\lambda-\mu^*)]$ holds.

As an example of this sort of procedure let me mention a calculation derived from the detection of extremely sharp peaks at ~ 500 meV in the IV curve from a tunneling experiment [8], which have been further investigated in Japan [9]. Such excitations are around energies proper for electrons, and may be thought to be produced by states of very polarizable electrons situated in the non-bonding localized oxygen P_z orbitals. Such states are assumed to act in a similar way as charged impurities in semiconductor, which greatly reduce the effective Coulomb repulsion through polarization terms: $\mu^*=U-2K$, where $K=e^2(1-1/\varepsilon)/R$, and U is the Hubbard repulsion. R is an effective measure of the radius of the polarized region around the impurity, and ε is the dielectric function. Using reasonable values, extracted from the experiments available, like $m^*=5m_0$, $\xi_0 \sim 30 \text{Å}$ and $a=3.78$ Å, the excited states were calculated to be of the order of the observed energies (~ 250 meV).

The temprature dependence of these excitations has been investigated [9] and they, most certainly, should be connected to the superconducting mechanism, since they disappear above 91.3 K in Y-Ba-Cu-O.

However, there is a word of warning, since the detailed description of a BCS pairing mechanism is much more laborious. There are beautiful and detailed calculations for predicting T_c using the Eliashberg equations [10].

Spanning the space of parameters in a rather systematic way, one can predict correct numbers for the La compounds but not for the Yttrium ones, and one is bound to think that the 1-2-3 materials belong to a new group of superconductors [11].

One of the main difficulties faced when adjusting the parameters of Y-Ba-Cu-O is the restriction imposed by the small isotope effect measured in the material. The isotope effect is predicted to be $T_c \sim M^{-\alpha}$, where $\alpha=1/2[1-\mu^*/(\lambda-\mu^*)]$, and therefore, the possible values of the parameters are greatly restricted.

In view of all these facts, most people are endured to think that, at least in these materials, phonons are not the only responsible for electron pairing.

II. NEW MECHANISMS

These early failures of having a satisfactory fitting of BCS predictions in the new high-T_c materials have produced a renewal of old theoretical ideas about possible superconducting states, arising from mechanisms different from conventional electron phonon pairing. Among these, the bipolaron models [12] have been mentioned.

However, they present difficult conceptual problems when one tries to predict high critical temperatures, since these models (in the strong coupling limit) predict T_c of the order of the energy of the boson, which lies below the BCS prediction. Indeed, in these models the electrons interact so strongly that they form pairs in real space with huge values of $|U|$.

One of the most exciting features of the new superconducting materials is the conspicuous low-dimensionality of the arrays in which the conducting ions lie. It was believed, from the very beginning, that these materials were the actual realization of the high temperature exciton superconductors [13]. Unfortunately there are problems when one considers electronic excitations as causing pairing. One important problem is the breakdown of the Migdal approximation. This approximation allows one to ignore vertex corrections, the lowest of which is

which in the case of phonons is of the order of $\hbar\omega_{exc}/E_F \sim 10^{-2}$. In these new materials E_F is very low and vertex corrections are larger than unity, therefore they cannot be neglected, and one has to be careful to restrict the exciton interaction to low momentum transfer events. This is an important point, as it is illustrated by the fact that if one ignores vertex corrections a plasmon should produce superconductivity in alkali metals like Na[14], which is not observed.

There are further problems with electronic excitations which arise from the very nature of the interaction. The electron-electron interaction can be written, quite generally, as

$$V(q,\omega) = \frac{4\pi e^2}{q^2 \epsilon(q,\omega)},$$

where $\epsilon(q,\omega) = 1 + 4\pi P_e + 4\pi P_{ph}$, contains information about the direct Coulomb repulsion and phonon interactions in the system. In particular

$$V(q,o) = \frac{\mu - \lambda}{N(o)},$$

where $N(o)$ is the density at the Fermi level. In general $\lambda < \mu$ and electrons repel each other, nevertheless, in many materials $\lambda > \mu$ due to local field corrections (unklapp processes), that arise from the fact that the local fields around the ions could be quite different from the average field sensed by extended electrons. Local field corrections are not adequate for non phonon mechanisms, because the volume of the unit cell occupied by the electrons is much larger than that in the case of ions, and it is hard to imagine a reason why there must be a difference in the fields felt by polarizable electrons and paired ones.

Another difficulty that arises from the separation of the electron gas is that the interaction vertex involves two identical particles and the exchange terms, which have opposite sign to the direct terms, dramatically reduce the effective strength of the interaction. Coupling of electrons through electrons must be achieved by a very delicate balance between the interactions involved.

Therefore, there is not much room to move if one is partisan of an electronic coupling. However, it is worthwhile studying in detail the possibilities of electronic pairing in these materials, and there is much work going on at the moment on plasmons and demons [15] [16], excitons

[17] [18] and polarization waves [19].

The problem of exchange and of possible magnetic interactions in these materials forces one to go back to the experiments and meditate. What is known for sure is 1) Charge is transported in pairs, 2) The materials present low-dimensional arrays of Cu-O atoms, 3) There is a gap of about $2\Delta/k_B T_c = 8$, which means very strong coupling and 4) There is antiferromagnetism in the La compounds and surely in the Y compounds as well.

Anderson has repeatedly stressed that the energy gap in this material is "inherently magnetic". Carrier transport in a Mott insulator is not "understood" yet because one has to solve the spin problem first. Consider, for instance, a Mott insulator in which the spins are arranged antiferromagnetically in the ion sites, and add a further electron. There is not a preferred site to which it can go and it is mobile, since it can tunnel to the neighbour sites. However this tunneling is restricted because, in order to occur, the spin of the neighbour has to be correct. This is the inspiration of applications of the Resonant Valence Bond theory [20] to these materials and could also be of the "magnetic polaron" (a hole trapped in a well in a Mott insulator) [21].

III. MAGNETIC THEORIES

One of the merits of the Resonant Valence Bond theory is that it opens vast horizons of thought to a great number of models, due to the interesting interplay between superconductivity an magnetic order. The basic physics starts with a single band Hubbard Model:

$$H_1 = U \sum_i n_{i\uparrow} n_{i\downarrow} - t \sum_{ij} c_i^+ c_j \qquad (1)$$

where $c_{i\sigma}^+ (c_{i\sigma})$ creates (anihilates) an electron on site i with spin σ. This can be canonically transformed to the superexchange picture, in which the number operators n are represented by spin operators. When applied to the Cu-O materials and Cu is stoichiometrically in a Cu^{++} state, the hopping part is zero and one is left with a Mott insulator that does not exhibit antiferromagnetism, but a new kind of state, which is a featureless singlet spin liquid of singlet pairs. Doping introduces interesting properties to the system: Acceptors (for instance Cu^{3+}) will drive the system to a macroscopic coherent state with broken symmetry. According to Anderson the basic nature of superconductivity is the condensation of holes in the presence of a Fermi gas of spinons. The RVB state in stoichiometry is insulator, in spite of the tremendous amount of pairing, since only local gauge symmetry is present and therefore no macroscopic coherent state can be attained.

There is some scepticism about the validity to describe these oxides with a single-band Hubbard model (equation(1)) [19]. This seems paradoxical, since the Hubbard model, Mott insulators and the superexchange mechanism, leading to antiferromagnetism, have been developed in order to understand the properties of transition metal oxides The band structure obtained with a self-consistent one-electron approach to (1) is as follows:

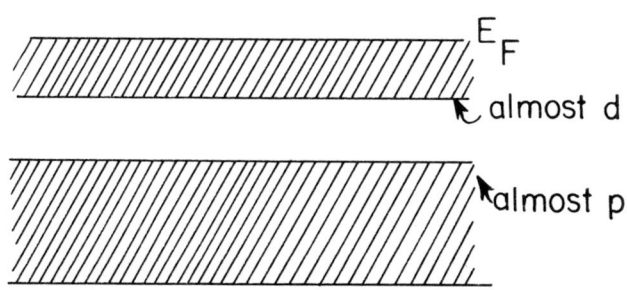

A set of filled oxygen bands and unfilled bands that are a covalent mixture of transition metal d orbitals and a small amount of oxygen p orbitals. The validity of this solution relies on the fact that the renormalized levels for oxygen p bands E_p and copper d bands E_d are well separated and therefore, the renormalized mean field hooping $\tilde{t} = t/(E_d-E_p)$ is small. As it happens these levels E_d and E_p are nearly degenerate in these oxides. Thus, restricting the attention to a Cu-like conduction band in these materials is not adequate.

Varma et al. [19] have shown these difficulties and suggested that superconductivity might arise from charge transfer resonances of the type

$$Cu^{3+} O^{2-} \rightleftarrows Cu^{2+} O^{-}$$

in which the transverse part, which appears as poles in the dielectric function, are responsible for superconductivity. These resonances do not couple to free electrons but to tightly bonded electrons.

All these theories need an electron gas slightly away from half-band filling, in which antiferromagnetic order is unlikely to appear. With Cu in a mixed valance state it is difficult to imagine long range magnetic order, as it is in the mixed valence state of the metallic rare earths.

A single-band Hubbard Model does not suffice to describe electrons in these materials, therefore there have been a number of variations on the theme that sound more reasonable to explain high-T_c superconductivity. In what follows there is a brief review of some of these ideas in chronological order.

1. Superexchange in the Single-Band Hubbard Model

Doniach et. al [22] use equation (1) in the strong correlation limit $t/U \ll 1$, which is canonically transformed in order to eliminate doubly occupied sites to lowest order in t/U. They use the "slave boson" picture, in which the charge and the spin of the fermions are separated $c_{i\sigma} = b_i^+ f_{i\sigma}$. Pairing occurs with increasing band filling via exchange of spin fluctuations on by localized (virtual), charge fluctuations. These are stabilized by the antiferromagnetic correlations in the underlying lattice. This theory predicts a rather short coherence lenght and therefore not only states near the Fermi surface pair together. In a mean field calculation one can show the coexistence of superconductivity and antiferromagnetism that persists even in the absence of long range order. The difficulty is that these effects should be observable by neutron scattering experiments as changes in the spin-lattice relaxation times, which of course are not observed yet.

2. Pairing through Local Spin Configurations

Emery [23] uses a simple extended Hubbard Hamiltonian in which off-diagonal correlations to nearest neighbours are considered in a square lattice model of the CuO_2 planes. Monte Carlo calculations by Hirsch [24] show that a ground state, that is an antiferromagnetic insulator, orders at finite temperature. Pictorically, when one dopes with holes one has the following

```
Cu↑  O   Cu↓  O    Cu↑
O         O ↑c      O
Cu↓  O   Cu↑  O↓  Cu↓
              a   b
O         O         O
Cu↑  O   Cu↓  O    Cu↑
```

Suppose that the states lie inside the gap and get localized into O p states. Consider a situation in which there are holes with opposite spins in oxygen positions b and c and then the exchange may be accomplished by an interchange between a-b followed by another a-c. This interaction is the primary cause of pairing. A virtue of this mechanism is that it is easy to obtain high values for T_c.

It is worthwhile to notice that it is possible to obtain an effective attractive interaction by rearranging a purely repulsive pairing force, in the strong coupling limit. There is also the possibility of having real bound pairs (spin polarons) for some values of the parameters used in the theory.

3. Antiferromagnetic Ordering and Pairing in 2D CuO_2.

Based on Emery's ideas, Hirsch[25] examines the Hamiltonian

$$H = \sum_{i,l} t_1 (d^+_{i\sigma} c_{l\sigma} + H.C.) + (\varepsilon-\mu) \sum_l c^+_{l\sigma} c_{l\sigma} + U\sum_i n_{i\uparrow} n_{i\downarrow} - \mu \sum_{i,\sigma} d^+_{i\sigma} d_{i\sigma}$$

where $n_{i\uparrow} = d_{i\uparrow}^+ d_{i\uparrow}$ is the number operator for holes in Cu sites and $c_{1\sigma}^+$ creates holes in oxygen sites, (i,1) are nearest neighbours arranged in a Cu-O square lattice. The effective hopping for a hole localized in a cage of O atoms surrounding a Cu site is $\tilde{t} = -t^2/(U-\varepsilon)$, where ε is the O level positions. When the hole propagates away from its site, it has to exchange spin with Cu sites. If exchange is such that $J_z/J_{xy} \gg 1$, it leaves a string of broken Cu-Cu bonds that costs an energy that increases linearly with distance, rather in the way quarks are confined in QCD. This provides an attractive interaction between holes. It is important to note that this does not happen in a linear chain. Pairing then is quite obvious and there is a Bose condensation, because the pair Hamiltonian is equivalent to a quantum XY model. Antiferromagnetic ordering is only needed to occur at distances of the order of the coherence distance of the pair. It is important to note that spin flip terms become important because they provide an independent mechanism for healing the defects produced by a single propagating hole, through the J_{xy} coupling.

4. Superconductivity from Antiferromagnetic Interactions

Parmenter [26] uses a Hamiltonian

$$H = H_1 + \frac{1}{2}\sum_{i,j} J_{i,j} \cdot S_i \cdot S_j + \frac{1}{2}\sum_{i,j,1} S_i \cdot (S_{j1} - \frac{1}{2}S_1 - \frac{1}{2}S_j)$$

where H_1 is equation (1), $S_{ix} + iS_{iy} = c_{i\uparrow}^+ c_{i\downarrow}$ and $S_{iz} = \frac{1}{2}(c_{i\uparrow}^+ c_{i\uparrow} - c_{i\downarrow}^+ c_{i\downarrow})$. Also S_{ij} is S_i but substituting c_{is} by $c_{j1s} = (c_{js} + c_{1s})/\sqrt{2}$. This Hamiltonian differs only slightly from the one by Hirsch.

Dividing the lattice into two square sublattices and bilinearizing with respect to the fermion operators Parmenter obtains an effective Hamiltonian that could be treated in the usual way, and obtains gap equations for the antiferromagnetic order parameter and the superconducting one, both as a function of J. Thus, within this approximation there is a possibility of the simultaneous presence of antiferromagnetic order and superconductivity, except in the half-filled band case.

The predictions of this theory are rather interesting, in the sense that pairing involves all the Brillouin zone and not only a thin shell near the Fermi level, as in BCS. However, the author does not show the implications of this fact over T_c.

5. Magnetic Interactions in One Dimensional Systems [27].

As could be deduced from the above descriptions, a Hubbard-like Hamiltonian is full of surprises and can be extremely useful when examining in detail the properties of the electron gas in a low dimensional system, in which not only the magnetic ordered states appear but also Charge Density Waves (CDW) and Spin Density Waves (SDW) are likely to be present. The very important issue of the coexistence of SDW and CSW with superconductivity has been analysed by Machida and Kato [28].

The purpose of the following ideas is to give a brief description of the fundamental properties of the Hubbard Hamiltonian when an antiferromagnetic interaction is enhanced between the conduction electrons. The simplest Hamiltonian is

$$H = H_1 + J \sum_{i,j} c^+_{i\sigma} c_{i\bar{\sigma}} c^+_{j\bar{\sigma}} c_{j\sigma} = H_1 + J \sum_{i,j} S^+_i \cdot S^-_j , \qquad (2)$$

where all the details of the electronic structure are assumed to be taken into account by H_1, that is to say, one could separate terms belonging to holes in Cu or O sites as in Ref. [25]. The main feature is that a particular group of terms has been isolated These terms have to do with hopping involving a spin flip, which is usually small in metals, because of Hund's rules. In the case of existing an antiferromagnetic interaction in the underlying lattice these terms become important. The main feature that arises from these terms can be investigated within the mean field approximation. It is found that the presence of the second term in equation (2) stabilizes a state with a gap, in which neither the charge or the spin fluctuate in space. The insulating state in the half-filled band case can be removed by doping and pairing appears. This can be seen by Fourier transforming equation (2), and the diagram for such an interaction is

```
        k+q ↑              k ↓
           ↖              ↗
             \          /
              \        /
               \~~q~~/
               /      \
              /        \
           ↗              ↖
        k ↑              k+q ↓
```

where the exitation q has a magnetic origin. It should be noticed that there is not an effective spin flip, but the momentum transfer between the pair gives the correct value of the spin. The properties of such a pairing are rather interesting. The density of states at the Fermi level becomes unimportant, since all electrons are involved, in principle; the coherence length is of the order of the lattice parameter and pairing in real space is possible.

However, there are problems when one tries to diagonalize the Hamiltonian, and a clear decoupling scheme is not yet available. Furthermore, there must be a physical way of justifiying an antiferromagnetic order in the lattice, that enhances the second terms in equation (2). Therefore, a detailed calculation of the dynamics of the lattice is needed before one could actually conclude that these ideas are relevant. Some hints are given by some new experiments:

1) Muon-spin-relaxation experiments [29] seem to show the existence of real pairs, rather than Cooper pairs.

2) Mössbauer spectroscopy [30] shows that there are high local fields around the Cu(I) sites.

3) Raman experiments [31] show extremely peculiar modes at ~335 cm^{-1} that behave strangely below T_c, and in which the oxygen motion is exagerately large.

4) The beautiful experiments presented here by Uchida [32], which show what could be called a phase diagram with doping, in which there is still an uncertainly about where the antiferromagnetic phase ends and superconductivity arises.

In conclusion, in spite of having at present quite a number of interesting ideas, there is not a definite theory that could explain high-T_c superconductors. It is quite clear that the key factors are the low dimensionality and the complicated antiferromagnetic interactions in the material. One should look optimistically for the near future. More detailed and complete calculations are in progress.

AKNOWLEDGEMENTS.

I am grateful to my colleages J. Tagüeña and Ch Wang for their encouragement and enthousiasm. Financial support from the programa Universitario de Superconductores de Alta Temperatura UNAM, is fully acknowledged.

REFERENCES.

1. G. J. Fisanick, S. Nakahara, S. Jin, T. F. Tiefel, R. C. Sherwood, M. Yan, R. Moore and R. B. van Dover (this volume)
2. D. Ríos-Jara (this volume)
3. B. Raveau, C. Michel, A. Maignan, M. Hervieu and J. Provost (this volume)
4. J. B. Goodenough and A. Manthiram (this volume)
5. K. Takita (this volume)
6. C. E. Gough et al., Nature 326. (1987)
7. J. Bardeen, L. N. Cooper and J. R. Schrieffer Phys. Rev. 106, 162 (1957)
8. R. Escudero, L. Rendón, T. Akachi, R. A. Barrio and J. Tagüeña-Martínez, Phys. Rev. B36, No. 7, 3910 (1987)
9. H. Enomoto, K. Gotoh, K. Ii, Y. Takano, N. Mori and H. Ozaki, paper AA 7.73, MRS 1987 Fall Meeting, Boston, Massachusetts.
10. G. M. Eliashberg, Soviet Phys JETP 11, 696 (1960)
11. J. P. Carbotte (this volume)
12. B. K. Chakraverty and J. Ranninger, Philos. Mag. B52, 669 (1985)
13. W. A. Little, Phys. Rev. 134, A 1416 (1964)
14. M. Grabowski and L. J. Sham, Phys. Rev. B29, 6132 (1984)
15. J. Ruvalds, Phys. Rev. B35, No. 16, 8869 (1987)
16. M. L. Cohen in "Novel Superconductivity", (ed. by S A. Wolf and V. Z. Kresin) Plenum, N.Y. (1987) p. 1095

17. W. A. Little in "Novel Superconductivity", op. cit. p. 341
18. J. Bardeen, D. M. Ginzberg and M. B. Salamon in "Novel Superconductivity" op. cit. p. 333
19. C. M. Varma, S. Schmitt-Rink and E. Abrahams in "Novel Superconductivity", op. cit. p. 335
20. P. W. Anderson in "Novel Superconductivity" op. cit. p. 295
21. R. B. Laughlin and C. B. Hanna in "Novel Superconductivity" op. cit. p. 553
22. S. Doniach, P. J. Hirschfeld, M. Inui and A. E. Ruckenstein in "Novel Superconductivity" op. cit. p. 395
23. V. J. Emery, Phys. Rev. Letters $\underline{58}$, No. 26, 2794 (1984)
24. J. E. Hirsch, Phys. Rev. $\underline{B31}$, 4403 (1985)
25. J. E. Hirsch, Phys. Rev Letters, $\underline{59}$, No. 2, 228 (1987)
26. R. H. Parmenter, Phys. Rev. Letters, $\underline{59}$, No. 8, 923 (1987)
27. R. A. Barrio, C. Wang, J. Tagueña-Martínez, D. Ríos-Jara, T. Akachi and R. Escudero, Proc. MRS Fall Meeting, Boston, Massachussetts (1987)
28. K. Machida and M. Kato, Phys. Rev. Letters $\underline{58}$, No. 19, 1986 (1987)
29. V. J. Emery, Proc. of the NATO Advanced Research Workshop on Organic and Inorganic Low-Dimensional Crystalline Materials, Spain (1987)
30. R. Gómez, S. Aburto, M. L. Marquina, M. Jiménez, V. Marquina, C. Quintanar, T. Akachi, R. Escudero, R. A. Barrio, and D. Ríos-Jara, Phys. Rev. $\underline{B36}$, 7226 (1987)
31. M. Cardona, R. Liu, C. Thomsen, M. Bauer, L. Genzel, W. König A. Wittlin, U. Amador, M. Barahona, F. Fernández, C. Otero and R. Saez (to be publihed in Solid State Commun.)
32. S. Uchida (this volume)

SUPERCONDUCTOR-NORMAL METAL INTERFACES IN HIGH-T_c SUPERCONDUCTORS

ROBERTO NICOLSKY
Instituto de Física, Universidade Federal do Rio de Janeiro, C.P.68528, Rio de Janeiro 21945, BRAZIL

ABSTRACT

The two types of Josephson junctions - tunnelling or metallic - and their current-voltage characteristics are reviewed. We discuss which general features can be associated with the dominant character of each Josephson junction and a network of them. A criterion for identifying this character is established and applied to analysing the experimental I-V curves of single crystals, point contacts and granular ceramics as well of high-T_c superconductors. The analysis indicates that the dominant character of the intrinsic Josephson junctions in these materials is metallic. This points out the important role played by the Andreev reflections in these copper oxides, which suggests a many-body Cooper pair ground state and a BCS-like superconductivity, no matter what mechanism can be responsible for the attractive pairing scattering processes in high temperatures.

INTRODUCTION

The recent discovery of the new superconductors with high (more than 30K) critical temperatures [1], [2] proposes a set of unusual problems in the field of superconductivity. These new materials form a novel class of superconductors: the Oxigen-deficient-metallic copper oxides. These copper oxides have many differences to the traditional superconductors, usually transition metals and their alloys. Thus, the first months of the investigation on these materials have been dominated by the search for identification of their structure and related topics as fabrication and characterization techniques. As a consequence, the structure has been well identified although there still are many open questions concerning this topic. One of these questions is just the nature of the non-superconducting interfaces present in those materials, including within the single crystals. It is well established by the high-resolution transmission electron microscopy the presence of inter-grain and intra-grain interfaces in those copper oxides. The importance of these interfaces is due to the resultant intrinsic Josephson effect, which can be experimentaly observed even in a sample of granular ceramic or single crystal. Both kinds of interfaces being intensively studied mostly from the structural point of view [3]. Otherwise, some important consequences of those interfaces in the transport properties [4] and magnetic behaviour [5] have been also discussed in the literature. However, very little attention has been paid to the experimental data on transport properties of these materials.

In this paper we analyse the nature of the interfaces which we can infer from the transport properties revealed in a non-equilibrium configuration, that is with a current larger than its critical one. In a bulk traditional superconductor this situation causes simply a transition to the normal state. However, in those new superconducting materials the overpass of the critical value of the current produces an intermediate state characteristic of the presence of Josephson junctions. Thus, we take the experimental data represented by a wide range of I-V curves obtained mainly with granular ceramics and point--contacts, but also with some single crystals. Initially, we briefly review the Josephson effect and how to obtain it. After this we analyse the typical I-V curves for different kinds of Josephson junctions. Based on this analysis we establish a criterion [6] for identifying from a particular I-V curve the dominant character of the Josephson junction which generate it: tunnelling or metallic. Applying this criterion to the experimental curves we can infer consistently that the nature of the interfaces in those superconducting ceramics or single crystals is superconductor - normal metal. However, the most important conclusion is that the superconductivity of those copper oxides has the Cooper pairing processes and a BCS - like ground state.

THE JOSEPHSON EFFECT

We start by emphasizing that the Josephson effect does not necessarily imply a tunnelling mechanism, as usually assumed. The Josephson effect, in a general sense, is a consequence of a phase coherence through an inhomogeneity of the pair potential, which is related to the energy gap of the excitation spectrum of the superconductor [7]. If this inhomogeneity implies a zero value of the pair potential in some part of the system, under certain conditions the Josephson effect will be a static phenomenum, called the DC Josephson effect, which consists of a DC current of Cooper pairs from one superconducting side to the other across the region which has a zero value of the pair potential, dragged by the phase difference between the superconducting sides. The conditions previously named are the width of the inhomogeneous region scaled to the coherence length of the superconductor, and the critical value of the current density. When this last critical parameter is overpassed the system reaches a non-equilibrium state and the AC Josephson effect takes place. The AC Josephson effect is the oscillation of the Cooper pairs between the superconducting sides of the system through the inhomogeneous region. The amplitude and the time average value of this AC current depend of the particular kind of system, but the oscillation frequency (Josephson frequency) is a quantum mechanical property which only depends on the time derivative of the phase difference between the superconducting sides. When the pair potential does not reach the zero value in the inhomogeneous region the DC Josephson effect is suppressed, but if the local critical current density is achieved the AC Josephson effect starts even if the whole system consists of superconductors. On the other hand, if some convenient voltage is applied across an inhomogeneous region an AC Josephson effect is immediately observed.

The superconducting systems which can generate the Josephson effect must be classified in two types, according to their suitability to one or to other model for the inhomogeneity of the pair potential. There are essentially two possible models for this inhomogeneity: the barrier and the well.

As we will see later, these models suggest two kinds of Josephson junctions respectively: the tunnelling Josephson junctions (TJJ) and the metallic (MJJ).

The most known but the less used of these systems is the TJJ. It consists of two superconductors separated by a thin insulating layer (≈ 10 Å) or a thicker semiconducting layer (the maximum width increases with the doping level).

The typical TJJ system is the sandwich, as showed in the Fig.1. The point contacts (Fig.1) can also represent the same TJJ when a thin oxide layer is developed between the superconductors (SIS junction). The model for TJJ system is illustrated in the Fig.2. The essential condition which must be satisfied for the model is that $\Delta_B \gg \Delta$, where Δ_B is the energy gap of the barrier and Δ is the pair potential of the superconductor.

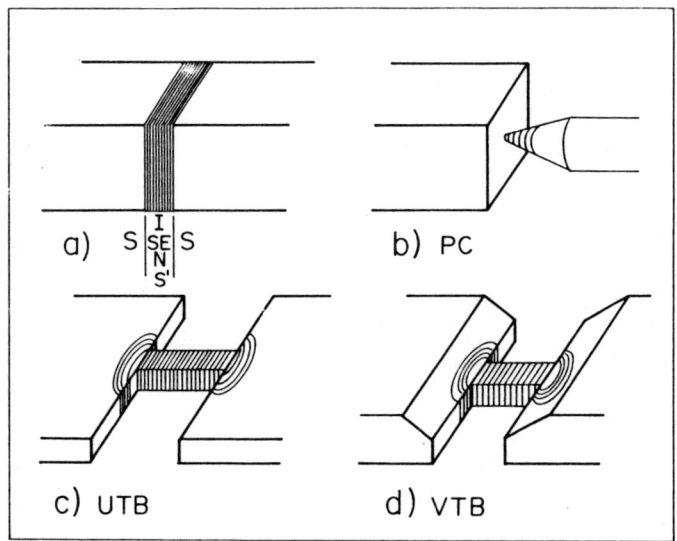

Figure 1. Examples of inhomogeneous systems. Sandwich (a), point contact (b), UTB - uniform thickness bridge (c) and VTB - variable thickness bridge (d).

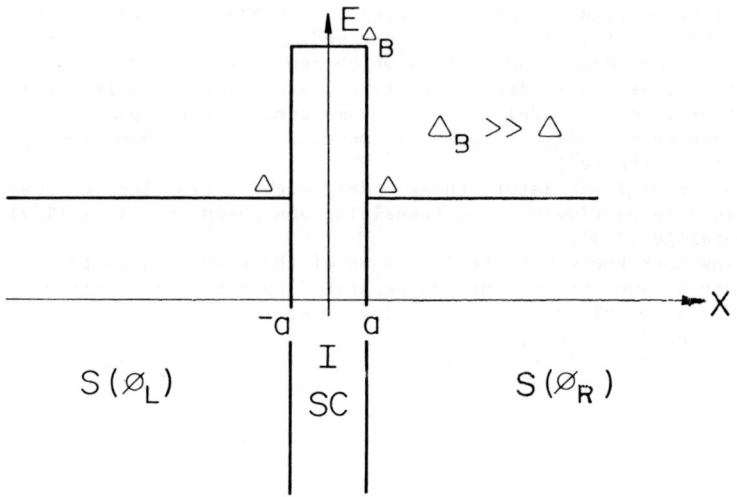

Figure 2. Energy model for a TJJ system.

To make a MJJ system, we have many possibilities as illustrated in the Fig.1. The sandwich case, or SNS (superconductor-normal metal-superconductor) junction, actually is not used due to its low resistance which causes a weak sensibility of the system for measurement.

The other MJJ systems are always strong constrictions of the cross section of the superconductors, which can be modeled geometrically [8]. In these MJJ systems when the current density in the constriction overpasses its local critical value the constriction is no longer in the superconducting state. Thus, the pair potential is depressed in the center of the constriction to a zero value, which occurs when the phase difference between the superconducting bulk sides reaches 2π and slips discontinously to zero. It is just the so called phase - slip - center [9].

After the phase - slip instant the pair potential grows up but does not retrieve anymore its equilibrium value in the center of the constriction [10], and a region around it remains always in the normal state [11], [12].

This non-equilibrium configuration produces an oscillation of the pair potential (AC Josephson effect) and develops a voltage across that normal region. The energy model for these MJJ systems is the quantum well instead of the barrier, as we see in the Fig.3. We will show later that this makes an enormous difference in the transport properties of the Josephson junctions, as we can show in the I-V curves for both ideal models.

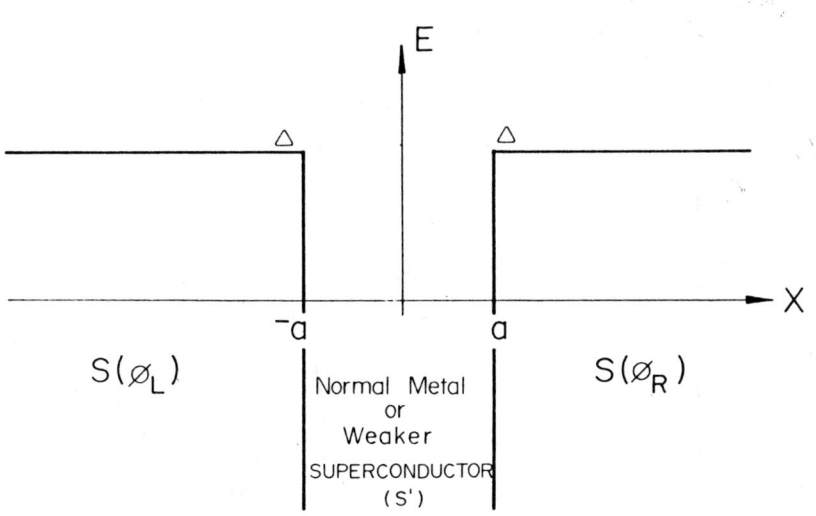

Figure 3. Energy model for a MJJ system.

I-V CURVES OF JOSEPHSON JUNCTIONS

The ideal TJJ system has a peculiar and well known I-V characteristic, as illustrated in the Fig. 4. We can observe the DC Josephson I_o current for zero voltage and the AC Josephson I_s current for voltages below $V_J = 2\Delta/e$, the voltage which restores the normal state in the system. The frequency is related to the time variation of the phase difference and the voltage by the well known Josephson relation [7], [13]

$$\frac{\partial}{\partial t}(\phi_R - \phi_L) = \omega_J = \frac{2eV}{\hbar} \qquad (1)$$

The temperature dependence of the DC Josephson I_o current [17] is

$$I_o = \frac{\pi S \Delta(T)}{2eR_N} \tanh \frac{\Delta(T)}{2kT} \qquad (2)$$

where S is the cross sectional area of the constriction.
For T near T_c the Eq. (2) gives a linear dependence with $(T_c - T)$

and it is easy to show that its maximum value is $\pi/4$ times the quasiparticle current for the breakdown voltage $2\Delta/e$. The alternating current is [17], [13].

$$I_S = I_o \sin(\phi_R - \phi_L) \tag{3}$$

and if the voltage is constant the sinus argument should be linear with time and its average value in a cycle should be zero.

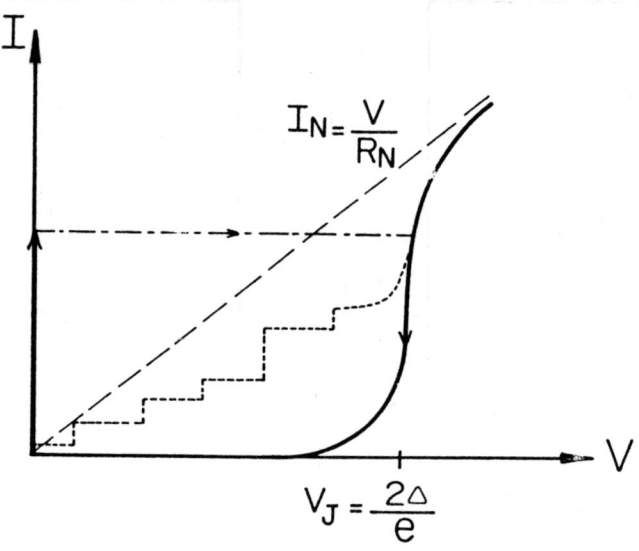

Figure 4. I-V curve of a TJJ system.

The alternating character of the AC Josephson current can be verified by the current enhancement when the junction is irradiated with microwaves. The absoption of the microwaves energy causes the appearence of the so called Shapiro steps, as shown in the Fig. 4 in dashed lines. The distance between the two consecutive steps is just $V = \frac{\hbar\omega}{2e}$, which confirms the pair structure of the carriers in the superconducting state. Finaly, the I-V curve of TJJ always exhibits a strong hysteresis, as indicated in the Fig. 4 by the arrows in the increasing and decreasing paths of the curve.

The ideal MJJ or SNS systems have an I-V curve much more complex, with a structure composed of several features. As each feature depends on its proper conditions [14], it is very rare to find all of them in the same curve. An exception is the classical work on VTBs by Octavio et al [15].

Fig. 5 shows the typical profile of a complete I-V curve for a MJJ system.

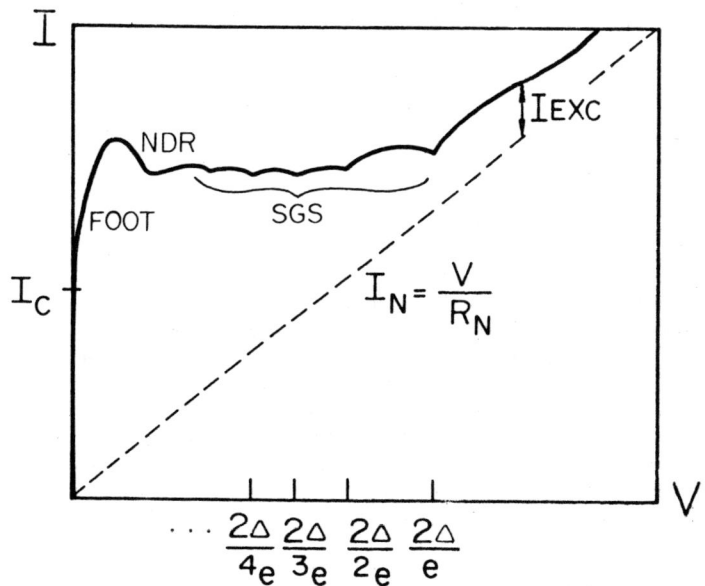

Figure 5. I-V curve of a MJJ system.

The features which characterize MJJ curve are:

- $V=0$: DC Josephson current I_c, below which we have a static regime;
- $V \ll 2\Delta/e$: foot, part of the curve with a differential resistance much smaller than the normal resistances ($\frac{\partial V}{\partial I} \ll R_N$) [15];
- $V \lesssim 2\Delta/e$: sub-harmonic energy gap structure, peaks which indicate the resonance of the voltage with the quasiparticle energy [16], [17];
- $V \gg 2\Delta/e$: I_{exc} (excess current), for high voltages the curve remains parallel to the normal current [16], [17];
- NDR: negative differential resistance, a novel effect proposed theoretically [18], [22] with some clear experimental evidence [15], [19],[20].

We see that the I-V curves of a MJJ are far more complex than the curves of the TJJ. Due to this, only recently it has been achieved a theory [21], [22] which expains the whole behaviour of MJJ systems, although partial explanations [10], [16], [17], were known previously. This theory describes the SNS junction dynamics under an applied fixed voltage with the time-dependent-Bogoliubov-de Gennes equations, taking into account the phase coherence across the system and the Josephson relation for the phase difference. The inelastic scattering by phonons, defects and impurities is also considered in the relaxation time approximation. The final result is the expression (1) of the Ref. [22] for the total current:

$$I = I_N + I_{AR} =$$

$$= \frac{V}{R_N} + \frac{eS}{m\ell N} \sum_{k_x} \sum_{k_y} \sum_{E_k \gtrless 0} \sum_{n \geq 0}^{\infty} \hbar k_{zF} f_0(E_k)(|A_n^+(E_k)|^2 - |A_n^-(E_k)|^2) \cdot$$

$$\exp - [(n2a + b - a)/e] \qquad (4)$$

where the symbols are explained in the Ref. [21], [22].

The current component I_{AR} corresponds to the contribution of the Andreev reflection mechanism [23] to the total current. We see that the sign of this component depends on the difference between the probability that a quasiparticle which moves dragged by the electric field ($|A_n^+|^2$) reaches n Andreev reflections and the probability of a quasiparticle which moves against the electric field drag ($|A_n^-|^2$) has the same number of Andreev reflections. However, due to the shift of the Fermi sphere in the presence of an electric field, when the summation is taken over all k states this difference is always positive, as we see in the Fig. 6.

From this we conclude that the Andreev reflection contribution is always non negative and the total current is larger, or at least equal, than the normal one. As far as we know, the Andreev reflection mechanism is the unique way to enhance the normal metal coductance above, and sometimes far above, the ohmic value (R_N^{-1}).

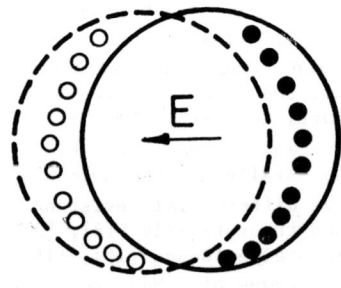

Figure 6. Shifted Fermi sphere in the presence of an electric field E.

Finally, we can observe that the I-V curves of MJJ usually do not exhibit hysteresis (only in pathological situations). The above I-V curves assume T far from T_c.

It is opportune to emphasize that the Andreev reflection mechanism is intrinsically associated to the many-body Cooper pair state, because it is a consequence of the kind of scattering which determines the pairing processes. The Cooper pair is formed in the k-space by the attractive scattering processes from the time-conjugate state (k', -k') to the state (k, -k), no matter which bosons are exchanged in this mechanism. Thus, according to the BCS theory, each time-conjugate state (k, -k) has a probability amplitude v_k to be occupied and u_k to be unoccupied. As a quasiparticle excitation k implies that we have definitely occupied the state k and unoccupied the state -k, a large number of Cooper pair scattering processes remains forbided by the Pauli principle and a minimum amount of energy Δ is required to insert a single quasiparticle into the superconductor. Thus, if a quasiparticle excitation, an electron for instance, with energy smaller than this minimum amount strikes the normal metal-superconductor interface it cannot penetrate in the superconducting side as a stable excitation and is reflected back to the normal metal side as a hole. This process due to the conservation of charges, energy and momenta, results in a pair formation in the superconducting condensate, as we see in Fig. 7. Of course, a hole quasiparticle excitation is, similarly, reflected back as an electron.

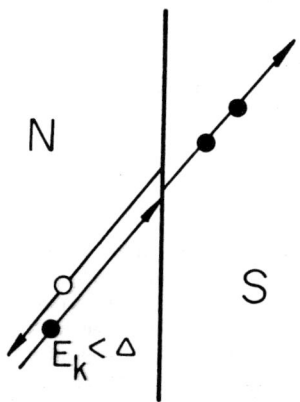

Figure 7. Andreev reflection mechanism: an electron strikes the N-S interface and is reflected back as a hole and a pair is formed in the superconducting condensate.

Due to the momentum conservation, the reflected quasiparticle excitation has an inverted group velocity and, in the absence of a magnetic field, moves back in the trajetory of the incident quasiparticle, the incident and the reflected, have the same contribution to the current because their charges, masses and group velocities have opposite signs. Thus, if a quasiparticle has one Andreev reflection its contribution to the current is twice a normal, if it has three its contribution is three times and for n Andreev reflection its contribution is n times the normal. This is the physical origin of that enhancement or the normal metal conductance due to the Andreev reflection mechanism [16], [17]. If the quasiparticle is dragged by an electric field it gains an amount of energy eV in each Andreev reflection and climbs up the well until its escape energy $E_k + neV > \Delta$ (n=number of Andreev reflections) and the quasiparticle can penetrate the superconductor as an imbalanced excitation causing the charge imbalance non-equilibrium phenomenum in the superconducting sides [23]. For low voltages and enough large inelastic mean free path a quasiparticle can reach a large number of Andreev reflections before achieving the escape energy and penetrating the superconductor. These additional contributions to the current are responsible for the low differential resistance at low voltages (foot) usually observed in MJJ current-voltage curves [14].

CRITERION FOR IDENTIFYING THE CHARACTER OF A JOSEPHSON JUNCTION

In the above description of the ideal Josephson junctions if we have to point out a more general property of each kind of junction, neglecting the details of the I-V curves, the constancy into the variety, we must observe these characteristics:

$$\text{TJJ:} \quad I < I_N = \frac{V}{R_N} \quad , \quad \text{for } V < V_J$$

$$\text{MJJ:} \quad I = I_N + I_{AR} > I_N = \frac{V}{R_N} \tag{5}$$

which are much more remarkable for low voltages. A simple comparison of the I-V curves of the Fig. 4 and 5 evidences this fact. In addition, we note that the TJJ presents hysteresis and the MJJ does not present it.

However, a real junction is only approximately an ideal one. How to interpret an I-V curve of a real junction? We can conclude it easily if we understand what is a real or imperfect Josephson junction. For instance, an imperfect TJJ means that the insulating layer of the junction has some metallic shorts and the imperfect MJJ has weak barriers in the interfaces. We can realize it as a superimposed arrangement of a barrier and a well. Thus, if the weak barriers of an imperfect MJJ grow up, the Andreev reflection component is gradually depressed [24], the whole system looses its metallic character and the current approaches the ohmic value. It is just the transition from metallic to tunnelling regimes [17], [25]. The inverse process can occur with an imperfect TJJ if its metallic shorts grow up. Fig. 8 illustrates these transitions processes in a junction under an applied fixed voltage and points out the appearance of the hysteresis when occurs the transition from metallic to tunnelling regime and its desappearance in the inverse process. Thus, we can treat a real junction as a composition of the two kinds of ideal junctions, MJJ and TJJ. The dominancy of one of them, i.e., the current value above or below the ohmic one, will define the character of the system.

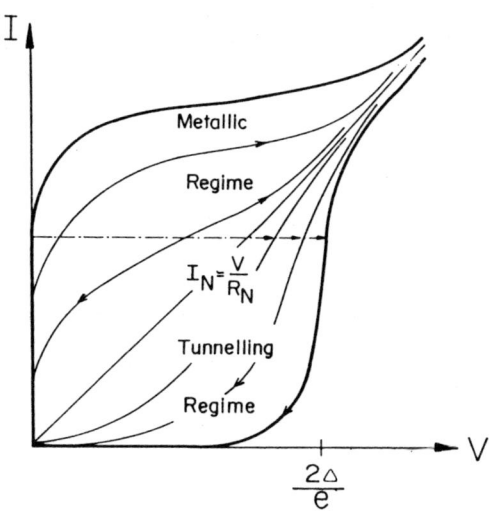

Figure 8. Transition from metallic to tunnelling character of a Josephson junction. The scales of the current axis are different and arbitrary for metallic and tunnelling regimes.

It has been analysed the transition of only one real or imperfect junction from one regime to the other. However, if we have a network of real Josephson junctions, how to treat it? We can simply generalize the transition of only one junction, representing the network of real junctions by a superposition of two networks of ideal TJJ and MJJ as well. For a given voltage below V_J, each MJJ of order i contributes to the current with two components, N_i and AR_i, and each TJJ does not contribute (its time average contribution is approximately zero at low voltages). Thus, the total current will be the summation over the i MJJ contributions only, and, depending on their quantity, can be larger, equal or smaller than the ohmic value. Thus, the character of the whole system can be metallic, normal or tunnelling, respectively. We can also produce the transition from one regime to the other changing the character of part of the junctions of the network. It is just what occurs with point contacts of those ceramic materials. With low pressure we have bad contacts and a tunnelling character with low current and hysteresis. Increasing the pressure we improve the contacts and the character can transit to the metallic one [26], [27]

with currents larger than the ohmic value and the hysteresis tending to
disappear. Finally, we are now able to define a general phenomenological
criterion for indentifying the character of some Josephson junction of
a network of Josephson junctions analysing the I-V curve produced by
the system. If the current is larger than the ohmic value, mainly at
very low voltages, and has no hysteresis above the critical current
the system of only one junction or a network of junctions has a metallic
character. It assures that the junction is dominantly MJJ or that at
least part of the network junctions are MJJ. However, it does not
assure the simultaneous presence of TJJ junctions. Only the transition
from one regime to the other can assure the presence of both types of
junctions. Similary, if the current is smaller than the ohmic value
which is always accompanied by hysteresis we have a dominant tunnelling
regime. Thus, this criterion is just the constancy into the variety,
as we said before. Now, we will see which consequences we can obtain
when we are able to identify the nature of the interfaces in high-T_c
superconductors applying the general criterion above.

APPLICATION OF THIS CRITERION TO HIGH-T_c SUPERCONDUCTORS

As these new superconductors are full of Josephson junctions, we
will see that the criterion above defined is very useful to identify
the dominant character of those junctions. The most interesting is
that this identification is not limited to itself, i.e., we can
extract very important and basic information about the kind of pairing
and the ground state of this high temperature superconductivity. We
can start with the I-V curves of single crystals of those copper oxides.
Surprisingly, the I-V curves are not of a bulk superconductor but of a
Josephson junctions network [28], [29]. We observe that above the
critical value of the current the voltage is developed , the current
keeps growing up and the curves are reversible, there is no histeresis.
Thus, following our criterion the character of the intra-grain
junctions is metallic in both copper oxides, of La and Y. It is well
established that in the Y compounds these junctions are formed at least
by twin boundaries between domains [4]. Thus, the metallic character
of I-V curves of single crystals of YBaCuO indicates strongly that these
twin interfaces must be of the SNS type. Other candidates to be normal
metals are the Y or rare earth planes due to the very small coherence
length in the c direction. Concerning the La compounds it is not clear
up to now what is the origin of its intra-grain Josephson junctions.

On the other hand, the I-V curves of point contacts, and break or
crack junctions, reveal the nature of the inter-grain junctions in these
materials [26], [30], [37]. With enough pressure we observe the same
metallic character in the current-voltage curves, mainly currents far
above the ohmic value at low voltages and the absence of hysteresis.
However, this is not so surprising because we can model a point contact
as a contact between superconducting grains, and this model is very
similar to a VTB, as presented in Fig. 2, a typical example of a MJJ
system. In some samples of point contacts at enough low temperature we
can observe even some typical characteristic feature of a metallic I-V
curve as the NDR [38], [39]. The same is observable in I-V curves

taken with granular ceramics [40], [46]. In this case we have a
network of inter and intra-grain junctions and the final result is a
pronounced metallic character of the current-voltage curves. Some
samples exhibit also some particular typical features of SNS junctions
as excess current [47]. More than the point contacts curves, the
metallic character of these curves is expected because a granular
ceramic is just a tri-dimensional network of VTBs, where the contacts
between the grains are the constrictions for the current flow. As a
conclusion we can say that the experimental data of current-voltage
characteristics of the copper oxide samples strongly indicate that
these material have inter and intra-grain intrinsic superconductor-
-normal metal interfaces. The most interesting conclusion is not about
the nature of the interfaces in these materials but what we can infer
about the essence of their superconductivity. The metallic character
or the I-V curves of single crystals, point contacts and even granular
ceramics, points out the important role played by the Andreev reflection
mechanism in the high temperature superconductors up to the T_c. This
strongly supports that the pairing processes are specificly the Cooper
pairing attractive scattering processes. This conclusion agrees with
experimental data of the flux trapping and Shapiro steps [49], [40],
which indicate that the carriers have a charge 2e.

The consequence is that the ground state of this "new"
superconductivity is a many-body Cooper pair configuration, which
means that it is a BCS-like superconductivity, no matter what kind of
mechanism (or bosons) is responsible for the pairing dynamics. Thus,
the mechanism actually is the main theoretical problem to be solved in
these copper oxide superconductors.

CONCLUSION

The application of the general criterion defined above to the
experimental data on transport properties of high-T_c superconductors
results in the identification of the nature of intra-grain and inter-
-grain interfaces as metallic. Thus, as a picture of these oxide
superconductors we can imagine an intrinsic superconductor-normal metal
multilayered structure. The pairing mechanism seems to be of the
Cooper type and the superconductivity of the BCS-like type, although
it is not clear which bosons are exchanged in the attractive pairing
scattering processes.

Otherwise, this work evidences the importance of the investigation
possibilities represented by the transport properties in superconductors,
mainly in non-equilibrium configurations where several aspects of the
ground state are revealed by the experimental data. This emphasizes
that more and more experiments are needed in transport properties and
dynamical processes as more complete I-V curves, charge imbalance,
tunnelling, quasiparticle injections, etc., which can evidence the
excitation spectrum of this superconductivity, the relaxation
processes and, possible, the pairing mechanism itself.

REFERENCES

[1]   J.G. Bednorz and K.A. Müller, Z. Physik B64, 189(1986).
[2]   M.K. Wu et al., Phys. Rev. Lett. 58, 908(1987).
[3]   See for instance the Proceedings of the Adriatico Reseach Conference on High Temperature Superconductors, S.Lundqvist et al. (ed.), Progress in High Temperature Superconductivity, Vol.1 (World Scientific, Singapore, 1987).
[4]   G. Deutscher and K.A. Müller, Phys. Rev. Lett. 59, 1745(1987).
[5]   M.M. Fang et al., preprint(1987).
[6]   R. Nicolsky, to be published.
[7]   M. Tinkham, Introduction to Supercondictivity (McGraw Hill, New York, 1975) pg. 192.
[8]   R. Nicolsky, M. Octavio and S. Frota Pessôa, J. Low Temp. Phys., 58, 11(1985) and references therein.
[9]   W.J. Skocpol, M.R. Beasley and M. Tinkham, J. Low Temp. Phys., 16, 145(1974).
[10] R. Nicolsky, Proc. of the LT-17, Wühl et al. (ed.), (North-Holland, Amsterdam, 1984) pg. 807.
[11] K.K. Likharev, Rev. Mod. Phys. 51, 101(1979).
[12] A. Barone and G. Paterno, Physics and Applications of Josephson Effect (John Wiley, New York, 1982).
[13] B.D. Josephson, Phys. Lett. 1, 251(1968).
[14] R. Nicolsky and R. Kümmel, to be published.
[15] M. Octavio, W.J. Skocpol and M. Tinkham, Phys. Rev., B17, 159(1978).
[16] T.M. Klapwijk, G.E. Blonder and M. Tinkham, Physica B+C 110, 1657 (1982) and references therein.
[17] G.E. Blonder, M. Tinkham and T.M. Klapwijk, Phys. Rev. B25, 4515 (1982), and references therein.
[18] R. Nicolsky and R. Kümmel, to be published.
[19] M. Octavio, Ph.D thesis, Harvard University, (1978).
[20] W. Klein et al., J. Low Temp. Phys., 61, 413(1985).
[21] R. Kümmel, B. Huckestein and R. Nicolsky, Jap. J. Appl. Phys. 26, Suppl. 26-3, 1471(1987).
[22] R. Kümmel, B. Huckestein and R. Nicolsky, to be published in Solid State Commun.(1988).
[23] R. Nicolsky and S. Frota Pessôa, J. Low Temp. Phys., 46, 107(1982) and references terein.
[24] P.A.M. Benistant, et al., Phys. Rev. B32, 335(1986).
[25] G.E. Blonder and M. Tinkham, Phys. Rev. B27, 112(1983).
[26] J. Niemeyer, M.R. Dietrich and C. Politis, Z. Physik B67, 155(1987).
[27] A.T. Golovashkin et al., communication to the ISEC'87 Conference, Tokyo, Japan(1987).
[28] K. Moriwaki, Y. Enomoto and T. Murakani, Jap. J. Appl. Phys. 26, Supl. 26-3, 1147(1987).
[29] R.C. Dynes, quoted by P.W. Anderson and Z.ou, Phys. Rev. Lett 60, 132(1988).
[30] J.S. Tsai, Y. Kubo and J. Tabuchi, Phys. Rev. Lett., 58, 1979(1987).
[31] A.Th.A.M. de Waele et al., Phys. Rev. B35, 8858(1987).
[32] J. Niemeyer et al., Z. Physik B69, 1(1987).
[33] T. Yamashita et al., Jap. J. Appl. Phys., 26, L635(1987).

[34] T. Yamashita et al., Jap. J. Appl. Phys., 26, L671(1987).
[35] T. Nishino et al., Jap. J. Appl. Phys., 26, L674(1987).
[36] J.S. Tsai et al., Jap. J. Appl. Phys., 26, Ç701(1987).
[37] T. Dowatsu et al., Jap. J. Appl. Phys., 26, L1148(1987).
[38] Y. Higashino et al., Jap. J. Appl. Phys., 26, L1211(1987).
[39] P. Leiderer et al., Z. Physik B67, 25(1987).
[40] A.Th.A.A. de Waele, private communication(1987).
[41] D. Esteve et al., Europhys Lett., 3, 1237(1987).
[42] J. Garcia et al., J. Mag. Mag. Mat., 69, 1225(1987).
[43] R. Escudero et al.,Proc. of the Conference on Novel Superconductivity, S.A. Wolf and V.Z. Kresin (ed.) (Plenum, New York, 1987), pg. 1011.
[44] K. Moriwaki et al., Jap. J. Appl. Phys., 26, L521(1987).
[45] I. Igushi et al., Jap. J. Appl. Physl, 26, L1021(1987).
[46] T. Komatsu et al., Jap. J. Appl. Phys., 26, L1148(1987).
[47] M. Wakata et al., Jap. J. Appl. Phys., 26, Suppl. 26-3, 1059(1987).
[48] Lu Li et al., preprint(1987).
[49] C.E, Gough et al., Nature 326, 855(1987).

CONTRIBUTED PAPERS

THE ORTHORHOMBIC TO TETRAGONAL PHASE TRANSITION IN THE $Er_{1-x}La_xBa_2Cu_3O_y$ SYSTEM

L. GOVEA[1], R. ESCUDERO[1], D. RIOS-JARA[1], C. PIÑA[2], C. WANG[1] AND R.A. BARRIO[1].

[1]Instituto de Investigaciones en Materiales, UNAM.
Apdo. Postal 70-360; 04510 México, D.F.
[2]División de estudios de Posgrado. Facultad de Química, UNAM. 04510 México, D.F.

ABSTRACT

We have studied the orthorhombic to tetragonal phase transition in the $Er_{1-x}La_xBa_2Cu_3O_y$ system. X-ray powder diffractometry was used to determine the dependence of the lattice parameters with x. A smooth increase in the lattice parameters was observed for increasing x values.

Critical superconducting temperatures (T_c), ranging from 90 K for $Er_1Ba_2Cu_3O_y$ to 48K for $La_1Ba_2Cu_3O_y$ were measured. Stepwise behaviour on the T_c versus x curve was observed, with drops in T_c at x≃.4 and 0.7. This stepwise behaviour is modeled by a percolation process associated to the ordering of the Er-La sublattice.

Samples of the $Er_{1-x}La_xBa_2Cu_3O_y$ were prepared by solid state reaction of the trivalent oxides of the rare-earths with $BaCO_3$ and CuO. The samples were reacted for 6 hrs in air at 900°C, reground, compacted under a load of about 4 tons. to form pellets; this process was repeated twice followed by sintering in air at 950°C for 6 hrs and slow cool. The pellets were oxygenated at 410°C for 2.5 hrs; except for the x=1 sample, in which 6 hrs of oxygenation were needed to get the highest value of T_c=48K. However, samples of $La_1Ba_2Cu_3O_y$ with highest T_c (81K) are reported by Maeda et al [1] and there is a suspicion that T_c must be of order of 100 K in an ideal sample [2].

X-ray powder diffraction patterns for the whole series of compounds are shown in Figure 1. For x values between 0 and 0.9, an orthorhombic structure was formed. For x=0 and 1 we obtained the same parameters as the reported for Er [3] and La [4]. A clear variation of the angular position of the peaks is observed, indicating a continuous increase in the lattice parameters as x changes from 0 to 1. This result indicates a continuous structural variation from the orthorhombic structure of Er 1:2:3 to the tetragonal structure of La 1:2:3 compounds.

The X-ray results and T_c measured by electrical resistance are summarized in Table 1.

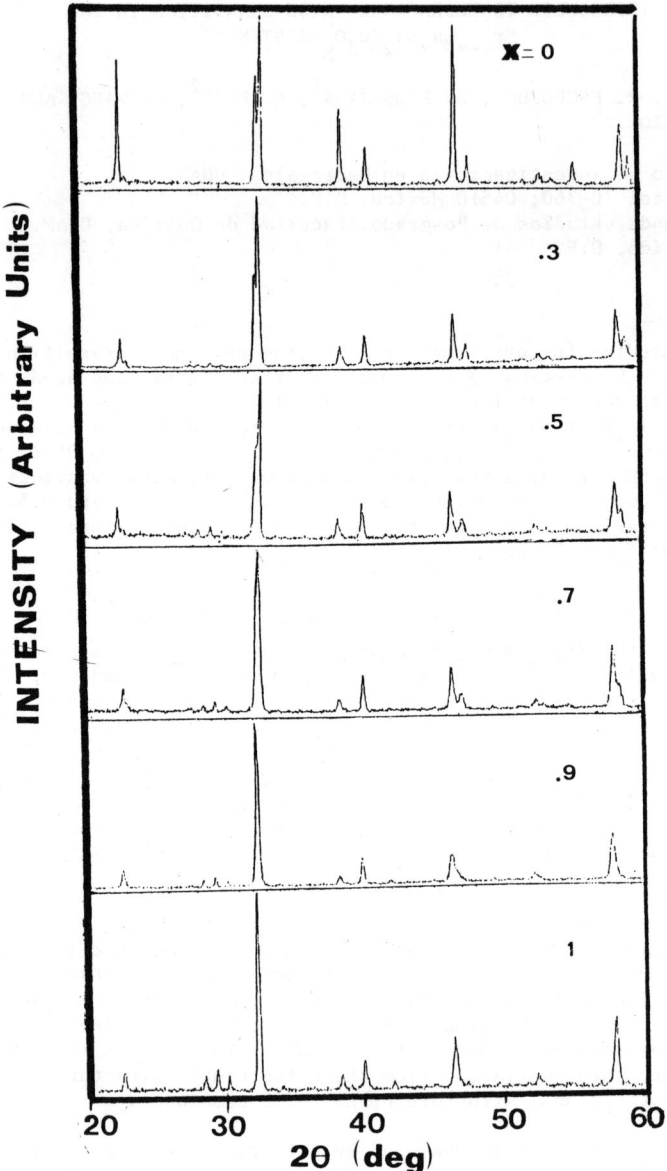

FIGURE 1. X-ray diffraction patterns of $Er_{1-x}La_xBa_2Cu_3O_y$ various values of La concentration x.

TABLE 1. Lattice Parameters (a,b, c; ± 0.005 Å) and Critical Temperature (T_c) for $Er_{1-x}La_xBa_2Cu_3O_y$ Compounds.

X	Lattice Parameters			T_c (K)
	a (Å)	b (Å)	c (Å)	
0.0	3 807	3.882	11.647	89.9
0.10	3.814	3.883	11.648	88.0
0.30	3.826	3.893	11.679	86.0
0.40	3.839	3.903	11.709	83.0
0.45	3.841	3.904	11.712	80.6
0.50	3.844	3 906	11.718	78.0
0.60	3.847	3.906	11.714	76.0
0.65	3.853	3.911	11.733	71.8
0.70	3.859	3.915	11.745	68.0
0.90	3.879	3.918	11.754	66.0
0.95	3.921	3.921	11.764	52.0
1.00	3.920	3.920	11.760	48.0

Figure 2 shows the variation of lattice parameters as a function of x determined from calibrated diffraction patterns. A continuous increase in a, b and c parameters is observed. No discontinuity at any value of x is observed, except probably at values greater than x = 0.9.

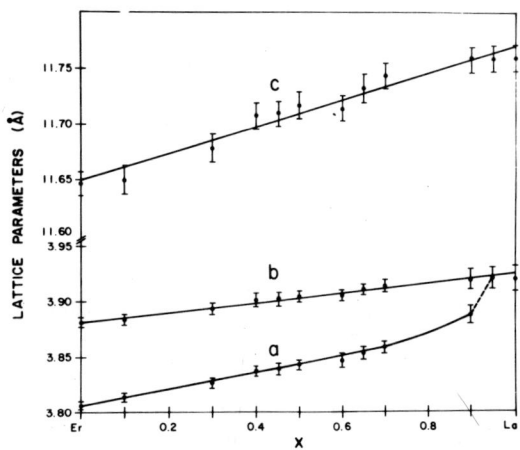

FIGURE 2

Variation of orthorhombic lattice parameters versus La concentration (x).

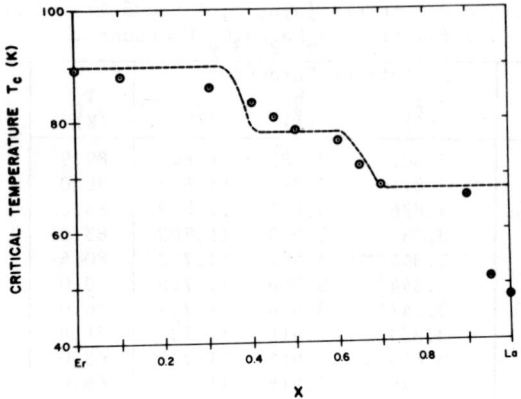

FIGURE 3 Variation of T_c versus La concentration (x).
Experimental points are shown in circles and the dotted
curve corresponds to the theoretical prediction of a
simple percolation model.

In Figure 3, the experimental results of the variation of T_c versus La concentration (x) are shown as dots. The curve presents changes at around 0.4, 0.7 and 1.0. The last one is due to structure transition from orthorhombic to tetragonal phase, which can be observed also in Fig. 2. The first two changes could be explained by a simple percolation model.

As one can notice in Figure 3, the T_c of the pure Er compound is about 90 K, while T_c for the compound with ninety per cent of La is about 66K. The variation of T_c as a function of x can be modeled by considering Er percolation in the Er-La sublattice. Let us consider a regular square lattice in which all sites can be randomly occupied by Er or La atoms (Fig. 4). Therefore two percolation thresholds should be present. The first one, at x=0.41, corresponds to first neighbours percolation in the square lattice [5]. If one considers perfect order at x=0.5, there should be a second neighbour percolation at x=0.705.

One notices that there is a plateau at around x=0.5 with an intermediate T_c of 78 K, which is telling us that ordering of the square lattice should be present. This agrees with the analysis of the parameters a and b, which shows a tendency towards a tetragonal lattice when x is increased from 0 to 1.

In Figure 3, there is a schematic theoretical prediction based on the above mentioned argument (dotted line). As it can be seen, the agreement is fairly good. It is important to note that the model is purely geometrical and does not take into account the complicated processes that relate the perturbations introduced by La-Er disorder with T_c (see Fig. 4). These latter most probably have to do with changes in the oxygen content of the sample, due to the inclusion of La. In fact, the low T_c

for the La compound tells us that it is difficult to obtain the optimum oxygen content in the range of oxygenation times used, as compared with the Er 1:2:3 compound. It is interesting to notice that the behaviour of T_c on the oxygenation in Y 1:2:3 superconductors [6] presents similar plateaux to the ones reported here. It is therefore assumed that the origin of both could be the same, namely, the oxygen stoichiometry. The beautiful part of the present discussion is that such a complicated interplay between T_c and oxygen content can be modeled by a purely geometrical change in the present experiment.

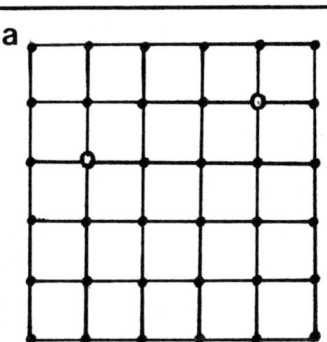

Let us consider a regular square lattice in which all sites can be randomly occupied by Er or La atoms. The Er first neighbour percolation limit is 0.41.

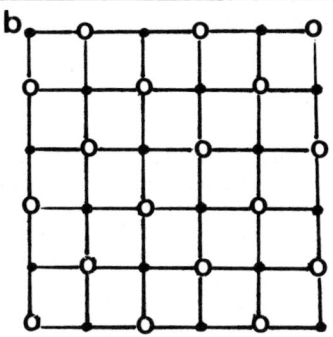

We assume that a diatomic square lattice is formed at x=0.5 since the analysis of parameters a and b shows a tendency towards a tetragonal lattice when x is increased from 0 to 1 and this sort of order is formed.

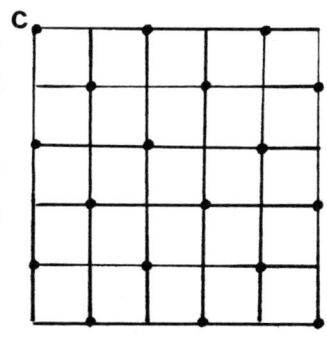

If one extracts the Er sublattice when x=0.5, it forms another regular square lattice of Er second neighbours. The corresponding percolation limit is 0.705.

FIGURE 4. Scheme showing the different steps on substituting La in an Er square lattice.

The smoothness of the curve in Fig. 3 can be thought as arising from La-Ba disorder, enhanced by the similar ionic radii of both ions, which obviously could disturb the effects predicted by the Er-La substitutional alloying alone. The presence of such a disorder is further supported by Fig. 1, in which clear changes in the relative intensity of the (00ℓ) reflexions are detected.

ACKNOWLEDGEMENTS

Financial support from the Programa Universitario de Investigación sobre Superconductores de Alta Temperatura de Transición, UNAM and CONACYT is fully acknowledged.

REFERENCES

1. A. Maeda, T. Yabe, K. Uchinokura, M. Izumi and S. Tanaka, Jpn. J. Appl. Phys. 26 (1987) L1550.
2. S. Uchida. To be published in Proceedings of the IX Winter Meeting on Low Temperature Physics. "High Temperature Superconductors". Morelos México (World Scientific) (1988).
3. T. Ishigaki, H. Asano and K. Takita, Jpn. J. Appl. Phys. 26 (1987) L987.
4. S. Lee, J.P. Golben, S.Y. Lee, X. Chen, Y. Song, T.W. Noh, R.D. McMichael, J.R. Garnes, D.L. Cox and B.R. Patton, submitted to Phys. Rev. Lett. (1987).
5. H.L. Frisch. J.M. Hammersley and D.J.A. Welsh, Phys. Rev. 126 (1962) 949.
6. R.J. Cava, B. Batlogg, C.H. Chen, E.A. Rietman, S.M. Zahurak and D. Werder, Nature 329 (1987).

SOLID SOLUTION IN THE SYSTEM $(Gd_yYb_{1-y})Ba_2Cu_3O_{7-x}$

G. PACHECO, MA. DE L. CHAVEZ, AND C. PIÑA.
DEPg. Facultad de Química, Cd. Universitaria
04510, México, D.F.

ABSTRACT

Crystal structures and different phases of high Tc oxide superconductor system $(Gd_yYb_{1-y}) Ba_2Cu_3O_{7-x}$ have been identified using X ray diffractometry and electron probe microanalysis. All the major diffraction peaks can be indexed in terms of orthorhombic phases. The identified phases have been confirmed by scanning backscattered electron images of the five samples.

The prospect of superconducting devices operating above the liquid nitrogen temperature is seen as significant technological development. Recently, materials of the general form R $Ba_2Cu_3O_{7-x}$ and with high critical temperature (Tc) have been found. R is a rare earth metal, such as Y, Nd, Sm, Eu, Gd, Dy, Ho, Er, Tm and Lu [1,2]. In this brief communications we report the crystal structure, microstructure and superconducting properties of the mixture with nominal composition (Gd_yYb_{1-x}) $Ba_2Cu_3O_{7-x}$ and y=0.0, 0.20, 0.50, 0.70, and 1.0 [3]. The samples were prepared by grinding, heating in air to 900°C for fifteen hours, pelleted, pressed and sintered at 1000°C for thereee hours. Then, after cooling, they were fired at 450°C for thereee hours in oxygen atmosphere. In Fig. 1 we compare the X ray diffraction lines for y=0.0, 0.5, and 1.0. All the major peaks can be identified in terms of the orthorombic phase assigned to the compund $YBa_2Cu_3O_{7-x}$ [4]. The small deviation of the (104) reflection is due to the difference in ionic radium of Gd and Yb and no indication of a second phase was found.

Resistance measurements, Fig. 2 show that the Gd and Yb content affects the high temperature resistivity of the samples, but does not affect their critical temperature at 87K.

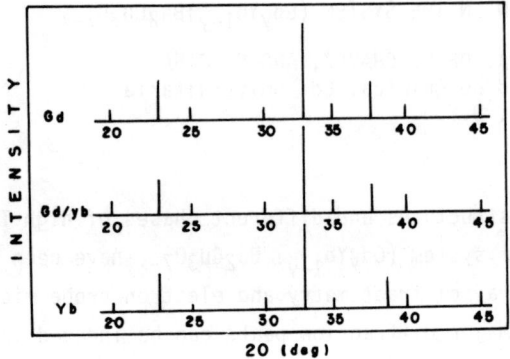

Fig.1

Fig.1 Principal characteristic X ray diffraction lines of $Gd_2Ba_2Cu_3O_{7-x}$, $Gd_{0.5}Yb_{0.5}Ba_2Cu_3O_{7-x}$, and $Yb Ba_2Cu_3O_{7-x}$. The simmetry is orthorhombic.

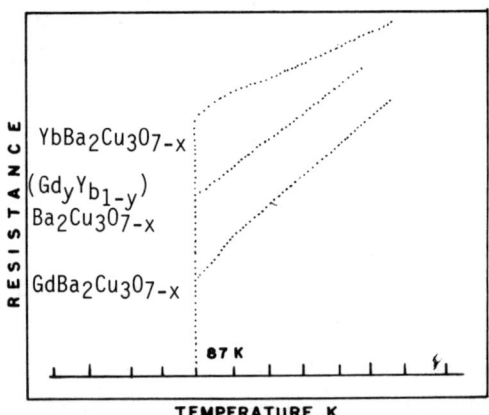

Fig. 2

Fig. 2 Shows the resistance-temperature characteristics of the samples $Yb Ba_2Cu_3O_{7-x}$, $(Gd_yYb_{1-y})Ba_2Cu_3O_{7-x}$, and $GdBa_2Cu_3O_{7-x}$.

In order to show that the nominal composition for he samples is actually the composition in the microcrystals, we performed electron probe microanalysis. Our scanning data are uniform throug out the samples and their averages are listed in Table I.

TABLE I

CHEMICAL COMPOSITION

Sample Y	%	Ba	Cu	Gd	Yb
1	w	43.42	32.55	24.05	-0-
	at	34.21	52.22	15.57	-0-
0.7	w	50.00	40.18	14.87	5.86
	at	59.00	52.87	20.30	4.00
0.5	w	49.89	38.61	10.17	13.14
	at	32.20	53.42	5.02	7.62
0.2	w	30.64	27.35	5.66	23.14
	at	24.72	47.35	4.08	16.80
0	w	43.40	32.28	-0-	11.64
	at	32.23	53.42	-0-	6.47

Since mechanical properties are determined by morphology, we also investigated this property by backscattering electron image for different values of y. In Fig. 3 we show this image for $y=0.50$.

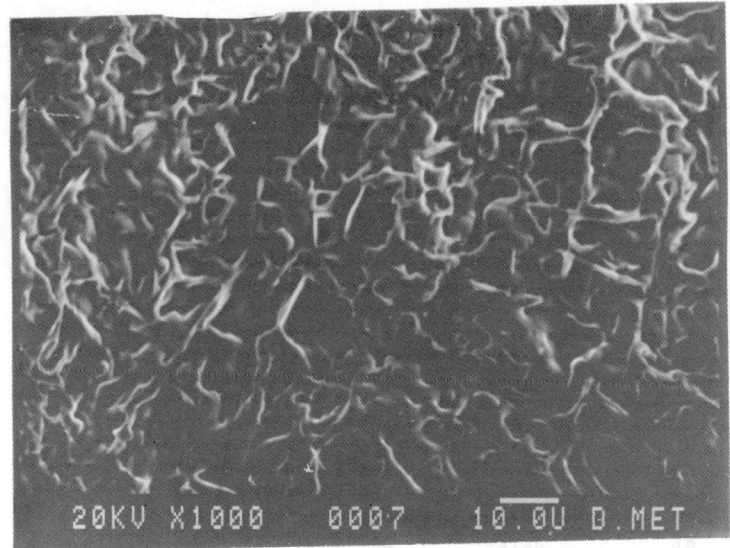

FIG. 3

Fig.3 Shows a typical backscatered electron image of sample $Gd_{0.5}Yb_{0.5}Ba_2Cu_3O_{7-x}$.

BIBLIOGRAPHY

1) D.W.Murphy, S. Sumshine, R.B. van Dover
   R.J. Cava, B. Battogg, S.M. Zahmreak and
   L.F. Schneemeyer; phys, Rev. Lett 58 (1987) 1888

2) P.H. Hor, R.L. Meng, Y.Q.Wang, L.Goo, Z.J. Hwang
   J. Bectilold, K. Forster and C.W. Chm. Phys, Rev. Lett 58
   (1987) 1891

3) J.O. Willis, z. Fisk, J.D. Thompson, S.W. Cheong
   R.M. Aiking, J.L. Smith and E. Zirngiebl
   Abstract Physic División. Los Alamos National
   Laboratory, Los Alamos, New México 87545.

4) F.Isumi, H. Asano, T. Ishigaki
   A.Ono and F.P.Okamura.
   Jpn J Appl Phys 26 (1987) L611

The authors would like to express their thanks to IQM F. Manriquez for measurements of microprobe and Dr. R. Escudero for superconductivity measurements.

SUPERCONDUCTIVITY OF Gd-X-Ba-Cu-OXIDES (X = Yb, La, Y)

M. de L. CHAVEZ, D. LOPEZ, G. PACHECO and C. PIÑA
División de Estudios de Posgrado, Facultad de Química,
Ciudad Universitaria, 04510 México, D.F.

ABSTRACT

The superconductivity of (Gd,X)-Ba-Cu-O compounds, where X=Yb,La,Y, and of possible solid solutions within the Gd-Ba-Cu-O system is reported. The compatibility relations in the Gd-Ba-Cu-O system have also been investigated and a tentative diagram is presented.

INTRODUCTION

The relations of compatibility in the system X-Ba-Cu-O, where X represents a rare earth component or an association of them, are studied with the objetive of establishing them and of identifying those phases which may exhibit superconductivity[1].

EXPERIMENTAL PROCEDURE

The compositions studied were prepared from reagent grade CuO, $BaCO_3$, Gd_2O_3, Yb_2O_3, and La_2O_3, which were weighed, blended, and fired at 900 - 950°C in an electric furnace. X-ray diffraction, with filtered CuK radiation, was used to identify the crystalline phases.

RESULTS AND DISCUSSION

For those compositions investigated in the system (Gd,X)-Ba-Cu-O, specifically $(Gd,X)Ba_2Cu_3O_z$, where X= Yb, La, Y and which were fired at 900°C, the measured temperatures of superconductivity (T_c) and the main characteristic diffraction peaks are indicated in Table I.

TABLE I

TEMPERATURE OF SUPERCONDUCTIVITY AND MAIN DIFFRACTION PEAKS OF SELECTED COMPOSITIONS

Composition	T_c (°K)	XRD peak d		
1 Gd Ba$_2$Cu$_3$O$_z$	87	3.87_{15}	2.73_x	2.23_{16}
2 (Gd Yb Y) Ba$_2$Cu$_3$O$_z$	-	3.85_{15}	2.73_x	2.23_{15}
3 (Gd La) Ba$_2$Cu$_3$O$_z$	67	3.85_{15}	2.74_x	2.23_{14}
4 (Gd Yb La) Ba$_2$Cu$_3$O$_z$	-	3.85_{15}	2.74_4	2.23_{15}

These compounds are isostructural with YBa$_2$Cu$_3$O$_z$ [2]. Samples 1 and 3 exhibit superconductivity at 87° and 67°K respectively. They both were sintered at 950°C for 4 hr and were further oxidized at 450°C during 4 hr. It is felt that the T_c could possibly be improved if the conditions of sinterization and oxidation were better known. Samples 2 and 4 in which part of the Gd was substituted for Yb, Y, and La failed to show any superconductivity under these conditions.

A systematic study of compositions within the system Gd_2O_3 - BaO - CuO allowed definition of the compatibility relations tentatively presented in Figure 1. Compatibility exists between BG - C, BG - BC, BG - *, and C - *. The symbol "*" is here used to designate the 123 type compound GdBa$_2$Cu$_3$O$_z$. In the binary system * - C, solid solution was detected near the 123 end member when the addition of up to 20 % CuO and reaction at 950°C failed to show any substantial displacement of the diffraction peaks. Ternary compositions within the small triangle BC - * - (* + 20% CuO) behaved similarly. Altogether an area of solid solution is depicted which may be of interest for developing superconductive materials. It is presently being evaluated by the authors. The temperatures of superconductivity recorded for those compositions near the 123 compound are indicated in Table II.

The system Gd_2O_3 - BaO - CuO presented in Figure 1 is tentatively proposed. Further studies are being carried using optical microscopy, SEM, and XRD coupled to O_2 measurements on materials sintered at 950°C and other temperatures as well.

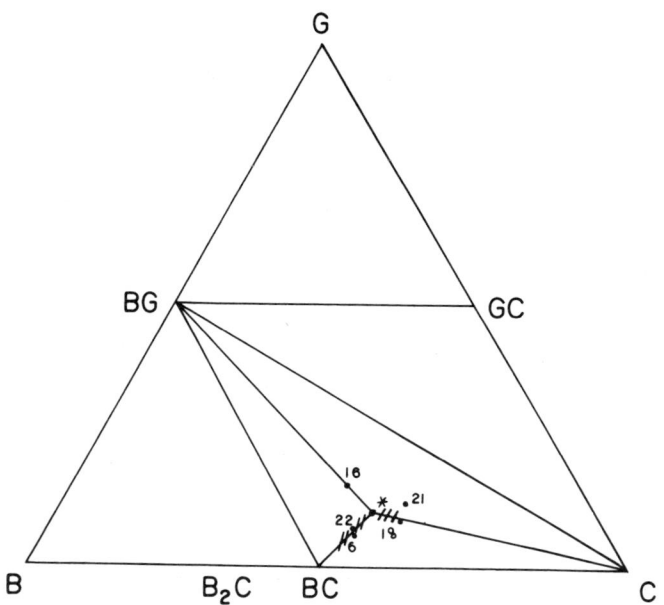

Fig. 1 Preliminary Subsolidus Phase Diagram
$Gd_2O_3 - Ba_2O - CuO$ at 900°C

/// SOLID SOLUTION, TRIANGLES BG–G–GC
and B–BG–BC need more experimentation

G Gd_2O_3, B BaO, C CuO, BG $BaO \cdot Gd_2O_3$, GC $Gd_2O_3 \cdot CuO$, BC $BaO \cdot CuO$, B_2C $2BaO \cdot CuO$, * $Gd_2O_3 \cdot 4BaO \cdot 6CuO$.

TABLE II
TEMPERATURE OF SUPERCONDUCTIVITY FOR COMPOSITIONS NEAR THE 123 COMPOUND

	Composition				T_c (°K)
	Gd_2O_3	BaO	CuO		
*	1	6	4	(123)	87
22	1	6	8		78
6	3	19	26		70
16	1	3	4		34
21	1	3	6		54
18	$GdBa_2Cu_3O_z$ + 20% CuO				83

CONCLUSIONS

The compounds $GdBa_2Cu_3O_z$ and $(Gd\ La)Ba_2Cu_3O_z$ exhibit superconductivity at 87°K and 67°K respectively. Similar compounds containing Yb and Y might have superconductivity if oxidating conditions could be improved. In the system $Gd_{1.5}$-BaO-CuO the 123 compound (*) presented conductivity at 87°K. Solid solutions within the systems BC - 123, C - 123, and BC - C - 123 were also found to be superconductive. These compositions are even more interesting in the sense that the rare earth elements are rarely pure. Thus, it could be that pure rare earths may not be needed in the preparation of superconductive materials, but a mixture of them.

The authors would like to express their gratitude to Dr. R. Escudero for the superconductivity measurements and to Dr. L. de Pablo for his suggestions throughout this work.

REFERENCES

1) L.F. Scheemeyer et al. Nature, vol. 328 13 August 1987.
2) J.O. Willis et al. Physics Div., Los Alamos National Lab 1988.

EVIDENCE OF STRUCTURAL CHANGES NEAR T_c IN A $YBa_2Cu_{3-x}Fe_xO_\delta$ SUPERCONDUCTOR.*

R. GOMEZ[1], S. ABURTO[1], V. MARQUINA[1], M.L. MARQUINA[1], M. JIMENEZ[1], C. QUINTANAR[1], T. AKACHI[2], R. ESCUDERO[2], R. A. BARRIO[2] AND D. RIOS-JARA[2].

[1] Facultad de Ciencias, UNAM. México D.F., 04510.
[2] Instituto de Investigación en Materiales, UNAM. México D.F., 04510.

ABSTRACT

In order to probe the local surroundings of the Cu sites in the $YBa_2Cu_3O_\delta$ superconductor, a series of transmission Mössbauer spectra at different temperatures were obtained from $YBa_2Cu_{3-x}O_\delta$ powdered samples. For $x = 0.125$, the spectra consist of three quadrupole doublets, two of which have Mössbauer parameters similar to the ones determined in a previous work [1]. The additional doublet obtained in this experiment has a larger quadrupole splitting than the previous ones but nearly the same isomer shift, indicating that the high local fields persist, although the Fe atoms responsible for the new doublet have a different environment. There are other two interesting features in this doublet: its intensity changes drastically with temperature and it presents a temperature dependent symmetry.

Mössbauer measurements have been reported recently in the Y-Ba-Cu-Fe-O [1-7], Gd-Ba-Cu-Fe-O [8] and the La-Sr-Cu-Sn-O [9] superconductor systems. The Mössbauer spectroscopy has been used in order to probe local properties around Cu sites. In a previous work [1] we reported the Mössbauer study of $YBa_2Cu_{2.9375}Fe_{0.0625}O_\delta$ superconductor. The spectra obtained at different temperatures showed two quadrupole doublets with parameters corresponding to low spin

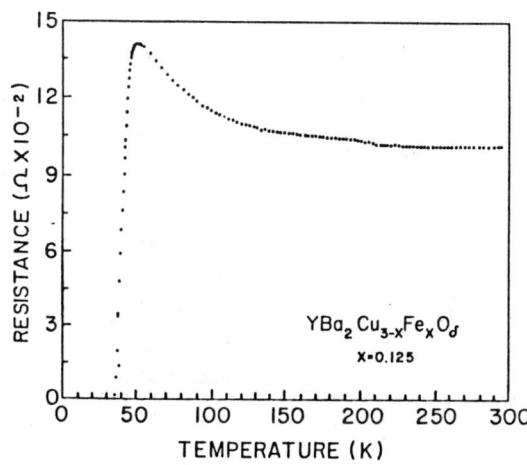

FIGURE 1
Resistance as a function of temperature for $x = 0.125$.

Fe. This experimental fact gives an indication of high local fields. Furthermore, the evidence obtained in this experiments imply that Fe atoms go to Cu(1) sites. The present work reports the Mössbauer study of $YBa_2Cu_{2.875}Fe_{0.125}O_\delta$ superconductor. The superconducting sample and the powder sample used for Mössbauer measurements were prepared under conditions similar to those described elsewhere [1].

Figure 1 shows the resistance versus temperature characteristic of the $x = 0.125$ sample. This curve presents a temperature onset of superconductivity at 50 K and fully superconducting behaviour at 35 K. It is interesting to note that the curve shows metallic like behaviour above 240 K, below which there is an increase in resistivity until the superconducting transition is reached.

Figure 2 exhibits the Mössbauer spectra obtained at 12, 60, 75, 90, 105 and 300 K. There are three quadrupole doublets, two of which have Mössbauer parameters similar to those reported previously (see Fig. 3 and 4). The new doublet (see Table 1) has a larger quadrupole splitting ($\approx$ 2.0 mm/s) and almost the same isomer shift ($\approx$ 0.0 mm/s, with respect to Fe). Therefore, we have once again the evidence of high local fields, although the Fe atoms associated with this doublet have a different environment.

A crude calculation of the electrical field gradient, based on a point charge model, at the center of an arrangement of two, four and six oxygen atoms reveals

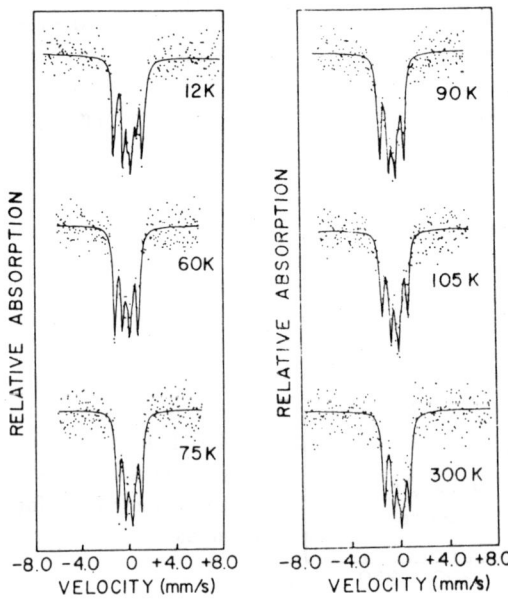

FIGURE 2
Mössbauer spectra of the sample for different temperatures.

that the Fe atoms go to the (three) Cu(1) sites. This is consistent with the results of our previous work [1] and with the results reported by Y.K. Tao et al. [10]. Besides, it is also possible to consider the Fe(III) substitution at the Cu(1) position with four and six oxygens and Fe(II) substitution at Cu(1) sites with two oxygens. The quadrupole splitting corresponding to the third doublet is temperature independent (see Fig. 4) in agreement with the Fe(II) behaviour [11] and with the Q value obtained (crude calculation). It is interesting to note that the absorption of this doublet shows an anomalous behaviour with temperature (see Fig. 5). This variations can only be explained assuming that the Fe atoms responsible for this doublet have a degree of mobility higher than the other Fe atoms. Actually this latter fact provides a further reason to assign only two oxygen neighbours to these Fe(II) atoms.

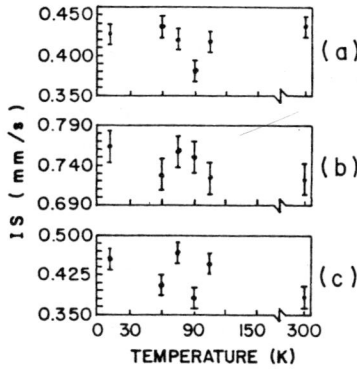

FIGURE 3
Isomer shifts for each doublet, as a function of temperature.

TABLE 1
Mössbauer parameters as a function of temperature for the $YBa_2Cu_{2.875}Fe_{0.125}O_6$ sample. A_1, A_2, A_3: absorptions. $\Gamma_1, \Gamma_2, \Gamma_3$: linewidths. $\Delta Q_1, \Delta Q_2, \Delta Q_3$: quadrupole splittings. IS_1, IS_2, IS_3 (SNP): isomer shifts with respect to sodium nitroprusside.

T(K)	A_1 $\times 10^{-3}$	IS_1(SNP) ± 0.01	ΔQ_1 ± 0.02	A_2 $\times 10^{-3}$	IS_2(SNP) ± 0.03	ΔQ_2 ± 0.04	A_3 $\times 10^{-3}$	IS_3(SNP) ± 0.01	ΔQ_3 ± 0.02
12	3.6	0.41	0.61	1.7	0.77	0.60	3.5	0.45	2.05
60	3.5	0.42	0.63	1.0	0.72	0.53	4.1	0.44	2.00
75	3.5	0.42	0.60	1.1	0.74	0.58	3.6	0.47	2.10
90	3.5	0.41	0.55	1.1	0.72	0.57	3.5	0.44	1.98
105	3.4	0.39	0.60	1.2	0.73	0.52	2.8	0.46	2.05
300	3.0	0.40	0.63	1.2	0.73	0.52	3.0	0.38	1.98

Γ_1 = 0.38 mm/s, Γ_2 = 0.28 mm/s, Γ_3 = 0.36 mm/s.

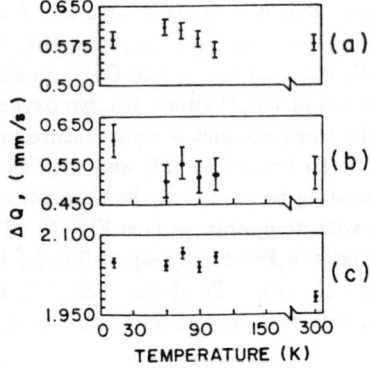

FIGURE 4
Quadrupole splitting as a function of temperature.

It is worthwhile emphasizing an another relevant fact related to the third doublet: it presents an asymmetry at 75 K. This asymmetric behaviour is also present at 60 and 45 K. However, at 12 K the asymmetry is inverted. The situation is completely different at higher temperatures, where the doublet is symmetric at 90, 105 and 300 K. It could be argued that the temperature dependence of the position of the lines could account for the observed changes in symmetry, however this is not seen, because the third doublet is rather isolated from the two others. This interesting behaviour could be attributed to the existence of weak magnetic internal fields. This statement is supported by preliminary calculations.

The authors wish to thank R. Gómez-Aiza, A. Calles and A. Salcido for their help in the computational work, and to G. González for preparing the sample.

* This work was partially supported by the Programa Universitario de Superconductores de Alta Temperatura de Transición and by the Consejo Nacional de Ciencia y Tecnología.

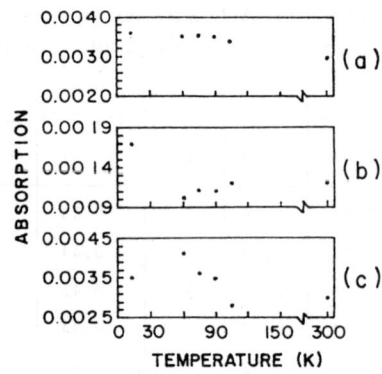

FIGURE 5
Absorption of each doublet, as a function of temperature.

REFERENCES

[1] R. Gómez, S. Aburto, M.L. Marquina, M. Jiménez, V. Marquina, C. Quintanar, T. Akachi, R. Escudero, R.A. Barrio, and D. Ríos-Jara. *Phys. Rev.* B36(13), 7226 (1987).

[2] E. Baggio Saitovitch, I. Souza Azevedo, R.B. Scorzelli, H. Saitovitch, S.F. da Cunha, A.P. Guimaraes, P.R. Silva, and A.Y. Takeuchi. Internal Report CBPF-NF-045/87. Centro Brasileiro de Pesquisas Físicas.

[3] Z.Q. Qiu *et al. J. Magn. Magn. Mater.* (to be published).

[4] X.Z. Zhou, M. Raudsepp, Q.A. Pankhurst, A.H. Morrish, Y.L. Luo, and I. Maartense. *Phys. Rev.* B36(13), 7230 (1987).

[5] E.R. Bauminger, M. Kowitt, I. Felner, and I. Nowik. Racah Institute of Physics. Jerusalem, Israel. Preprint (1987).

[6] E. Mattievich, L.F. Moreira, M.F. da Silveira, R.F.R. Pereira, H.S. de Amorim, M.R. Amaral Jr., and E. Meyer. ICAME. Melbourne, Australia (1987).

[7] Q.A. Pankhurst, A.H. Morrish, X.Z. Zhou, and I. Maartense. ICAME. Melbourne, Australia (1987).

[8] H. Tang, Z.Q. Qiu, Y.W. Du, Gang Xiao, C.L. Chien, and J.C. Walker. *Phys. Rev.* B36(7), 4018 (1987).

[9] J. Giapintzakis, J.M. Matykiewicz, C.W. Kimball, A.E. Dwight, B.D. Dunlap, M. Slaski, and F.Y. Fradin. *Phys. Lett.* A121,H 307 (1987).

[10] Y.K. Tao, J.S. Swinnea, A. Manthiram, J.S. Kim, J.B. Goodenough, and H. Steinfink. Material Research Society Fall Meetting. Boston, USA (1987).

[11] N.N. Greenwood, and T.C. Gibbe. Mössbauer Spectroscopy. Chapman and Hall, London (1971).

THERMOPOWER OF $Y_{(1-x)}Pr_xBa_2Cu_3O_{7-\delta}$ $(0 \leq x \leq 1)$

A.P. GONCALVES, I.C. SANTOS, E.B. LOPES, R.T. HENRIQUES, M. ALMEIDA
AND L. ALCACER.
Departamento de Química, ICEN-LNETI, P-2686 SACAVEM CODEX, PORTUGAL

ABSTRACT

The thermopower of the ceramic oxides of the series $Y_{1-x}Pr_xBa_2Cu_3O_{7-\delta}$ ($0 \leq x \leq 1$) is reported. Substitution of Praseodymium for Yttrium gradually changes the cell parameters but the structure remains orthorhombic up to full substitution, superconductivity being however supressed for x more than 0.6. The results are interpreted on the basis of increasing band filling due to the Pr(IV) extra electrons, and they stress the importance of correlation effects in these oxides.

INTRODUCTION

The ceramic oxides of the series $Y_{(1-x)}Pr_xBa_2Cu_3O_{7-\delta}$ ($0 \leq x \leq 1$) were synthesized by the solid state reaction of Y_2O_3 (99.99%, Ventron), Pr_2O_3 (99.9%, Pechiney-St.Gobain), $BaCO_3$ (99.5%, Carlo Erba RPE-ACS) and CuO (99.999%, Ventron), in the appropriate proportions. Pellets of intimate mixtures of the starting materials were first fired in air at 850°C for 12 hours and then slowly cooled in the furnace. These pellets were then milled, repressed at 8Kbar, re-heated at 850°C for 12 hours in air, heat treated under pure oxygen for 16 hours at 600°C and then cooled to room temperature at a rate of 30°C/h.

The materials with the different compositions were characterized by X-ray diffraction and resistivity. The X-ray diffraction can be indexed to the orthorhombic $YBa_2Cu_3O_{7-\delta}$ structure with a slight expansion of the cell as x increases, the structure of $PrBa_2Cu_3O_{7-\delta}$ being still orthorhombic. For $x \leq 0.5$ the superconductivity transition temperature decreases as x increases and the materials with $x > 0.5$ are semiconductors. A more detailed description of these results will be published elsewhere.

The thermopower was measured by a slow a.c. (~ 0.01 Hz) technique previously described [1,2]. Elongated samples with dimensions 4x1x0.5 mm^3, and with evaporated silver contacts at the ends, were mounted with platinum paint between two 25μm gold foils connected to two quartz ther mal reservoirs. The measurements were performed in vacuum and the temperature gradients were of the order of 1 to 2 K and measured using gold 0.7% at. Fe vs chromel or copper vs constantan thermocouples depending on the temperature range. The absolute thermopower was corrected for the absolute thermopower of the gold leads [3].

RESULTS AND DISCUSSION

The absolute thermopower data of $YBa_2Cu_3O_{7-\delta}$ is plotted in Fig.1. The results are similar to some of those previously reported [4] but with values about 50% smaller. They are however significantly different from those of other authors [5,6,7,8]. Results on $YBa_2Cu_3O_{7-\delta}$ from samples of our laboratory with different thermal treatments show that the observed differences are related to the oxygen content. The 1:2:3 single phase samples submitted to long annealing in pure oxygen always give the values reported here. Larger values were found in samples imperfectly reacted or annealed.

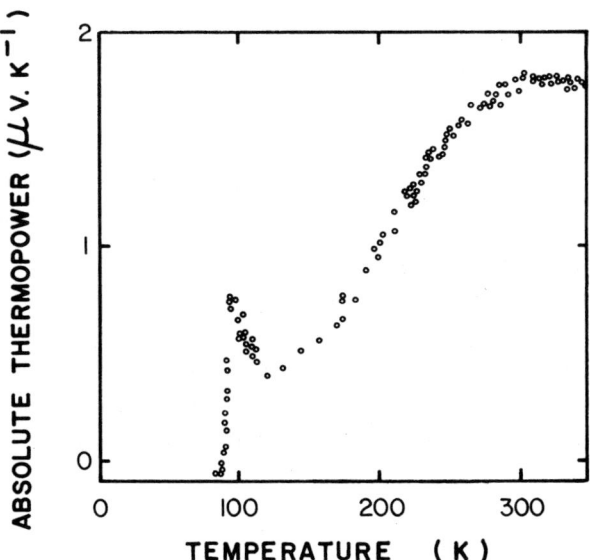

Fig.1. Absolute thermopower of $YBa_2Cu_3O_{7-\delta}$

These small values of S are typical of good metals but, in this case S is positive for T>Tc (hole like) and not linear with T as expected. Such nonlinearity has been attributed to phonon drag effects, denoting strong electron-phonon interactions in these materials [4,6,8]. Such interpretation is not adequate, however, namely the temperature independent values in the range 280-350K and the effects of substitution of Pr for Y as reported in this paper. The peak immediately after Tc is a remarkable feature which has not been explained so far. Fig. 2 shows the absolute thermopower as a function of the temperature for the different compositions of the series $Y_{(1-x)}Pr_xBa_2Cu_3O_{7-\delta}$, $0 \leq x \leq 1$.

For $x \leq 0.5$, S drops to zero at the transition temperature observed in conductivity. Inspite of metallic resistivity above Tc, again, the thermopower is not linear with T. The substitution of Pr for Y has drastic effects on the thermopower which increases with x and has negative dS/dT in the high temperature ranges. For $x > 0.5$ the materials are semiconductors and the thermopower would be expected to be linear with 1/T. Its behaviour is, however, more consistent with the picture of correlated electrons in a narrow band. In these materials the electronic conduction most certainly involves hopping. The substitution of Pr for Y primarily increases the band filling and consequently S. The decrease of S, for $x \geq 0.6$, approaching zero upon cooling below 100K is indicative of conduction via a continuum of states around the Fermi level as expected in a partially filled band even within the framework of correlated electrons in a narrow band [9,10].

In conclusion, the thermopower results, again indicate that the electronic properties of these materials are most probably dominated by electron correlation effects in relatively narrow bands.

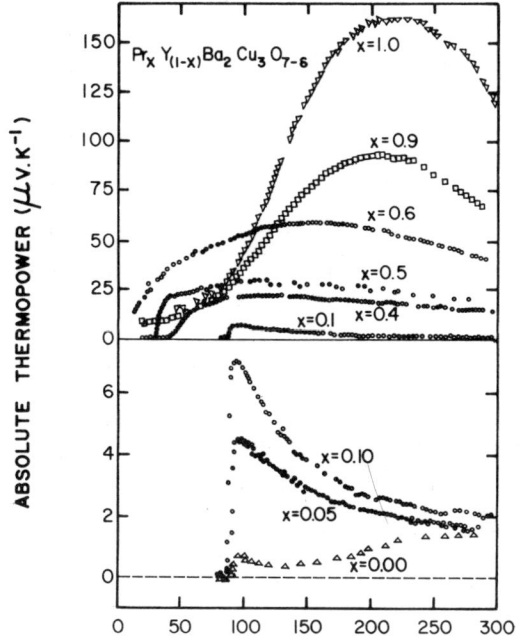

Fig. 2. Absolute thermopower of $Pr_xY_{(1-x)}Ba_2Cu_3O_{7-\delta}$ for $0 \leq x \leq 1$.

REFERENCES

1. P.M. Chaikin and J.F. Kwak, Rev. Sci. Instrum. 46, 218 (1975).

2. M. Almeida, L. Alcacer and S. Oostra, Phys. Rev. B:30, 2839 (1984).

3. R.P. Huebner, Phys. Rev. 135, A1281 (1964).

4. A. Mawdsley, H.J. Trodhal, J. Tallon, J. Sarfati and A.B. Kaiser, Nature 328, 233 (1987).

5. Z.G. Khim, S.C. Lee, J.H. Lee, B.J. Suh, Y.W. Park, C. Park, I.S. Yu and J.C. Park, Phys. Rev. B 36, 2305 (1987).

6. R. Yaozhong, H. Xuelong, Z. Yong, Q. Yitai, C. Zuyao, W. Ruiping and Z. Oirui, Solid State Commun. 64, 467 (1987).

7. S.W. Cheong, S.E. Brown, Z. Fisk, R.S. Knok, J.D. Thompson, E. Zirngrebl, G. Gruner, D.E. Peterson, G.L. Wells, R.B. Schwartz and J.R. Cooper, Phys. Rev. B, 36, 3913 (1987).

8. J.T. Chen, C.J. McEwan, L.E. Wenger and E.M. Logothetis, Phys. Rev. B. 35, 7124 (1987).

9. J.F. Kwak and G. Beni, Phys. Rev. B.13, 652 (1976).

10. J. Ihle and T. Eifrig, Phys. Stat. Solidi, 91, 135 (1979).

CRYSTAL FIELD INTERACTION AND MAGNETIC ORDER IN GdBa$_2$Cu$_3$O$_7$

M.T. CAUSA[1], C. FAINSTEIN[1], G. NIEVA[1], R. SANCHEZ[1], L.B. STEREN[1], M. TOVAR[1]
R. ZYSLER[1], D.C. VIER[2], S. SCHULTZ[2], S.B. OSEROFF[3], Z. FISK[4] AND J.L. SMITH[4]

[1]Centro Atómico Bariloche and Instituto Balseiro, 8400 Bariloche, ARGENTINA.
[2]University of California San Diego, La Jolla CA 92093, U.S.A.
[3]San Diego State University, San Diego CA 92182, U.S.A.
[4]Los Alamos National Laboratory, Los Alamos NM 87545, U.S.A.

We report experimental data on the crystal field interaction of Gd ions with their environment in Gd$_x$Eu$_{1-x}$Ba$_2$Cu$_3$O$_7$. The ESR spectrum and a Schottky anomaly in the specific heat of dilute samples indicate the existence of a crystal field splitting of the $^8S_{7/2}$ ground state of the Gd^{3+} ions of about 1.5K. Since the single ion energies involved are of the same order of magnitude as the energies associated with the magnetic ordering of the Gd moments in GdBa$_2$Cu$_3$O$_7$ (T_N = 2.24K) we analyze the effects of the crystal field interaction on the magnetic transition.

Many of the recently discovered superconducting oxides, ABa$_2$Cu$_3$O$_7$, (with A = Y or rare earths) show coexistence of superconductivity and magnetic order at low temperatures [1,2]. It is then of interest to characterize the type of magnetic order present in these materials and to study the interrelation of this order with the superconducting properties.

We report here a calorimetric and electron spin resonance (ESR) study of the crystal field interaction of Gd^{3+} ions in the ABa$_2$Cu$_3$O$_7$ structure, and we analyze its effects on the magnetic order of the GdBa$_2$Cu$_3$O$_7$ compound. We have measured pure GdBa$_2$Cu$_3$O$_7$ and dilute samples of Gd$_x$Eu$_{1-x}$Ba$_2$Cu$_3$O$_7$ with 0.005<x<0.05, where we have chosen EuBa$_2$Cu$_3$O$_7$ as a non-magnetic host in order to separate the single ion interactions of Gd atoms.

The samples were prepared by sintering thoroughly mixed powders of Eu$_2$O$_3$ Gd$_2$O$_3$, BaCO$_3$ and CuO, in appropriate concentrations. The raw materials were allowed to react in air for 20 hours at 980°C. The samples were then ground, pressed into pellets, heated in an oxygen atmosphere at 985°C, and slowly cooled to room temperature in the same atmosphere.

Measurements of the specific heat were made down to 0.45K using a semi-adiabatic calorimeter. The specific heat of GdBa$_2$Cu$_3$O$_7$ presents a lambda type anomaly peaked at 2.24K, in agreement with previous reports [1]. In the case of dilute samples, the magnetic contribution to the measured specific heat was obtained after subtraction of the specific heat of the EuBa$_2$Cu$_3$O$_7$ host.

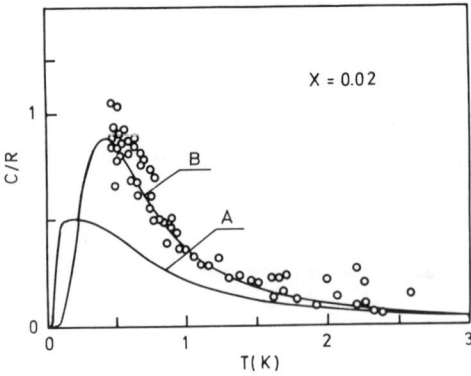

Fig.1. Specific heat of Gd$_{0.02}$Eu$_{0.98}$Ba$_2$Cu$_3$O$_7$.

A low temperature upturn was observed for samples of low concentration as shown in Fig.1 for x = 0.02. We have associated this feature with the high temperature tail of a Schottky anomaly due to the crystal field splitting of the ground state of Gd^{3+} ions, based on the results of ESR measurements as we discuss below.

The ESR spectra of powdered polycrystalline samples with $0.005 < x < 0.05$ were measured at 9 GHz and 35 GHz from 2K to room temperature. At the lowest temperatures a spectrum with resolved fine structure was obtained as shown in Fig.2 for x = 0.05. This spectrum has a central line around g = 2 and satellite lines at both sides.

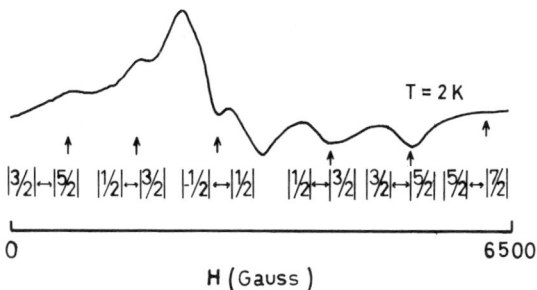

Fig.2. ESR spectrum for $Gd_{0.05}Eu_{0.95}Ba_2Cu_3O_7$ measured at 9 GHz.

The spectrum of powdered samples is expected to show a superposition of lines arising from all the possible orientations of the particles relative to the applied magnetic field. Theoretical simulations have been reported [3,4] for Gd^{3+} for two cases: cubic and axial symmetries. In the first case four separate satellite line have been observed experimentally and six in the case of axial symmetry. The number of separate lines observed in our samples and their relative spacing allowed us to describe the spectrum with a simple effective Hamiltonian:

$$\mathcal{H} = (D/3) \{ 3S_z^2 - S(S+1) \} \qquad (1)$$

Although this Hamiltonian is appropriate for axial or tetragonal (C_{4v}) symmetry and the Gd sites in $ABa_2Cu_3O_7$ have orthorhombic (C_{2v}) symmetry, the deviation from tetragonal symmetry is very small and the proposed Hamiltonian is expected to give a reasonable description of the experimental results. By fitting the position of the ESR lines to this Hamiltonian we have obtained a value of $|D/k_B| = 0.13K$. This value of the crystal field parameter would give rise to a total crystal field splitting of about 1.5K and a Schottky anomaly in the specific heat as shown in Fig.1 for a positive (curve A) and a negative (curve B) value for D.

When the Gd magnetic moments order in the concentrated system $GdBa_2Cu_3O_7$ the crystal field interaction causes anisotropy in the magnetic properties. A negative value of D favors, at T = 0K, ordering of the magnetic moments parallel to the symmetry axis of the crystal field interaction, and a positive value of D favors ordering in the plane perpendicular to it. The negative value of D suggested by the interpretation of our specific heat data in terms of the simple crystal field Hamiltonian of Eq.1, would indicate magnetic order along the c-axis. However, classical dipole-dipole interactions should also be considered, not only as a source of magnetic anisotropy, but rather as a possible cause for the magnetic order itself. In this case antiferromagnetic order is expected [5] with the Gd moments forming ferromagnetic chains alligned along the a-axis. Mössbauer spectroscopy of ^{155}Gd ions should in principle provide experimental evidence concerning the relative orientation of the magnetic moments with respect to the symmetry axes of the electric field gradient at the Gd site, which in turn is related to the crystal field interaction. Unfortunately, a comparison with existing experimental data is difficult because the interpretations given by different authors [6,7] are not unique.

The shape of the specific heat anomaly as a function of temperature is also expected to depend on the crystal field field interaction. In cases where this interaction is much larger than the magnetic energies involved in the ordering process, its main effect is to partially remove the degeneracy of the free ion ground state. This is the case of the $ABa_2Cu_3O_7$ compounds with A = Dy or Er, where only a Kramers doublet remains populated at the temperatures where magnetic order takes place [8]. In the case of A = Gd, and because Gd^{3+} is an S-state ion, the crystal field interaction is much weaker and its effects are noticeable around T_N. Our experimental data for $GdBa_2Cu_3O_7$ are shown in Fig.3, in comparison with the specific heat of $GdVO_4$, a typical antiferromagnetic compound [9] with similar Neel temperature (T_N = 2.5K). At high temperatures a large specific heat tail indicates that short-range magnetic order is important well above T_N, and at low temperatures a shoulder is observed, which is characteristic of many magnetic compounds containing Gd atoms [10]. This shoulder is primarily due to the high degeneracy (8-fold) of the ground state of Gd^{3+} ions [10]. In our case, the shoulder appears around 1K, where thermal energies are of the same order of magnitude as the crystal field interaction.

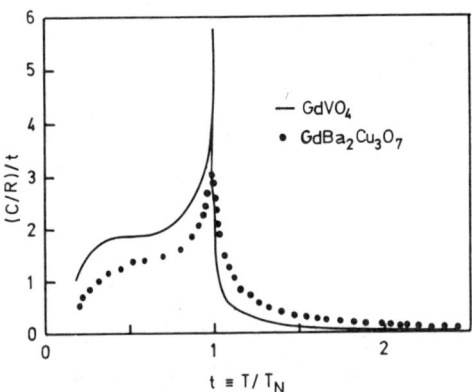

Fig.3. Specific heat for $GdBa_2Cu_3O_7$. For comparison it is also shown the specific heat of the antiferromagnetic compound $GdVO_4$.

In order to analyze the effects of this interaction on the specific heat temperature dependence, we have carried out a mean field calculation of the magnetic transition including an axial crystal field term, as given by Eq.1. It was observed in this model that at temperatures below $\sim T_N/2$, the energy levels are almost temperature independent and equally spaced in absence of crystal field interaction. This gives rise to the low temperature shoulder in the specific heat, as shown in Fig.4, where we display C/T vs. temperature in order to visually enhance the presence of the shoulder. Our mean field calculation shows that for our experimental value of D, only minor changes in the overall shape of the specific heat are expected, with a shift of the transition temperature of about 15%, as shown in Fig.4.

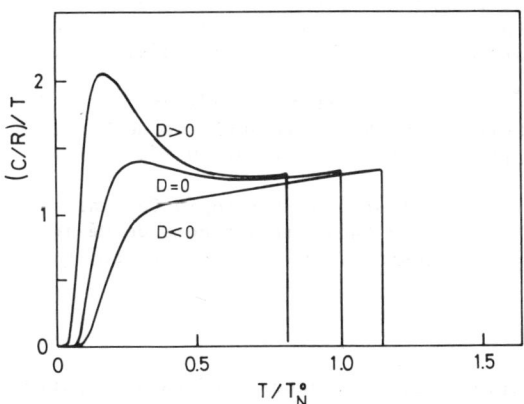

Fig.4. Mean field calculation of the magnetic anomaly of the specific heat for different values of the crystal field parameter D. The molecular field parameter has been chosen to give the measured Neel temperature for $D = 0$.

In conclusion, we have obtained a coherent description for the ESR and specific heat data of dilute samples, using a simple crystal field Hamiltonian with axial symmetry. From this simple model it follows that the crystal field parameter should be negative in order to fit the specific heat data. This is also consistent with the shape of the low temperature shoulder observed for the concentrated system $GdBa_2Cu_3O_7$.

We acknowledge partial support by CONICET (Argentina).

References:

1. Z.Fisk, J.D.Thompson, E.Zirngiebl, J.L.Smith and S-W Cheong, Solid State Commun. (1987).
2. M.T.Causa, S.M.Dutrús, C.Fainstein, G.Nieva, H.R.Salva, R.Sánchez, M.Tovar and R.Zysler, J. Modern Phys. B1, 989 (1987).
3. P.Urban and D.Seipler, J. Phys. F7, 1589 (1977).
4. D.Seipler and T.Plefka, J. Phys. F8, 969 (1978).
5. V.Massida, 72nd. Mtg. Argentine Physical Soc., Bariloche (1987).
6. E.E.Alp, L.Soderholm, G.K.Shenoy, D.G.Hinks, D.W.Capone II, K.Zhang, and B.D.Dunlap, submitted to Phys. Rev. Letts. (1987).
7. G.Czjzek, H.J.Borneman, C.Meyer, D.Ewert and B.Renker, Int. Conf. on Applications of the Mossbauer Effect, Melbourne, Australia (1987).
8. B.W.Lee, J.M.Ferreira, Y.Dalichaouch, M.S.Torikachvilii, K.N.Yang and M.B.Maple, submitted to Phys. Rev.B (1987).
9. J.D.Cashion, A.H.Cooke, T.L.Thorp and M.R.Wells, Proc. Roy. Soc. London A318, 473 (1970).
10. R.W.Hill, J.Cosier and D.A.Hukin, J. Phys. C16, 2871 (1983).

SOME REMARKS ON EPR STUDIES OF Y-Ba-Cu-O COMPOUNDS

G.Aguilar, H.Murrieta+, J.Ramírez*, T.Akachi,
R.A.Barrio and R.Escudero

Instituto de Investigaciones en Materiales
Universidad Nacional Autónoma de México
Apartado Postal 70-360, 04510 México, D.F.

+ Instituto de Física
Universidad Nacional Autónoma de México
Apartado Postal 20-364, 01000 México, D.F.

* Instituto de Fisiología Celular
Universidad Nacional Autónoma de México
Apartado Postal 70-600, 04510 México, D.F.

ABSTRACT

Electron paramagnetic resonance studies were performed on green, tetragonal and orthorhombic phases of the Y-Ba-Cu-O compound. The spectrum of the green phase consists of a very intense signal due to Cu^{2+} ions in a high symmetry site, while the tetragonal phase presents an asymmetric spectrum assigned to Cu^{2+} in Cu(II) sites. The spectrum of the orthorhombic superconducting phase consists of two superimposed signals due to Cu^{2+} ions in Cu(I) and Cu(II) sites, and another very broad and weak signal attributed to the presence of O^- ions.

The samples used in this study were characterized by X-ray powder diffraction patterns and by resistance measurements as a function of temperature. The critical temperature of the superconducting orthorhombic samples was 90 K.

Electron paramagnetic resonance (EPR) experiments were carried out using a reflection X-band Brucker ESR200D spectrometer. The g-values were determined using an HP frequency counter and a proton NMR gaussmeter.

An EPR spectrum obtained from a green phase, Y_2BaCuO_5, samples is shown in Fig. 1, and consists of a very intense symmetrical signal at about 3400 gauss. The line width, measured peak to peak, was 260 gauss and the g-value was 2.100 ± 0.001. In the case of Y_2BaCuO_5 only Cu^{2+} ions are present[1], in a five-fold O^{2-} coordination. This fact could explain the large intensity of the signal observed in this phase, relative to that of the superconducting $YBa_2Cu_3O_{7-\delta}$ phase, discussed below. The symmetrical line shape could be due to the high symmetry site occupied by the copper ions in this compound, where the Cu^{2+} ions are located almost at the center of a square base pyramid.

The non-superconducting $YBa_2Cu_3O_{7-\delta}$ tetragonal samples exhibit an

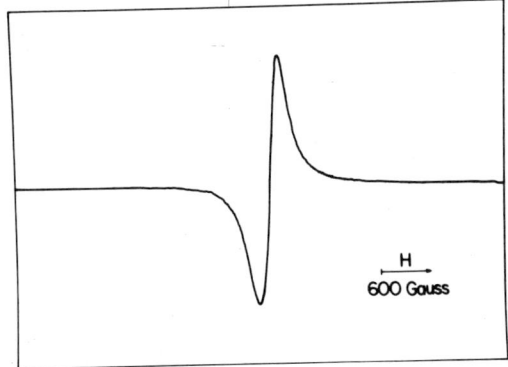

FIG. 1. Electron paramagnetic resonance spectrum for a Y_2BaCuO_5 sample (green phase), taken at room temperature.

asymmetrical EPR signal with an average line width of 210 gauss and a g-value of 2.076 ± 0.001 (Figure 2). In the tetragonal phase Cu(I) sites are occupied by monovalent copper ions, while Cu(II) sites are occupied by Cu^{2+} ions [2]. In this phase, the oxygen coordination in Cu(II) sites is such that the O^{2-} ion, located along the x-axis, has a longer bonding distance than that of the oxygen ions situated in the base of the pyramid[3]; on the other hand, the Cu^{2+} ions in the pyramid occupy a less symmetrical situation, in comparison to the green phase. These facts might explain the asymmetry of the line shape in the tetragonal phase.

In the superconducting $YBa_2Cu_3O_{7-\delta}$ samples, the EPR spectrum also presents an asymmetric signal (Figure 3), although the line width is about 60 gauss less than that of the tetragonal phase and also the "peak to peak" slope, of the first derivative EPR line, is more stepped. Now, in the orthorhombic phase, both the Cu(II) and Cu(I) ions could be in a divalent state[2]; the bonding distances for Cu(II) sites are quite similar to those of the tetragonal phase[4], while the Cu(I) ions are in a four-fold coordination, with shorter bonding distances. These facts suggest that the spectrum of the orthorhombic phase consists of the superposition of the signal coming from the Cu(II) sites, producing the same kind of asymmetrical line-shape, and a

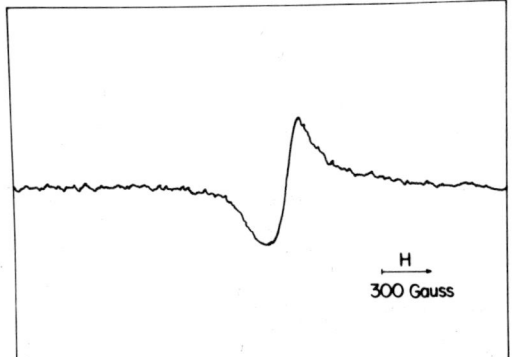

FIG. 2. Electron paramagnetic resonance spectrum for a non-superconducting $YBa_2Cu_3O_{7-\delta}$ sample (tetragonal phase), taken at room temperature.

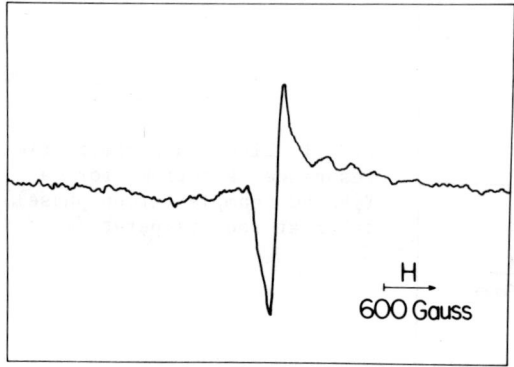

FIG. 3. Electron paramagnetic resonance spectrum for a superconducting $YBa_2Cu_3O_{7-\delta}$ sample (orthorhombic phase), taken at room temperature.

slightly splitted line ($g_{\parallel}$ = 2.109 ± 0.001, $g_{\perp}$ = 2.055 ± 0.001) produced by the tetragonal four-fold coordinated Cu(I) ions. It should be mentioned here that, under the same experimental conditions, the intensity of the tetragonal phase signal is smaller than that of the orthorhombic phase, a fact consistent with the presence of divalent copper ions in sites I and II.

An interesting result in these studies, is that in the well oxygenated superconducting samples only, a further weak and fairly broad line appears. This signal can be appreciated in figure 3 as a deviation from the base line. The width of the observed line being at least 900 gauss. In samples with a defficient oxygen treatment, which have T_c <70 K, it was not possible to detect such a signal. The origin of this signal could be attributed to holes in the oxygen sites, producing O^- ions; which have an effective spin of 1/2. Similar broad signals appear in alkali halides with a high density of color centers where they are attributed to these defects [5]. It is expected that these O^- defects have a large mobility in the lattice. Therefore the interaction of these defects with other type of impurities, as well as dipole-diple interactions among them, could be the origin of the observed large broadening. It has been mentioned in the literature [6] that there might be charge transfer fluctuations in mixed valence states of the type $(Cu^{3+} O^{2-}) \longrightarrow (Cu^{2+} O^-)$, which could be the origin of the above mentioned defects.

To conclude, EPR experiments are valuable to explore the local defect surroundings of Cu ions, and relate them with the superconducting properties of these ceramics. In the present study the detection of a signal that could be attributed to valence fluctuation defects is of paramount importance. Further studies in this direction are certainly needed.

ACKNOWLEDGEMENTS

The authors are thankful to UAM-I for allowing the use of their EPR facilities. Also thanks are given to the Programa Universitario de Investigación en Superconductores de Alta Temperatura de Transición of the UNAM for financial support.

REFERENCES

1. R.M.Hazen, L.W.Finger, R.J.Angel, C.T.Prewitt, N.L.Ross, H.K.Mao, C.G.Hadidiacos, P.H.Hor, R.L.Meng and C.W.Chu, Phys. Rev. B35, 7238 (1987).

2. H.Oyanagi, H.Ihara, T.Matsubara, M.Tokumoto, T.Matsushita, M.Hirabayashi, K.Murata, N.Terada, T.Yao, H.Iwasaki and Y.Kimura, Jpn. J. Appl. Phys. 26, L1561 (1987).

3. F.Izumi, H.Asano, T.Ishigaki, E.Takayama-Muromachi, Y.Uchida and N.Watanabe, Jpn. J. Appl. Phys. 26, L1214 (1987).

4. T.Kajitani, K.Oh-ishi, M.Kikuchi, Y.Syono and M.Hirabayashi, Jpn. J. Appl. Phys. 26 L1144 (1987).

5. J.H.Schulman and W.D. Compton "Color Centers in Solids" (Pergamon Press, New York, 1962), p.66.

6. C.M.Varma, S.Schmitt-Rink and E.Abraham, in Novel Superconductivity, edited by S.A.Wolf and V.Z.Kresin (Plenum Press, New York, 1987), p.355.

MECHANICAL CHARACTERIZATION OF $YBa_2Cu_3O_{7-x}$

L. MARTINEZ, J.L. ALBARRAN, S. VALDES AND J. FUENTES.
Institute of Physics, University of Mexico, P.O. Box 139-B, 62191
Cuernavaca, Morelos, MEXICO.

ABSTRACT

We report micro and macro hardness measurements made on a polycrystalline sample of $YBa_2Cu_3O_{7-x}$. Microhardness indentations in the interior of grains revealed that this material has, in many cases, enough ductility to allow the formation of clear diamond marks. The distribution of the measurements made on many individual grains of a polished surface shows indications of an anisotropic hardness response. The macrohardness indentations, which covered many grains, induced intergranular fracture.

The application of the new ceramic superconductors for the development of devices is limited mainly because of low critical current densities and poor mechanical properties [1]. Progress has been made in improving the critical current density of the superconductor $YBa_2Cu_3O_{7-x}$ [2]. We have not found reports of the plastic behavior of this compound which is generally considered brittle.

The superconductor $YBa_2Cu_3O_{7-x}$, and its family of rare earth substitutions, has attracted unprecedent attention [3]. It is a superconductor which below the transition temperature, between 90 and 95 K, has four important features: zero resistivity; Meissner effect; reproducibility and relative stability [4]. There are many recent reports of discoveries of other ceramic compounds claiming critical temperatures ranging approximately between 150 and 240 K [5]. However, none of these materials has shown to have at the same time the four features mentioned before [6].

The polycrystalline samples of $YBa_2Cu_3O_{7-x}$ are usually composed of heavily twinned grains of faceted quadrangular shape and of pores which may account up to 25% of the volume of the sample [7,8]. The mechanical characterization of $YBa_2Cu_3O_{7-x}$ has to include attempts to isolate the possible sources of its brittleness. The poor granular adhesion and the intergranular mode of its failure are suggestive that grain boundaries may be playing a critical role.

In the present work we employ hardness measuring techniques to test and compare the mechanical response of both the interior of the grains and large areas covering many grains of a metallographic section.

A mixture of yttrium (Y_2O_3) and copper (CuO) oxides and barium carbonate ($BaCO_3$) was prepared according to the stoichiomety of $YBa_2Cu_3O_{7-x}$. The mixture was ground in an agate mortar during 1.2 ks, annealed at 1163 K for 86.4 ks and slowly cooled; ground and annealed a second time in the same conditions; compacted to the form of a pellet of 12 mm diameter and 5 mm thick; annealed for sintering at 1193 K during 10.8 ks and slowly cooled. Since coarser grains were desired to facilitate the microhardness indentations the sample was further annealed for 86.4 ks at 1173 K and slowly cooled [7]. In all the cases the cooling rate was not faster than 0.1 K/s.

To confirm that the right phase was formed the four probe test of superconductivity was performed in a closed circuit helium liquefactor. The current used was 50 mA and the voltage and temperature were monitored employing a data acquisition unit. Since a chromel-alumel thermocouple was used the temperature error band is about ± 3 K.

The pellet was ground using coarse sand paper to remove surface irregularities and obtain a surface representative of the bulk of the material. This surface was polished using alumina powder up to 0.3 μm. The grains and voids are revealed to the optical microscope without any chemical etching.

The hardness inside the grains was measured using a Vickers microhardness indenter mounted in an optical microscope. Loads between 10 and 60 grams were applied. We made 30 indentatios for the loads of 20 to 60 grams and 80 for the load of 10. Surface Rockwell N indentations were made to measure the collective response of many grains of the material and further analyzed in a scanning electron microscope.

The four probe test indicated that the critical temperature of this sample was 92 K. Magnetic repulsion at 77 K was also exhibited by the sample.

In the figure 1 the typical optical micrograph of the orthorrombic twinned phase of $YBa_2Cu_3O_{7-x}$ is shown. The porosity of the sample is 24% determined calculating the area of pores and dividing by the total area. The grain size, using the line intercept method, is 14 μm. Many of the grains are faceted and rectangular and some of them have two or more twinned domains in different directions.

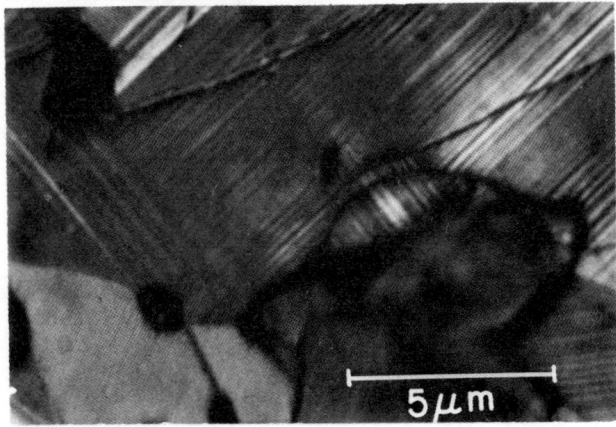

Fig. 1. Micrograph showing the heavily twinned grains of the orthorrombic phase of $YBa_2Cu_3O_{7-x}$.

The response of the grains to the micro hardness Vickers indentation in general was to yield and form the typical diamond mark shown in Figure 2a. Some of the indentations induced the fracture in the grain and many developed cracks at the diamond corners, specially when they were close to the grain boundary (Figure 2b).

The microhardness Vickers is a decreasing function of the applied load as depicted in Figure 3. The microhardness values are about 25% smaller than those of a medium carbon steel and much smaller than several metallic oxides. The microhardness values are scattered as indicated by the error bars in Figure 3. The distribution of the microhardness values obtained employing the loads of 10 and 60 grams are depicted in Figures 4a and c respectively. The forms of these distributions suggest that they might be composed of two distributions each one having its average value and standard deviation. Since crystallographic orientations are likely to be randomly distributed in this polycrystalline sample it is conceivable that the hardness values may be dependent of grain orientation.

Fig. 2a. Diamond mark in a grain of $YBa_2Cu_3O_{7-x}$.

Fig. 2b. Cracks developed at the corners of the diamond. Fracture induced by one indentation.

Using standard numerical techniques we fitted the histograms of Figure 4 with two normal distributions for each one, as it is depicted, and obtained the average microhardness and the standard deviation of each one.

The distributions of microhardness values obtained using the loads of 20, 30, 40 and 50 grams are not so clearly splitted and look similar to the one for 30 grams shown in Figure 3b.

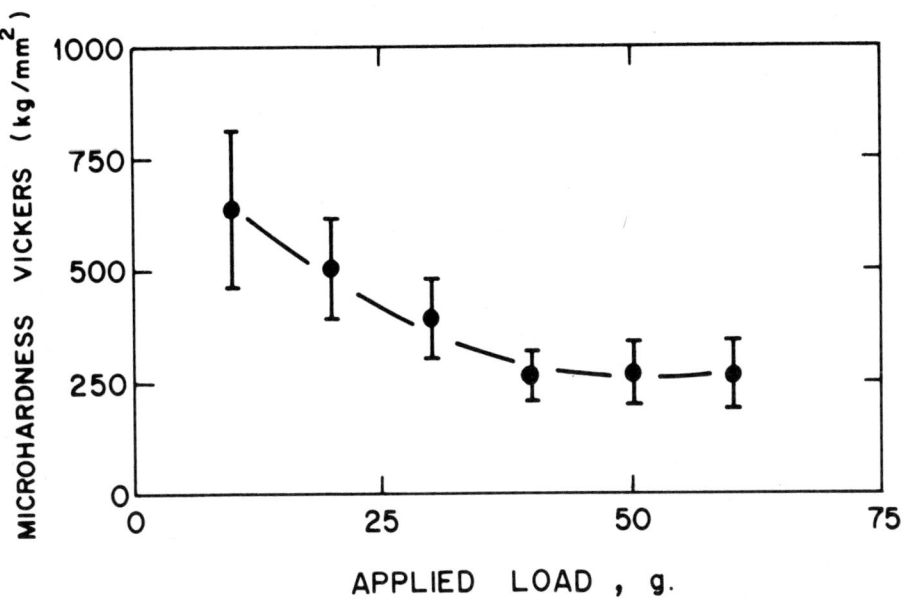

Fig. 3. The microhardness Vickers is a function of the applied load. The error bars indicate the scattering of the measurements.

Unfortunately it is not possible for us to determine the crystallographic orientation of each grain where a hardness indentation was made. Hardness measurements made on single crystals are necessary to confirm the possible annisotropy of the microhardness response of this material.

The average surface Rockwell N hardness is 40 similar to one of a medium carbon steel. The indentation shown in Figure 5a, was made of thousands or millions of intergranular cracks. This is shown in more detail in Figure 5b. The crack shown in 5b evolved along the grain boundaries joining the voids.

In summary we can conclude that the microhardness marks in the grains $YBa_2Cu_3O_{7-x}$ show that they are capable of been deformed plastically to some extent. The cracks developed at the corners of the indentations are frequently associated to the proximity of grain boundaries. The surface Rockwell N indentations were formed by many intergranular failures and cracks were developed following paths

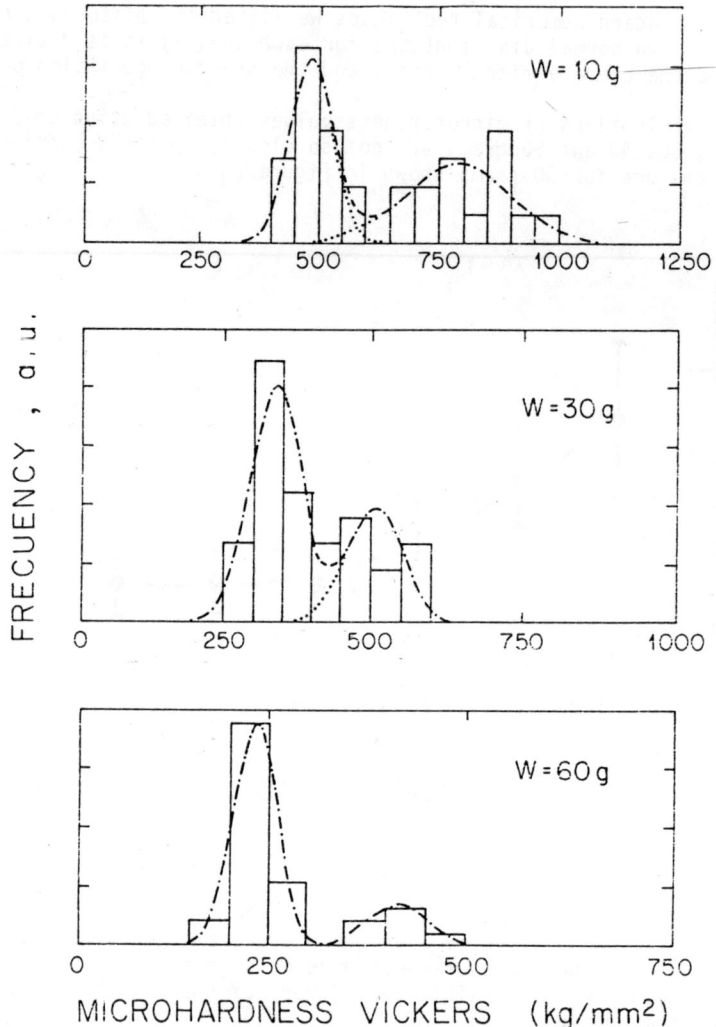

Fig. 4. Dispersion of the hardness measurements for applied loads. (a) is for 10 grams; (b) for 30 and (c) for 60.

connecting voids. The bulk mechanical behavior of the polycrystalline samples is more likely to be defined by the grain boundary poor adhesion and by the presence of the voids. Apparently the reserve of plasticity of the interior of the grains has not opportunity to perform. The distribution of hardness values has indications of a possible anisotropic mechanical response of single crystals.

Fig. 5a. Surface Rockwell N indentation.

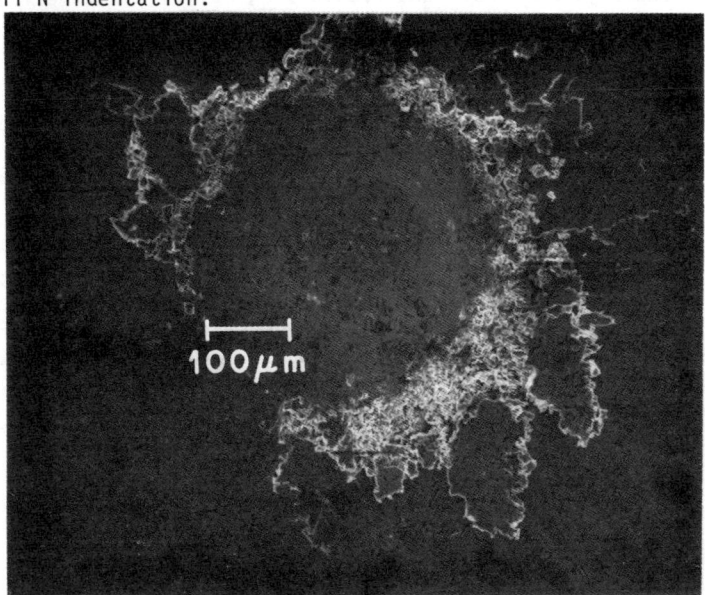

Fig. 5b. Notice the crack that evolved joining the voids along the grain boundaries.

This work was supported by the Programa Universitario de Superconductividad of the Universidad Nacional Autonoma de Mexico. The useful comments of J.H. Schneibel are acknowledged. Thanks are given to Alfredo Sanchez, Monica Lopez and Anselmo Gonzalez for their technical support.

References

1. W.C. Shumany, Adv. Mat. and Proc., 132, No. 3, 44 (1987).
2. P. Chaudari, R.H. Koch, R.B. Laiwobitz, T.R. McGuire and J. Gambino, Phys.Rev.Lett., 58, 2684 (1987).
3. C.W. Chu, P.H. Hor, R.L. Meng, L. Gao, Z.J. Huang and Y.Q. Wang, Phys.Rev.Lett., 58, 405 (1987).
4. C.W. Chu, Special Adriatico Conference on High T_c Superconductivity, Trieste, Italy, July (1987).
5. S.R. Ovshinsky, R.T. Young, D.D. Allred, G. De Maggio and G.A. Van der Leaden, Phys.Rev.Lett., 58, 2579 (1987).
6. P.M. Grant, personal communication.
7. L. Martinez, L. Vazquez, J.L. Albarran, J. Fuentes, E. Carrillo, A. Mendoza, E. Orozco, J.G. Perez Ramirez, R. Perez, L. Cota, J.L. Boldu, and M. Jose-Yacaman, Mat. Sci. and Eng., in press.
8. R. Perez, J.G. Perez Ramirez, J. Reyes-Gazga, F. Ruiz, J. Fuentes, J.L. Albarran, E. Carrillo, E. Orozco, L. Cota, A. Mendoza, J.L. Boldu, L. Martinez y M. Jose Yacaman, Int. Jour. of Mod. Phys., in press.

STUDIES ON THE ELECTRONIC STRUCTURE OF $YBa_2Cu_3O_{7-x}$

CARMEN DE TERESA[1], PABLO DE LA MORA[2] AND JAIME KELLER[1]
1 Depto. de Física y Química Teórica, División de Ciencias Básicas, Fac. de Química, U.N.A.M., Apartado 70-528, 04510 México, D.F.
2 Depto. de Física, Fac. de Ciencias, U.N.A.M., Ciudad Universitaria, 04510 México, D.F.

ABSTRACT

The electronic structure of the $YBa_2Cu_3O_{7-x}$ system is studied using a tight binding calculation with an Extended Hückel parametrization. This method is used for calculations on a cluster and on a 2D extended solid.

The density of states, the distribution of charges of the wave functions of the different atoms and the band structure, as well as the composition of the wave functions are discussed and analyzed as a function of the number of atoms in the cluster or in the unit cell.

The effect of vacancies on the electronic structure of the material is studied.

INTRODUCTION.

A tight-binding approach to the calculation of the electronic structure, specially parametrized (as in fact the Cu-O bond has been extensively studied, see for example Hay, Thibeault and Hoffmann [1]), to reproduce with higher accuracy the states near the Fermi level, allows the systematic study of the relevant or more sigificant contributions to the electronic states that should be "active" in the transport and electronically driven properties.

In this paper the Cu(I) plane was chosen as a reference because from standard band structure calculations [2] it is thought that superconductivity happens mainly on that plane. The first stage consisted of a study on this plane, later on other atomic layers were added in order to see how the electronic structure is being built up, and to see how this affects the Cu(I) layer.

Two types of calculations were done; one on 2D extended solids with periodic-boundary conditions where the band structures were obtained, the other on clusters in order to study the effect of local chemical bonding and the influence of defects in the electronic structure of the material. In all studies the geometry used was as in the reported crystalline structure.

RESULTS.

The band structure of the Cu(I) plane shows the splitting in bonding and antibonding states (Figure 1). The charge on the Band at the Fermi level is distributed mainly in Cu(I): s(0.3), $d_{x^2-y^2}$ (1.1), d_{z^2} (0.4) and O: P_y(0.2). When the oxygens of the barium planes are added (Figure 2) the degeneracies are removed and the Fermi-energy band shifts up forming a gap of $\simeq$ 1.4 eV. The composition of this band remains almost the same but it is important that a small contribution of the O(II): p_z(0.1) appears.

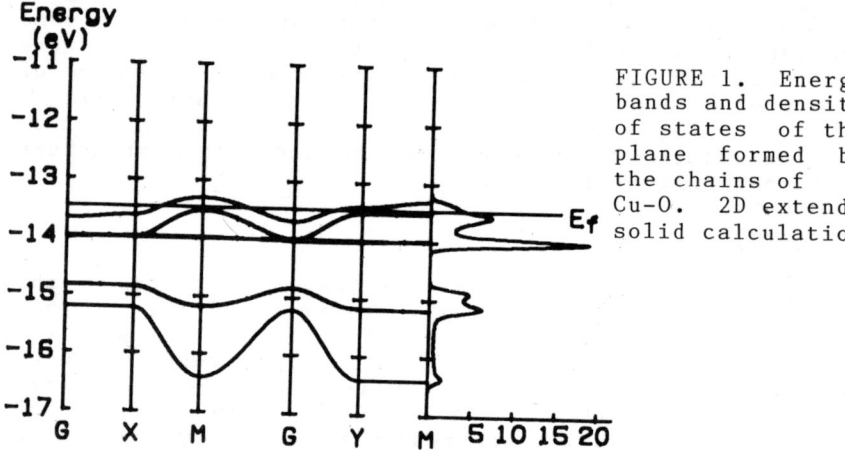

FIGURE 1. Energy bands and density of states of the plane formed by the chains of Cu-O. 2D extended solid calculation.

Cu(II) contributes to the electronic structure considerably, increasing the number of bands. But the composition of the band at the Fermi level remains the same. A new Cu(II)-O(III) band shows above the Fermi level. We are currently analyzing the conditions under which this band could appear at E_f.

The system formed by the Ba(-)-Cu(I)-Ba(+) planes was studied (Figure 3) both with band structure and with a cluster-type calculation. On adding the barium atoms an additional peak is formed on the density of states at the Fermi energy. Otherwise the barium orbital contributions appear only on bands above the Fermi level. The cluster chosen was big enough so that a density of states could be built. This density of states shows the same features as the above calculation. Comparing

Figures 2b and 3b, cluster calculations, it is seen that the enhancement of localization introduced by the finite size of the cluster, splits the peaks at the Fermi level as a result of the Ba^{++} electrostatic influence.

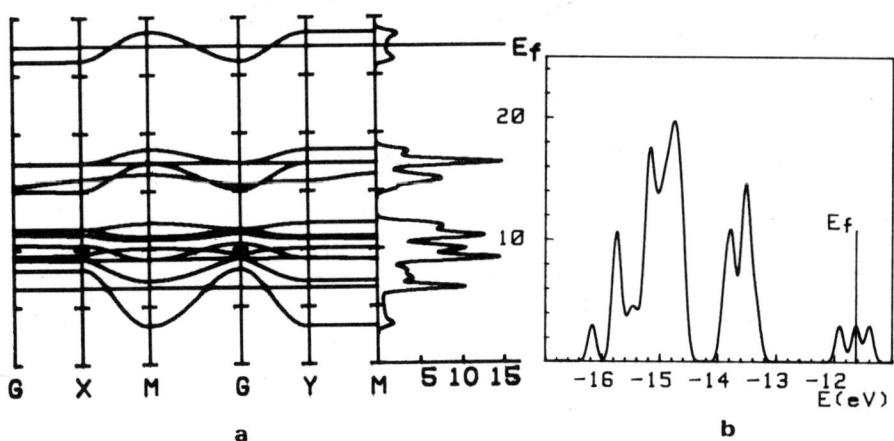

FIGURE 2. Same as Fig. 1, but with the adjacent oxygens that form kites. a) 2D extended solid. b) Cluster.

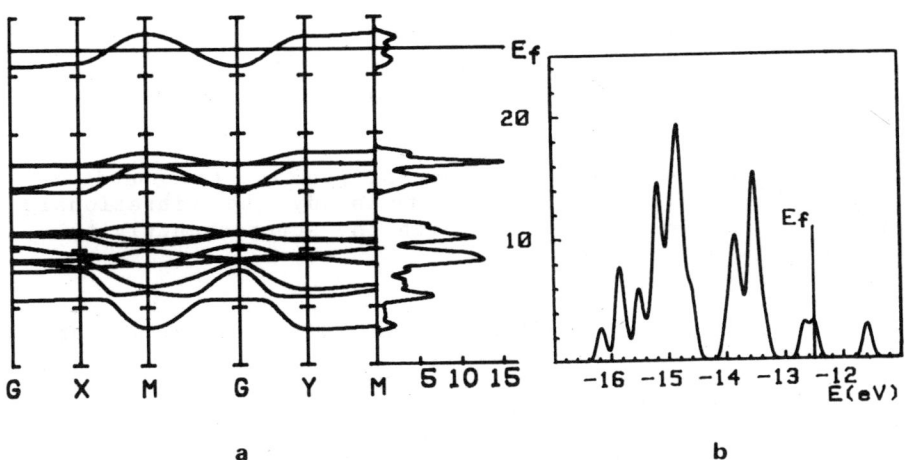

FIGURE 3. Same as fig. 2 with upper and lower barium planes.

The O(I) atoms around the Cu(I) atom were removed from the barium-free cluster to study the effect of this kind of vacancies on the crystal. A small peak appears in the gap below the Fermi level due to the decrease in the antibonding of Cu(I) atoms.

DISCUSSION.

The size of the superconducting-crystal unit cell makes the calculations of the band structure with traditional solid-state methods complicated. The above results show that a method like Extended Hückel is useful to give a qualitative description of this kind of system. The main features of the density of states agree with other first-principle calculations. It allows the study of configurations containing local detailed description of the system as was shown above.

It is interesting to remark that even if a simple "kite" CuO_4 cluster of Cu(I), O(I) and O(II) atoms is computed, the main features mentioned above appear, with the antibonding HOMO being Cu(I) $d_{x^2-y^2}$ (0.16) d_{z^2} (0.5), O(I) P_y (0.1), O(II) p_z (0.09) composition. In all cases studied the charge of the Cu ions was not larger than +1.7, it does not seem to be possible that Cu ions might have any Cu^{+++} contributions in these materials.

These calculations confirm that if the electrons near the Fermi level are the relevant states, the superconductivity of the material should be due to the states formed by Cu(I) and its surroundings oxygens [3].

The knowledge of the composition of the band at the Fermi level will be used to study the vibrational states and their relationship with the superconductivity.

REFERENCES.

1. P. J. Hay, J.C. Thibeault and R. Hoffmann, J. Am. Chem. Soc. 97, 4884 (1975).
2. L.F. Mattheiss, Phys. Rev. Lett. 58, 102B (1987). H. Krakauer and W.E. Pickett (to be published).
3. J. D. Jorgensen et al. (submitted for publication to Phys. Rev. (1987)).

ANALYSIS OF JAHN-TELLER DISTORTIONS IN $YBa_2Cu_3O_7$ COMPOUND

A. CALLES[1], E. YEPEZ[1,2], A. SALCIDO[1], J.J. CASTRO[3] AND A. CABRERA[1]

1 Facultad de Ciencias, UNAM, Apdo. Post. 70-646
04510 México D.F., MEXICO.
2 On sabatical leave from ESFM, IPN. (supported in part by COFAA).
3 Depto. de Física, CINVESTAV, IPN, Apdo. Post. 14-740,
México D.F., MEXICO.

ABSTRACT

With the purpose of making a theoretical prediction of the importance the Jahn-Teller effect has in the high Tc superconducting perovskite compounds, we present a group theoretical analysis that allows to predict the systems where this effect may play an important role.
We also present a comparative analysis of the calculated phonon spectra coming from a Jahn-Teller deformed structure and a non-deformed one.

Introduction.

The discovery of high Tc superconductors [1] has brought up a lot of theoretical work trying to explain the possible mechanisms governing this new phenomenon, since the classical BCS [2] theory has found difficulties to explain all the properties of the new superconductors.
In BCS theory the formation of Cooper pairs depends on the number of carriers contributing to the superconducting state and on the interaction of this electrons with the phonons. This dependence is given by the relation

$$T_c \simeq \theta_D \exp(-1/\lambda) \qquad (1)$$

$$\lambda \simeq N(E_F) \, V_{ep} \qquad (2)$$

where θ_D is the Debye temperature, $N(E_F)$ is the density of states at the Fermi energy (number of carriers involved in superconductivity), and Vep is the electron-phonon interaction.

Hence, within BCS theory, higher Tc might be possible if the electron-phonon interaction and the density of states at the Fermi level could be further enhanced.

Bednorz and Müller [1] suggested the polaron formation involving the Jahn-Teller effect [3] as a possible mechanism for enhancing the electron-phonon interaction.

In the present work we study the superconducting structure of $YBa_2Cu_3O_7$ within the Jahn-Teller framework finding the situations where this effect exist.

The Jahn-Teller effect exists if an atom in a crystal possesing a degenerate electronic state moves from its original position giving rise to a structural distortion of the lattice lowering its symmetry and splitting the degenerancy. Hence in order to have the Jahn-Teller effect in a crystal we need (a) to have degenerate electronic states within a specific coordination, defined by the neighbours, and (b) the existence of a non-symmetrical vibrational normal mode, corresponding to the structure defining the coordination, which will couple with the degenerate electrons. The electronic states must be classified according to the coordination point symmetry group.

Coordination defined by the unit cell around Yttrium atom.

The structure of $YBa_2Cu_3O_7$ has as the point group associated to the coordination around Y the Pmmm (D_{2h}) group. This group does not have degenerate irreducible representations. Therefore around Y and Ba we do not have the conditions to have a Jahn-Teller effect.

Coordination defined by O(1) around Cu(1).

On the other hand the coordination around Cu(1) (see Fig. 1) in the plane defined by the vectors b and c is the D_{4h} group. This group has four one dimensional irreducible representations and one two-fold degenerate. The Cu(1) can enter with valence 2^+ or 3^+ which means respectively one or two holes in the d shell.

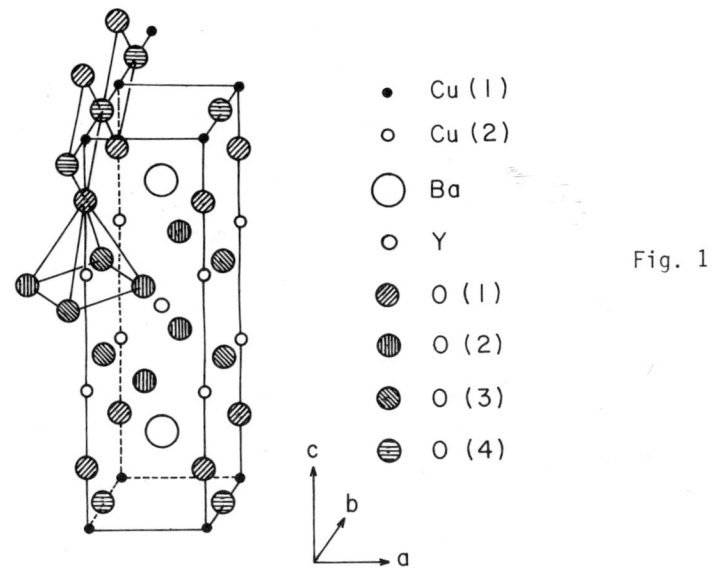

Fig. 1

- Cu (1)
- Cu (2)
- Ba
- Y
- O (1)
- O (2)
- O (3)
- O (4)

The two-fold degenerate representation is contained in the decomposition of the $\ell = 2$ representation of O(3) in D_{4h} [4], therefore the electronic state within the square defined by the oxygen of the crystalline field is two-fold degenerate. And since we have two-fold degenerate normal modes of vibration, the structure along the Cu(1) in the b-c plane must change from square to rhombic and depending on the intensity this could be static or dynamic. Hence the Jahn-Teller theorem predicts that the structure of $YBa_2Cu_3O_7$ must be slightly deformed in the b-c plane around the chains of Cu(1). This result is in agreement with the structure determined by neutron diffraction data [5] for the orthorhombic phase, therefore being the Jahn-Teller effect a plausible explanation for the experimentally observed distortion.

Coordination defined by O(2) and O(4) around Yttrium atom.

If we look at Y^{3+} in the coordination defined by the eight oxygens (O(2) and O(4)) the associated point group is D_{4h} and now the possibility of having the Jahn-Teller effect depends on which electronic states are empty (d or s).

Phonon density of states.

With the purpose of observing some changes in the physical properties induced by the Jahn-Teller effect we calculated the phonon density of states for a structure without deformation and for a Jahn-Teller deformed structure in the b-c plane. (For details of the calculation, see ref. [7] in this volume). Since, as we mention before this coincides with the experimentally determined one, we took for our calculation the actual structure. The results found are shown in Fig. 2 (a) y (b), and we compare them with the experimental density of states determined by Rhyne et al. [6] through neutron scattering measurments (see Fig. 2 (c)).

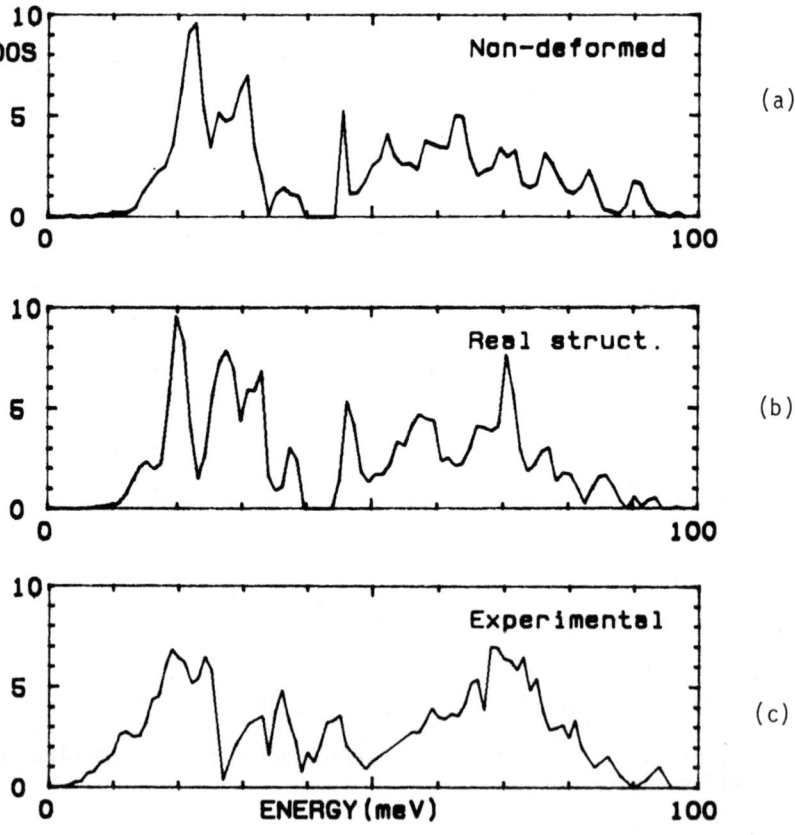

Fig. 2

As can be seen from these figures the phonon density of states calculated from the actual structure (which corresponds to our Jahn-Teller distorted structure) predicts the main features of the experimental density, whereas the one calculated from the ideal structure (without deformation) compares less favourable. Notice for example, how the initial shoulder found at 15 meV has almost disappeared and the peak at 70 meV has completely vanished and also how the non-deformed structure tends to smooth out the structure between 50 and 100 meV.

The results presented give an evidence of the importance the Jahn-Teller effect may have in the superconducting compound $YBa_2Cu_3O_7$

We acknowledge valuable discussions with R. Escudero and R. Barrio from I.I.M. - U.N.A.M.

References

[1]. J.G. Bednorz and K.A. Müller, Z. Phys. B 64, 189 (1986)
C.W. Chu, D.H. Hor, P.L. Meng, L. Gao, Z.J. Huang and Y.Q. Wang Phys. Rev. Lett. 58, 405 (1987)
[2]. J. Bardeen, L.N. Cooper and J.R. Schrieffwe, Phys. Rev. 108, 1175 (1957)
[3]. A. Jahn and E. Teller, Proc. Roy. Soc. A161, 220 (1937)
[4]. E.P. Wigner, Group Theory, Academic Press, N.Y. (1959)
[5]. F. Izumi, et al. Jpn. J. Appl. Phys. 26 (1987) L649.
[6]. J.J. Rhyne et al. Phys. Rev. B 36 (1987) 2294.
[7]. A. Calles, E. Yépez, A. Salcido, J.J. Castro and A. Cabrera. Normal Modes of Vibration in High Tc Superconductors. A Pictorial View. Included in this volume.

STUDY OF THE SUPERCONDUCTORS CERAMICS OF THE TYPE $Yb_2Ba_4Cu_6O_x$, $YbGdBa_4Cu_6O_x$ AND $Gd_2Ba_4Cu_6O_x$.

C. PIÑA[1], A. MONTOYA[2], P. BOSCH[2], R. ESCUDERO[3].

1 División de Estudios de Posgrado, Facultad de Química, UNAM. 04510. México, D. F.

2 Instituto Mexicano del Petróleo. México, D. F.

3 Instituto de Investigación en Materiales. UNAM. Apdo. Postal 70-360, 04510. México, D. F.

ABSTRACT.

We present the results of a comparative study of the system: $RE_2Ba_4Cu_6O_x$ (RE=Yb, YbGd, Gd). Lattice parameters, radial distribution function and calculated power X-ray spectra, indicate that the compounds are structurally similars.

INTRODUCTION.

In the past few months several new superconductors: $RE_1Ba_2Cu_3O_x$ (RE=Y, Yb, Gd, Er, Ho, Tm,..) have been reported [1-7]. The materials have perovskite- related structures. Several groups have studied the oxygen stoichiometry in these systems and their structures and superconducting properties [8-9].

In this report we present the results of a comparative study for the systems: $Yb_2Ba_4Cu_6O_x$, $YbGdBa_4Cu_6O_x$ and $Gd_2Ba_4Cu_6O_x$ prepared under the same conditions. The relative sizes of metal ions of rare earths and their electronic structure are different, but the results of the lattice parameters, radial distribution function and calculated and experimental patterns powder diffractograms indicate that the three compounds are structurally similar, however, the mixed case was found to be very close to the $Yb_2Ba_4Cu_6O_x$ phase.

SAMPLE PREPARATION, T_c MEASUREMENTS AND CRYSTAL STRUCTURE.

Conventional solid state methods were used to prepare $RE_2Ba_4Cu_6O_x$ (RE=Yb, YbGd, Gd). Appropiate amounts of RE_2O_3, $BaCO_3$ and CuO were mixed by milling and calcined overnigth at 900°C. Pressed pellets were then sintered at 950°C and reheated for 3 hours in oxygen at 450°C. The compounds of (a) $Yb_2Ba_4Cu_6O_x$, (b) $YbGdBa_4Cu_6O_x$ and (c) $Gd_2Ba_4Cu_6O_x$ shows a superconducting transition at 85°K, 87°K and 75°K repectively. Fig. 1, shows the resistance versus temperature characteristics.

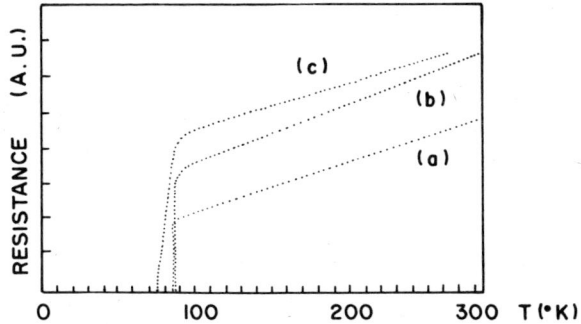

Fig. 1. Resistance vs Temperature of (a), (b) and (c).

The diffractograms were obtained with a copper anode X-ray tube coupled to a Siemmens goniometer. The monochromator provided a Cu K_α radiation. The crystalline compounds were identified using J.P.C.D.S. indexes in the conventional way.

The samples were mixed with $\alpha\text{-}Al_2O_3$ as internal standard. The 2θ positions of the diffractions peaks were corrected and exact values of the cell parameters were determined, also calculated powder diffraction patterns were obtained, varying the atomic positions and their ocupation factor but maintaining the symmetry and space group. The calculations were carried out with the program Lazy Pulverix [10].

The diffractograms were obtained for the radial distribution function, with a molybdenum anticathode X-ray tube (Mo K_α) which allowed high values of the angular parameter to be obtained ($s = 4\pi \operatorname{sen}\theta / \lambda$), where θ is the diffraction angle and λ the wavelength. The intensities were recorded at intervals of $2\theta = 1/8°$, from $2\theta = 4°$ to $2\theta = 130°$ and formed the input data to the program Radiale [11].

The three oxides show the same kind of diffractogram which correspond to a defect perovskite with an orthorhombic structure similar to $Y_{0.4}Ba_{0.6}CuO_{2.22}$[12].(Table I).

Table I. d-spacings of the $Yb_2Ba_4Cu_6O_x$, $YbGdBa_4Cu_6O_x$, and $GdBa_4Cu_6O_x$ compounds compared with Syono Model [12].

	$Y_{0.4}Ba_{0.6}CuO_{2.22}$	$Yb_2Ba_4Cu_6O_x$	$YbGdBa_4Cu_6O_x$	$Gd_2Ba_4Cu_6O_x$
hkl	d_{calc}	d_{hkl}	d_{hkl}	d_{hkl}
001	11.67	11.63	11.01	11.19
003	3.889	-	3.89	-
100	3.818	-	-	3.83
012	3.235	3.79	-	-
013	2.750	2.90	2.74	-
103	2.724	-	2.72	2.70
112	2.468	2.68	-	-
104	2.318	2.30	2.33	2.31
113	2.231	2.19	2.22	2.21
020	1.9441	1.94	1.94	1.93
200	1.9089	1.92	1.90	1.90
116	1.5827	-	1.66	-
123	1.5825	-	1.58	-
213	1.5681	-	1.56	-

The resolved peaks corresponding to (013) and (110, 103) reflections (d=2.74Å and d=2.72Å) show that the symetry is orthorhombic, cell parameters were determined using the peaks (003), (013) and (110). The results are summarized in Table II. As expected the cell parameters of the Gd phase are larger than those of the Yb phase (ionic radius of Gd is larger than Yb) also we finds that a and c orthorhombic cell parameters of mixed phase (YbGd) are intermediate, but closer to those of the Yb phase. The b parameter, however, is shorter in the mixed phase than in the other two compounds. We conclude that the Gd is in a solid solution in the Yb phase structure.

Table II. Cell parameters of the compounds compared with Syono [12] and Beno [13] compounds.

Compound	a_o(Å)	b_o(Å)	c_o(Å)	V (Å)
$Yb_2Ba_4Cu_6O_x$	3.815	3.891	11.654	172.86
$YbGdBa_4Cu_6O_x$	3.822	3.880	11.665	172.98
$Gd_2Ba_4Cu_6O_x$	3.845	3.909	11.695	175.77
$YBa_2Cu_3O_{7\pm\epsilon}$	3.823	3.886	11.680	173.55
$Y_{0.4}Ba_{0.6}CuO_{2.22}$	3.818	3.888	11.667	173.19

The general features of the three radial distribution functions are the same, showing that the structure is similar in the three compounds. In Table III, the interatomic distances found in the superconducting materials are compared. The peaks found at radii larger than 6Å in the mixed phase (r=6.45Å, 8.46Å and 9.94Å) are located between the positions of the Yb and Gd phases, nevertheless the structure show different characteristics as the distances, 2.33Å, 3.64Å, 4.73Å and 5.33Å, are not found in the other two compounds.

Table III. Interatomic distances (Å) of the compounds.

$Yb_2Ba_4Cu_6O_x$	$YbGdBa_4Cu_6O_x$	$Gd_2Ba_4Cu_6O_x$
2.41	2.33	2.35
3.58	3.64	3.57
4.40	4.73	4.67
5.38	5.33	5.44
6.39	6.45	6.50
8.44	8.46	8.53
9.91	9.94	9.97

In conclusion we found that the three compounds present the same crystalline structure, but at atomic level there are some small differences which could be important to explain the absorption of oxygen in the Cu (I) chains.

REFERENCES.

1. Takayama, E. et al. High T_c Superconductor $YBa_2Cu_3O_7$ Oxygen Content vs T_c Relation. Japanese Jour.of Appl. Phys. 26(7) L1156-58 (1987).
2. Asano, H. et al. Crystal Structure of the High T_c Superconductor $YbBa_2Cu_3O_{7\pm\xi}$ Jpn. Jour. of Appl. Phys. 26(6) L1064-65 (1987).
3. Takabatake, T. et al. Superconductivity in Metamorphic Phases of $ErBa_2Cu_3O_{7\pm\xi}$ Jpn. Jour. of Appl. Phys. 26(7) L1231-32 (1987).
4. Ishigaki, T. et al. Crystal Structure of the High T_c Superconductor $ErBa_2Cu_3O_{7\pm\xi}$ Jpn. Jour. of Appl. Phys. 26(6) L987-88 (1987).
5. Kawasaki, et al. High T_c YbBaCuO Thin films Deposited on Sintered YSZ Substrates by Sputtering. Jpn. Jour. of Appl. Phys. 26(5) L738-40 (1987).
6. Asano, H. et al. Crystal Structure of the $HoBa_2Cu_3O_x$ Superconductor

Determined by X-ray Powder Diffraction. Jpn. Jour. of Appl. Phys. $\underline{26}$(5) L714-15 (1987).
7. Ishigaki,T. et al. Crystal Structure of the High T_c Superconductor $TmBa_2Cu_3O_{7\pm\xi}$. Jpn. Jour. of Appl. Phys. $\underline{26}$(7) L1226-27 (1987).
8. Rao, C. N. et al. Structure Property Relations in High T_c Oxide Superconductors. Jpn. Jour. of Appl. Phys. $\underline{26}$(6) L882-84 (1987).
9. Raveau, B. et al. Orthorhombic Superconducting $YBa_2Cu_3O_{7\pm\xi}$. Com. Letters $\underline{4}$(2) 205-10 (1987).
10. Yvon K., Jeitschko W. and Parthé,E. Jour, Appl. Cryst. $\underline{10}$, 73-74 (1977).
11. Magini, M. and Cabrini, A. Jour. Appl. Cryst. $\underline{5}$, 14 (1972).
12. Syono, Y. et al. X-Ray and Electron Microscopic Study of a High Temperature Superconductor $Y_{0.4}Ba_{0.6}CuO_{2.22}$ Jpn. Jour. of Appl. Phys. $\underline{26}$(4) L498-501 (1987).
13. Beno, M. A et al. Structure of the Single Phase High Temperature Superconductor $YBa_2Cu_3O_{7\pm\xi}$. Appl. Phys. Lett. $\underline{51}$(1) 57-60 (1987).

VIBRONIC STATES AS THE POSSIBLE ORIGIN OF HIGH T_c SUPERCONDUCTIVITY.

JAIME KELLER
Depto. de Física y Química Teórica, División de Ciencias Básicas
Fac. de Química, U.N.A.M., Apartado 70-528, México, D.F., 04510.

I. ABSTRACT.

It is suggested that vibronic states, arising from the coupling of the electrostatic field of the positive ions with the Cu-O bonds, may generate high T_c superconductivity and a theory is constructed based on that idea.

II. VIBRONIC STATES IN THE $YBa_2Cu_3O_7$ HIGH T_c SUPERCONDUCTORS.

The observation of Müller and Bednorz [1] that an oxide of lanthanum and copper became superconducting, at temperatures up to 35K, when a fraction of the lanthanum was replaced by barium and the report by Wu et al. [2] which ultimately led to superconductivity definitely being set up 15K to 20K above liquid nitrogen temperature, in a series of new compounds, with formula $TBa_2Cu_3O_{7-\delta}$ (T = ytrium or a trivalent rare earth) with $0.0<\delta<0.5$ (the perovskite structure would have the, oxygen rich, formula $YBa_2Cu_3O_9$, the stoichiometric compound $YBa_2Cu_3O_{6.5}$) started what has been the largest response of the material sciences community to a single discovery. Here we discuss that a special type of vibronic state could be at the origin of this behaviour.

As reported in this same volume (de Teresa, de la Mora and Keller [3]) we have concluded that from electronic band structure calculations of the High T_c SC $Y_1Ba_2Cu_3O_{7-\delta}$ the electronic states which must enter into the superconducting state are those at the Fermi level, of a set of copper (1) (d_{z^2}, $d_{x^2-y^2}$)-oxygen (1) and (4) (p_y, p_z) states which are within the highest occupied subband.

The ytrium atomic orbitals do not contribute to the states at the Fermi level as they lie 2-4 eV below it and the bonding bands between copper and oxygen atoms are also located 3-4 eV below the Fermi level.

The antibonding band at the Fermi level presents a dispersion relation (in the direction of the Cu-O chain bonds) with two maxima in the density of states separating a minima where the Fermi level sits.

The subbands around the Fermi level, separated by some 50-500 meV, have Cu(1) to O(1) and Cu(1) to O(4) antibonding character in different ratios. Below the Fermi level the Cu(1) to O(1) character is dominant, whereas above the Fermi level the proportion of Cu(1) to O(4) character is increased. This is important for the discussion below because when the Ba^{++} ion is displaced towards the O(4) atoms, this atom in turn moves away from the Cu(1) and the Cu(1) to O(1) states get stronger weight. The opposite being the case when the Ba^{++} ions are displaced away from the $O(4)$ atoms. Of course the actual normal mode is such that we can think of the $O(4)$ atoms as being the ones which oscillate because of their lower mass ratio to the Ba^{++} ions.

We have found that the Ba^{++} atoms can move inside the Cu-O cages in a potential which is either a double well with a shallow bump in the middle or a symmetry breaking (Mexican hat) potential such that the amplitude of the oscillations of the Ba^{++} ions is large (the actual normal mode being better described as a displacement of the $O(4)$ atoms). As a result the interaction is strong and a vibronic local state, that is a state where the adiabatic approximation breakes down, can set in with the formation of local electron pairing $\phi_i(\underline{r}_1-\underline{r}_i,\underline{r}_2-\underline{r}_i)$. The pairs are not of the Cooper type as they are not paired in k-space but only locally, neither are they superconducting states. The existence of this local strong correlation can be responsible for the special features observed for the resistivity and susceptibility in some samples behaviour that the high T_c superconductors present even at room temperature or higher. When the material is cooled to low temperatures, the localized boson states can condense into extended pairs

$$\psi_s(\underline{k}) = \sum_i a_{\underline{k}}^i \phi_i(\underline{r}_1,\underline{r}_2) \qquad (1)$$

with a reduction of electron-electron repulsion and of kinetic energy for a range of $\underline{k}$ where "bonding" between the vibronic states may occur, which in turn may contribute to form a superconducting state. We call the model generated by this physical considerations a "two step approach to superconductivity" (Keller [4]).

III. A TWO STEP APPROACH TO HIGH T_c SUPERCONDUCTIVITY.

The BCS theory constructs the superconducting state as that which minimizes the energy simultaneously with the introduction of the electron-phonon interaction, after a diagonalization of the Hamiltonian of the system. From the considerations of the previous section we see that a more general theory can be developed where the local pairing at temperature T_1 and the establishment of extended states out of local vibronic states and their condensation into a true superconducting state at temperature T_2 do not occur simultaneously. Formally we can also have the two steps occurring simultaneously. The T_c will be equal to the lowest T_i.

In our case the first step corresponds to the formation of vibronic states $\phi_i^v(\underline{r}_1-\underline{r}_i,\uparrow;\underline{r}_2-\underline{r}_i,\downarrow)$ where a local pairing occurs. If H_i is the Hamiltonian for a vibronic state referred to a local space origin $\underline{r}_i$ then

$$H_i \phi_i^v = \varepsilon \phi_i^v \quad (i) = \text{(regions of the crystal around } \underline{r}_i) \qquad (2)$$

The second step corresponds to the formation of a set of extended states

$$\psi_s(\underline{k}) = A(\underline{k}) \sum_i e^{i\underline{k}\cdot\underline{r}_i} \phi_i^v(\underline{r}_1-\underline{r}_i,\uparrow;\underline{r}_2-\underline{r}_i,\downarrow) \qquad (3)$$

which will contribute to a (correlated) superconducting (SC) state of the system, each $\psi_s(\underline{k})$ being a solution of a quasiparticle wave equation, $A(\underline{k})$ contains the effect of correlation resulting in an effective attractive interaction between superconducting pairs,

$$\psi = \pi_{\underline{k}} \psi_s(\underline{k}), \quad E = \langle\psi|\hat{H}_s|\psi\rangle$$

$$\hat{H}_s = \sum_i \hat{H}_i + \sum_{ij} \hat{H}_{ij} \text{ or } (\hat{H}_s = \sum_i H_i V_i^+ V_i + \sum_{ij} H_{ij} V_i^+ V_j) \quad (4)$$

where the $\hat{H}_{ij}$ correspond to the hopping interaction between vibronic states in sites i and j which will lower the kinetic energy of the states. As a first approximation $\hat{H}_s$ is the Hamiltonian corresponding to a collection of (extended normal) modes, where the "active" local modes have been added up to an extended vibration. V_i^+ creates a vibronic state around site i.

The Hamiltonian $\hat{H}_s$ can be transformed to extended normal variables λ, using a Bogoliubov-Tyablikov transformation, to obtain:

$$\hat{H} = \sum_\lambda \varepsilon_\lambda b_\lambda^+ b_\lambda - \sum_{\lambda'\lambda} V_{\lambda\lambda'} b_{\lambda'}^+ b_\lambda \quad (5)$$

with $V_{\lambda\lambda'}$ positive for those λ where correlation stabilizes the SC state, which has the form of a BCS Hamiltonian but where the parameters and the wave function will be largely different.

There will then exist several characteristic temperatures in the model: T_1 the critical temperature for the formation of a local vibronic state. T_2 the critical temperature for the formation of the extended states where the oscillations in one region of the lattice do not interfere with the formation of the vibronic state in a second region, a third characteristic temperature T_3 would correspond to that at which enough pairs exist for the system to condense into a SC state.

The presence of defects in the Cu-O chains in the lattice will reduce T_2 or T_3 and then the probability of the extended superconducting state being formed. However, if annealing or thermal cycling of the material occurs, the defects could be concentrated at random but in alternate oxygen defective Cu-O chains only, with an effective doubling of the $\underline{a}$ vector of the lattice. The Cu-O chain with less defects could then stablish the extended vibronic states at a higher T_2 and the critical concentration of pairs reached at a higher T_c. The introduction of chemically inactive atoms in the defective chains would stabilize the lattice.

The b_λ^+ are boson operators, but they do not correspond to <u>independent</u> quasiparticle operators, in fact they exist as bosons formed from fermion states, then the maximum numbers of quasiparticles in the b_λ^+ states is fixed from the availability of electron like quasiparticles from which they are constructed, via the vibronic interaction. This is similar to the case of the BCS boson operators.

V. DISCUSSION.

We have presented several aspects of a novel approach to high T_c superconductivity. In it at temperatures below T_0 and T_1 the localized or extended boson states ϕ_i will start to form, causing changes in the electric and magnetic properties. This transition is not of 1st. order as the states ϕ_i will gradually be produced on cooling down the sample. When the density of ϕ_i states exceeds a limit given both by percolation arguments and by the strength of the $\phi_i \rightarrow \phi_j$ interaction, the system will condense into a true superconducting state at a temperature T_2. At T_c = smaller (T_1, T_2) the transition will be of first order as a result of it being a cooperative phenomena similar to the gas-liquid phase transition.

REFERENCES.

1. J.G. Bednorz and K.A. Müller, Z. Phys. B64, 189-93 (1986).
2. M.K. Wu, J.R. Ashburn, C.J. Tong, P.H. Hor, R.L. Meng, L. Gao, Z.J. Huang, X.Q. Wang and C.W. Chu, Phys. Rev. Lett. 58, 908 (1987).
3. De Teresa C., de la Mora P. and Keller J., this volume.
4. J. Keller, paper presented at the XI Int. Workshop on Condensed Matter Theories (Oulu, Finland, July 26th-August 3, 1987) and at the panel on high T_c Superconductivity, Mérida, México, November 1987, to be published in Revista Méx. de Física, special issue on high T_c superconductivity, 1988.

ACKNOWLEDGEMENTS.

This work was supported in part by the "Unión Química en Materia Condensada" Project, Clave PCEXCNA-022702, CONACYT and by the "Proyecto Universitario Interdisciplinario de Superconductores de Alta Temperatura", UNAM. The technical assistance of Mrs. Irma Aragón is gratefully acknowledged as well as the many very fruitful discussions with Carmen de Teresa and Pablo de la Mora.

AUTHOR INDEX

ABURTO, S. 251
AGUILAR, G. 264
AHN, B. T. 38
AKACHI, T. 251,264
ALBARRAN, J. L. 161,268
ALCACER, L. 256
ALMEIDA, M. 256
BARRIO, R.A. 206,237,251,264
BEYERS, R. 38
BOSCH, P. 284
CABRERA, A. 167,279
CALLES, A. 167,279
CARBOTTE, J. P. 196
CARRILLO, E. 161
CASTRO, J. J. 167,279
CAUSA, M. T. 260
CHAVEZ, M. DE L. 243,247
CIVALE, L. 102
CRUZ, J. 154
DECCA, R. 102
DE LA CRUZ, F. 102
DE LA MORA, P. 275
DEL ANGEL, P. 172
DE LOZANNE, A. L. 82
DE TERESA, C. 275
DOMINGUEZ, J. M. 172
D'OVIDIO, C. 102
ENGLER, E. M. 38
ESCUDERO, R. 54,237,251,264,284
ESPARZA, D. A. 102
FAINSTEIN, C. 260
FALCONY, C. 172
FISANIK, G. J. 88
FISK, Z. 260
FUENTES, J. 161,268
GOMEZ, R. W. 113,251
GONCALVES, A. P. 256
GOODENOUGH, B. 18
GOVEA, L. 237
GRANT, P. M. 38
GÜR, T. M. 38
GUZMAN, O. 172
HENRIQUES, R. T. 256
HERVIEU, M. 5
HUANG, C. Y. 28
HUGGINS, R. A. 38

JACOWITZ, R. D. 38
JIMENEZ, M. 251
JIN, S. 88
JOSE, J. V. 123
JOSE-YACAMAN, M. 154
KELLER, J. 275,289
LEE, V. Y. 38
LIM, G. 38
LOPES, E. B. 256
LOPEZ, D. 247
MAIGNAN, A. 5
MANTHIRAM, A. 18
MARQUINA, M. L. 251
MARQUINA, V. 251
MARTINEZ, L. 161,268
MENDOZA, A. 161
MICHEL, C. 5
MONTOYA, A. 172,284
MOORE, R. 88
MURRIETA, H. 264
NAKAHARA, S. 88
NICOLSKY, R. 219
NIEVA, G. 260
OMAÑA, J. 172
OROZCO, E. 161
OSEGUERA, U. 154
OSEROFF, S.B. 260
OSQUIGUIL, E. 102
PACHECO, G. 161,243,247
PARKIN, S. S. P. 38
PEREZ, R. 154
PEREZ-RAMIREZ, J. G. 154
PIÑA, C. 237,243,247,284
POLITIS, C. 44
PROVOST, J. 5
QUINTANAR, C. 251
RAMIREZ, J. 264
RAMIREZ, M. L. 38
RAVEAU, B. 5
REYES-GASGA, J. 154
RIOS-JARA, D. 135,237,251
ROBLEDO, A. 145
ROCHE, K. P. 38
SAFAR, H. 102
SALCIDO, A. 167,279
SANCHEZ, R. 260
SANTOS, I. C. 256
SCHLUTER, M. 183
SCHULTZ, S. 260
SHERWOOD, R. C. 88

SMITH, J. L.   260
STEREN, L. B.   260
TAKITA, K.   73
TIEFEL, T. F.   88
TOVAR, M.   260
UCHIDA, S.   63
VALDES, S.   268
VAN DOVER, R. B. 88
VAREA, C.   145
VAZQUEZ, J. E. 38
VIER, D. C.   260
WANG, C.   237
YAN, M.   88
YEPEZ, E.   167,279
ZYSLER, R.   260